Sources in the History of Mathematics and Physical Sciences

12

Sources in the History of Mathematics and Physical Sciences

Vol. 1: G.J. Toomer (Ed.)
Diocles on Burning Mirrors: The Arabic Translation of the Lost Greek Original, Edited, with English Translation and Commentary by G.J. Toomer

Vol. 2: A. Hermann, K.V. Meyenn, V.F. Weisskopf (Eds.)
Wolfgang Pauli: Scientific Correspondence I: 1919–1929

Vol. 3: J. Sesiano
Books IV to VII of Diophantus' *Arithmetica:* **In the Arabic Translation Attributed to Qustā ibn Lūqā**

Vol. 4: P.J. Federico
Descartes on Polyhedra: A Study of the *De Solidorum Elementis*

Vol. 5: O. Neugebauer
Astronomical Cuneiform Texts

Vol. 6: K. von Meyenn, A. Hermann, V.F. Weisskopf (Eds.)
Wolfgang Pauli: Scientific Correspondence II: 1930–1939

Vol. 7: J.P. Hogendijk
Ibn Al-Haytham's *Completion of the Conics*

Vol. 8: A. Jones
Pappus of Alexandria Book 7 of the *Collection*

Vol. 9: G.J. Toomer (Ed.)
Appollonius *Conics* **Books V to VII: The Arabic Translation of the Lost Greek Original in the Version of the Banū Mūsā,** Edited, with English Translation and Commentary by G.J. Toomer

Vol. 10: K. Andersen
Brook Taylor's Role in the History of Linear Perspective

Vol. 11: K. von Meyenn (Ed.)
Wolfgang Pauli: Scientific Correspondence III: 1940–1949

Vol. 12: F.J. Ragep
Naṣīr al-Dīn al-Ṭūsī's *Memoir on Astronomy*
(al-Tadhkira fī ᶜilm al-hay'a)

F.J. Ragep

Naṣīr al-Dīn al-Ṭūsī's *Memoir on Astronomy*

(al-Tadhkira fī ᶜilm al-hay'a)

Volume I
Introduction, Edition, and Translation

With 96 Illustrations

Springer Science+Business Media, LLC

F.J. Ragep
Department of the History of Science
University of Oklahoma
601 Elm Street #622
Norman, OK 73019 USA

Library of Congress Cataloging-in-Publication Data
Ṭūsī, Naṣīr al-Dīn Muḥammad ibn Muḥammad, 1201–1274.
[Tadhkirah fī ᶜilm al-hay 'ah. English]
Naṣīr al-Dīn al-Ṭūsī's Memoir on astronomy = al-tadhkira fī ᶜilm al-hay 'a /[commentary by] F.J. Ragep
p. cm. —(Sources in the history of mathematics and physical sciences ; v. 12)
Includes bibliographical references and index.
ISBN 978-1-4757-2243-7
1. Astronomy, Arab—Early works to 1800. 2. Ṭūsī, Naṣīr al-Dīn Muḥammad ibn Muḥammad, 1201–1274. 3. Astronomers—Iran—Biography. I. Ragep, F.J. II. Title. III. Title: Tadhkira fī ᶜilm al-hay 'a. IV. Series.
QB23.T87 1993
520—dc20 93-10445

Originally published by Springer-Verlag New York, Inc. in 1993
Softcover reprint of the hardcover 1st edition 1993

Production managed by Hal Henglein; manufacturing supervised by Vincent R. Scelta.
Camera-ready copy prepared by the author.

ISBN 978-1-4757-2243-7 ISBN 978-1-4757-2241-3 (eBook)
DOI 10.1007/978-1-4757-2241-3

to Anwar to Lina

who endured

this most recalcitrant

of siblings

كم كتاب كتبته بيدي
سوف تبلا يدي ويبقى الكتاب
كانت الدنيا لقوم غيرنا رحلوا منها وخلوها لنا
سوف نسكنها ونرحل بعدهم
ونخليها كما خليت لنا

How many books have I written with my hand!
My hand shall wither but the book will remain.
Once the world was to another people
But they have since departed
Leaving it with us.
We too shall dwell in it
But we also shall follow them
Leaving it as it was left for us.

Anon., Damascus, Ẓāhiriyya MS 4871, f. 36a

Preface

I was introduced to Ṭūsī and his *Tadhkira* some 19 years ago. That first meeting was neither happy nor auspicious. My graduate student notes from the time indicate a certain level of confusion and frustration; I seem to have had trouble with such words as *tadwīr* (epicycle), which was not to be found in my standard dictionary, and with the concept of solid-sphere astronomy, which, when found, was pooh-poohed in the standard sources. I had another, even more decisive reaction: boredom. Only the end of the term brought relief, and I was grateful to be on to other, more exciting aspects of the history of science.

A few years later, I found myself, thanks to fellowships from Fulbright-Hays and the American Research Center in Egypt, happily immersed in the manuscript collections of Damascus, Aleppo, and Cairo. Though I had intended to work on a topic in the history of mathematics, I was drawn, perhaps inevitably, to a certain type of astronomical writing falling under the rubric of *hay'a*. At first this fascination was based on sheer numbers; that so many medieval scientists could have written on such a subject must mean something, I told myself. (I was in a sociological mode at the time.) As I began to read, or rather try to read, these manuscripts, some of daunting size, I became more and more engaged in a world of mostly forgotten scientists, many from a period that modern scholarship had deemed, with the hubris that only modernism can muster, both invisible *and* unworthy. But these late medieval astronomers of "decline" seemed to me to be saying interesting things, and significantly they themselves thought they were saying interesting things as they spoke to one another over geographical and chronological distance. And two names kept recurring with astonishing frequency in these works: Naṣīr al-Dīn al-Ṭūsī and his *Tadhkira fī ᶜilm al-hay'a*. I had come full circle.

To write this book, I have incurred enormous debts to both institutions and individuals, and it gives me great pleasure to be able at long last to express my gratitude in print. They are, of course, neither responsible for the opinions expressed nor for the remaining shortcomings. The National Endowment for the Humanities generously awarded me a research grant (RL-20578-84) that allowed me to work without interruption on the edition and translation. Another grant, this one from the National Science Foundation (SES-8618656), was for research on trepidation, and I have incorporated some of the results in the commentary. The Department of the History of Science at Harvard University appointed me a postdoctoral fellow on two occasions and provided me with the necessary facilities to take full advantage of those grants. A year at the Society for the Humanities at Cornell was enormously stimulating and gave me the opportunity both to learn and to test my ideas with a very talented group of non-historians of science. Travel funds from the Department of the History of Science at the University of Oklahoma allowed me to check and recheck countless details and footnotes during preparation of the final copy.

During the many years of research and writing, I have benefited from my acquaintance and friendship with a number of extraordinary individuals, far more than I had ever expected to meet in a lifetime. In Cairo, I had the great fortune to have Edward S. Kennedy and David King as next-door neighbors *(yaᶜnī)*, and they were, despite some scepticism concerning my project, always ready and able to provide advice and guidance on matters great and small. Also in Cairo, Aḥmad Harīdī was unceasingly patient in teaching this *khwāja* about the beauty and intricacy of Arabic and Arabic paleography. Back home, Aron Zysow, my next-carrel neighbor at Widener Library, shared his very considerable knowledge of Islam and Islamic history, Wheeler Thackston offered advice and assistance on all things Persian, and Marina Tolmacheva graciously looked over the maps and made several helpful suggestions. Over many long, sometimes difficult years, Raine Daston, Mollie Palchik, and Noel Swerdlow provided inspiration and encouragement and were always there when needed most.

A. I. Sabra, *shaykhunā al-ra'īs*, was the one who introduced me to the *Tadhkira* those many years ago and, for reasons known only to himself, thought that I could be entrusted with its study. His teachings, methodology, and inspiration are such an integral part of this work that it would be less than elegant to provide a list; let me simply say that it was he who made *hay'a* such an important part of my intellectual vocabulary.

D. E. Pingree, *ḥakīm extraordinaire*, has been unceasing in his efforts to help me broaden my horizons and use what he calls common sense, which, in his version, is quite uncommon. I very much benefited from his critique and suggestions on an earlier draft of a chapter concerning the tradition of the *Tadhkira*, and his "common sense" led me to uncover the relationship between the various versions of the text.

G. J. Toomer, φιλόπονος τε ὁμοῦ καὶ φιλαλήθης, bestowed upon this book a degree of attention that went far beyond what the designation editor normally calls for. He saved me from innumerable blunders of detail and interpretation, and his stern kindness made me feel that I should and could meet his soaring standards.

S. P. Ragep, friend and co-worker of some twenty years, drafted and photographed the figures and collated them with those in the manuscripts, prepared the concordance of manuscripts, typed the Arabic text and apparatuses (in multiple versions), offered advice on both style and substance, much of which was adopted, helped prepare the index, and somehow managed in addition to have a career and the dedicatees. But above all, she helped me find words and meaning where once there was only a cacophony of silence.

F.J.R.

Table of Contents

Volume Two

Volume One

Part I

General Introduction

Part II

Edition and Translation

Part III

Commentary Figures

Part I

General Introduction

§1. Naṣīr al-Dīn al-Ṭūsī

A. Life

1. Introduction

Abū Jaᶜfar Muḥammad b. Muḥammad b. al-Ḥasan Naṣīr al-Dīn al-Ṭūsī was born on Saturday (at dawn according to Riḍawī[1]), 11 Jumādā I, 597 H./17 February 1201 A.D. in Ṭūs or its environs[2] and died in Baghdad (at sunset according to Baḥrānī[3]), 18 Dhū al-Ḥijja 672 H./25 June 1274 A.D. In a remarkable century that began with the death of Chu Hsi and further witnessed the birth, death or both of such luminaries as Roger Bacon, Ibn ᶜArabī, Maimonides, Gregory Chioniades, Ibn Taymiyya, Thomas Aquinas and Levi ben Gerson, Ṭūsī's breadth and depth of learning stand out. Even in his lifetime he was something of a legend and acquired the honorific title of *khwāja* (distinguished scholar and teacher);[4] later he came to be referred to by the rather more impressive *ustādh al-bashar* (teacher of mankind) and *al-muᶜallim al-thālith* (the third teacher, *i.e.* after Aristotle and Fārābī).[5] After his death, Ṭūsī's influence, which can only be described as monumental, continued in fields as diverse as ethics, natural philosophy, mathematics, Sufism, astronomy, *kalām* (dialectical theology), *fiqh* (law), and logic.

A full-scale political and intellectual biography of Naṣīr al-Dīn, though much needed, would be out of place here; my much more limited aim is simply to provide a chronology of his life and to give a sense of why he maintained a

---

[1] *Aḥwāl*, p. 3.

[2] Naṣīr al-Dīn gives his place of origin as Ṭūs in the introduction to the *Īlkhānī Zīj*; see Boyle [1963], pp. 246–247. Ṭūs may indicate either the district or city in Khurāsān; the city itself, near modern Mashhad, was destroyed in 1389 A.D. ("Ṭūs," *EI*[1], 4: 974).

[3] *Lu'lu'at al-Baḥrayn*, p. 246.

[4] Bar Hebraeus (1226–1286), who knew Naṣīr al-Dīn at Marāgha, calls him "*Khwāja Naṣīr*" (*Chronography*, 1: 451).

[5] Riḍawī, *Aḥwāl*, p. 3.

lifelong interest in astronomy. But even this circumscribed goal must inevitably lead us into difficult questions of the typicalness of his education, the role of Hellenism in his worldview, the importance of patronage on the quantity and content of his scientific production, and the impact of his changing religious beliefs on his science. In order to approach these questions and, in general, to organize the details of Ṭūsī's multifaceted life, I shall make a basic assumption, namely that Ṭūsī was committed throughout his life, though in varying degrees, to what we may broadly call a Hellenistic attitude. This does not mean that he abandoned Islam (though he was accused of such); it does mean that he was heir to the intellectual traditions of late Greek antiquity that had begun to be "naturalized"[6] within an Islamic context by such personages as Kindī (fl. 3rd/9th c.), Fārābī (fl. 4th/10th c.) and Ibn Sīnā (fl. 5th/11th c.). Ṭūsī's contribution to this tradition was to continue the rapprochement of Hellenism and the Islamic disciplines. But more fundamentally, Ṭūsī's Hellenism led him to pursue knowledge for its own sake; thus we would be deeply mistaken to view his pursuit of the sciences as being primarily for the purpose of serving the practical needs of the Islamic community or his patrons. Only by situating Ṭūsī within this "ancient" tradition can we hope to comprehend his lifelong love and devotion to astronomy, a commitment that should be understood as epistemological rather than utilitarian.[7]

2. Education and Early Life

How did one come to be educated in this Hellenistic tradition over and above the more customary religious one? Although Ṭūsī's education may not be typical, it does provide some insight into the factors that would lead a young student to pursue the ancient sciences and the manner in which he might go about such a task. Over the course of his life, Naṣīr al-Dīn would travel widely to satisfy his craving to be educated in virtually all fields of learning; his first instruction, however, occurred in his own home under the tutelage of his father, Wajīh al-Dīn Muḥammad b. al-Ḥasan al-Ṭūsī. Wajīh al-Dīn could trace his own education through a line of Shīᶜite scholars back to ᶜAlam al-Hudā al-Murtaḍā (355–436/966–1044), the *naqīb* (head) of the ᶜAlīds in Baghdad and defender of

[6] This term was coined by Sabra [1987a].

[7] This *should* be obvious; however the insistence by numerous modern commentators that Islamic scientists could never rise to the Greek view that knowledge was to be pursued for its own sake compels one to state the obvious.

the Imāmī (Twelver) Shī^{c}ites[1] against the attacks of the Mu^{c}tazilite[2] Judge ^{c}Abd al-Jabbār (died 415/1024);[3] he continued this lineage by teaching his son *sharī^{c}a* (law) through the reading of al-Murtaḍā's works. The young Naṣīr al-Dīn also studied with his father's maternal uncle, Naṣīr al-Dīn ^{c}Abdallāh b. Ḥamza al-Ṭūsī, another eminent Imāmī scholar, and his own maternal uncle Nūr al-Dīn ^{c}Alī b. Muḥammad al-Shī^{c}ī.[4] But there was evidently more to Naṣīr al-Dīn's family than a simple adherence to exoteric Shī^{c}ite dogma. In his so-called autobiography, Ṭūsī tells us that his father, someone who had seen the world [*jahān-dīdah*], encouraged the young boy to study different sciences and listen to the masters of various sects and opinions.[5] This seemingly broad-minded attitude may have been indirectly inspired by Tāj al-Dīn al-Shahrastānī, who had been a teacher of Naṣīr al-Dīn ^{c}Abdallāh b. Ḥamza, who had in turn taught his nephew, *i.e.* Ṭūsī's father. Tāj al-Dīn had what Madelung calls "crypto-Ismā^{c}īlī thought,"[6] and this may have given the otherwise rather strait-laced Twelver family a somewhat more "worldly" view; at least it may have been part of the widened horizon that would lead the young Naṣīr al-Dīn to give both philosophy and perhaps the heretical Ismā^{c}īlī doctrine a hearing. At any rate, Ṭūsī took his father up on his suggestion and studied the branches of philosophy (*ḥikma*) and especially mathematics with a certain Kamāl al-Dīn Muḥammad al-Ḥāsib, a student of Afḍal al-Dīn al-Kāshī.[7]

After this initial period of his education, Naṣīr al-Dīn apparently traveled to Nīsābūr, a major city in Khurāsān located to the west of Ṭūs, to study with the noted physician Quṭb al-Dīn al-Miṣrī and with the polymath Farīd al-Dīn Dāmādh. Both provide us with evidence that he was actively pursuing further education in the ancient sciences and philosophy.

Quṭb al-Dīn, though originally from the Maghrib, had studied for some time in Egypt (whence the appellation *al-Miṣrī* [the Egyptian]), and then had traveled

---

1 Shī^{c}a is the designation of a large minority group in Islam (the majority being called Sunnīs) that are united in their belief in the legitimacy of the claim of ^{c}Alī, the son-in-law of the prophet Muḥammad, and his family to the caliphate. Though there were (and are) several sects that can be called Shī^{c}ite, the two we shall be concerned with in this chapter are the Imāmīs (or Twelvers) and the Ismā^{c}īlīs (or Seveners), the two disagreeing on the identity (whether twelfth or seventh) of the last, and at present hidden, Imām. Today the Imāmīs, by virtue of their majority status in Iran, are the largest Shī^{c}ite sect. The Ismā^{c}īlīs, who are now led by the Āghā Khān, are a relatively small group that is no longer the major political force it was during the Middle Ages.

2 The Mu^{c}tazilites were rationalist theologians who held a predominant position during the early centuries of ^{c}Abbāsid rule.

3 Brockelmann, "al-Murtaḍā," *EI*[1], 3: 736.

4 Ibn al-Fuwaṭī, vol. V (in *Oriental College Magazine*, suppl. 1939), p. 177; Riḍawī, *Aḥwāl*, pp. 158–161.

5 Ṭūsī, *Sayr*, p. 38.

6 Madelung [1985], pp. 87–89.

7 Ṭūsī, *Sayr*, p. 38.

eastward to study with the well-known *mutakallim* (dialectical theologian) Fakhr al-Dīn al-Rāzī (ca. 543–606/1149–1209). Eventually he came to reside in Nīsābūr where Ṭūsī studied medicine and *ḥikma* (presumably philosophy) under him.[8] Quṭb al-Dīn wrote a commentary on the *Kulliyyāt* (which is the part dealing with general topics) of Ibn Sīnā's medical work, the *Qānūn*, in which he is reported to have been rather critical of the author.[9]

Farīd al-Dīn Dāmādh was well-versed in philosophy, *uṣūl* (jurisprudence), logic, and medicine.[10] He also had studied with Fakhr al-Dīn al-Rāzī and, of even more interest, could trace his pedagogical lineage back to Ibn Sīnā.[11] Presumably then, Naṣīr al-Dīn studied the works of his famous Persian predecessor with Farīd al-Dīn.[12]

We can, with a fair degree of confidence, place Naṣīr al-Dīn in Nīsābūr sometime after 610/1213, when he would have been a mere lad of twelve or thirteen, and before 618/1221, when the armies of Chingiz Khān completed their devastation of Khurāsān. Indeed, Ibn abī Uṣaybiᶜa informs us that Quṭb al-Dīn al-Miṣrī was one of the victims of the final Mongol onslaught on Nīsābūr that occurred in that year.[13]

Perhaps because of the precarious situation in his native Khurāsān, or perhaps simply out of a desire to further his studies, Naṣīr al-Dīn traveled to Iraq[14] to study with the Shīᶜite legal scholar Muᶜīn al-Dīn Sālim b. Badrān al-Miṣrī[15] and with the extraordinary Kamāl al-Dīn ibn Yūnus (551–639/1156–1242), whose expertise ranged from Qur'ān exegesis to the *Almagest*.[16]

We know from several sources that Ṭūsī studied with this Muᶜīn al-Dīn; in addition, there is extant an *ijāza* (a license allowing the recipient to teach a given work) granted by Muᶜīn al-Dīn to Naṣīr al-Dīn in 619/1222.[17] Little more is

---

[8] Riḍawī, *Aḥwāl*, pp. 167–169.

[9] Ibn abī Uṣaybiᶜa, *ᶜUyūn*, 2: 30.

[10] Ibn al-Fuwaṭī, vol. IV (Damascus, 1962–65), 3: 458–459 and 466.

[11] Shūshtarī gives the names of four intermediaries (*Majālis*, 2: 203 [7th *Majlis*]).

[12] Apparently Zand has claimed that Ṭūsī studied Ibn Sīnā's *Al-Ishārāt wa-'l-tanbīhāt* with Farīd al-Dīn. I do not have access to Zand's article and have been unable to confirm this independently; I here depend on Siddiqi [1963], 1: 564–565.

[13] Ibn abī Uṣaybiᶜa, *ᶜUyūn*, 2: 30; Juwaynī, *History*, 1: 176–178.

[14] On this early trip to Iraq (Baghdad and/or Mosul), see Siddiqi [1963], p. 565 (quoting Zand) and Riḍawī, *Aḥwāl*, p. 179. In the *Sayr* (p. 41), Ṭūsī refers to a trip from Iraq to Khurāsān.

[15] I am basing Muᶜīn al-Dīn's residence in Baghdad on the citation by Siddiqi of the Zand article.

[16] Ibn Khallikān, *Wafayāt*, 5: 311–313 (English trans. 3: 467–468); see also Ibn abī Uṣaybiᶜa, *ᶜUyūn*, 1: 306–308.

[17] This was for Ṭūsī's successful completion of the study of the third part and most of the second of a work on Shīᶜite law by ᶜIzz al-Dīn Ḥamza b. Zuhra, who had been Muᶜīn al-Dīn's teacher. Ibn Zuhra himself was apparently a mainstream Twelver Shīᶜite. (I am indebted to A. Zysow for this information.) We also know that Naṣīr al-Dīn collated the third part of this work, completing it in 624/1227 (Riḍawī, *Aḥwāl*, pp. 163–166).

known about Mu^cīn al-Dīn, however. That he or his ancestors came from Egypt (or perhaps simply passed through as we saw above with Quṭb al-Dīn al-Miṣrī) is clear from his name. Ṣafadī and Kutubī both inform us[18] that he was also *al-Mu^ctazilī*[19] and *al-Rāfiḍī*.[20] Despite the little we do know about him, he does seem to have played an important part in Naṣīr al-Dīn's education. Ibn Kathīr (ca. 700–74/1300–73), for example, states that his basic education came from Mu^cīn al-Dīn who, he continues, influenced Ṭūsī to such an extent that he undermined his religious belief *(afsada i^ctiqādahu)*.[21] Ibn Kathīr, who held strong Ḥanbalī biases,[22] may have somewhat exaggerated the influence of Mu^cīn al-Dīn who was, as far as he was concerned, twice cursed—once with Shī^cism and a second time with Mu^ctazilism. We should keep in mind, however, that his influence is also implied in the account of Ṣafadī/Kutubī by the fact that he is referred to twice. Furthermore he is one of only two of Naṣīr al-Dīn's teachers who there receive specific mention.

The other is Kamāl al-Dīn ibn Yūnus.[23] In all probability, Naṣīr al-Dīn studied astronomy and mathematics with Kamāl al-Dīn who was noted for his

---

[18] Ṣafadī, *Wāfī*, 1: 179; Kutubī, *Fawāt*, 3: 246. The biography of Ṭūsī given in these two works is virtually identical; Iḥsān ^cAbbās in the introduction to his edition of Kutubī would seem to explain this not uncommon occurrence by asserting that Kutubī copied much of his work from Ṣafadī (1: 4–5). Though both men were contemporaries and indeed died in the same year (764/1363), Ṣafadī seems to have completed his work first.

[19] There is a possibility that *al-Mu^ctazilī* may be a misreading of *al-Māzinī*, which is given by Mu^cīn al-Dīn in the *ījāza* referred to above. However, we should note that *al-Mu^ctazilī* is also recorded by Ibn Kathīr.

[20] During this period, the Imāmī Shī^cites had accepted some of the doctrines of the Mu^ctazilites; thus it is not surprising that one person could be considered to be a member of both groups despite the earlier animosity between them. (This is a rather complicated issue; for some indication of the relationship between Mu^ctazilism and Shī^cism, see Madelung [1970] and Momen [1985], pp. 76–82.) *Al-Rāfiḍī* may identify him as belonging to a certain Shī^cite sect that formed in opposition to Zayd b. ^cAlī's (d. 122/740) refusal to allow the denunciation of the Companions of the Prophet, or, more probably, it may simply identify him as a Shī^cite. (For this latter use of *Rāfiḍa* by the opponents of Shī^cism, see Madelung [1970], p. 3; for the former usage, see Aḥmad Amīn, *Ḍuḥā*, 3: 136 and Lane, *Lexicon*, 3: 1121.)

[21] Ibn Kathīr, *Bidāya*, 13: 267–268.

[22] The Ḥanbalīs were hostile to speculative theology as well as to legal approaches that did not adhere to a strict reading of the Qur'ān and the traditions of the Prophet.

[23] Because of his great fame—at least among his contemporaries—one would have expected that many of Naṣīr al-Dīn's biographers would have recorded that he studied with the great scholar of Mosul. Surprisingly, we find this information only in Ṣafadī/Kutubī, who rely on the authority of al-Shams, the son of Mu'ayyad al-Dīn al-^cUrḍī (Ṣafadī, *Wāfī*, 1: 181; Kutubī, *Fawāt*, 3: 249.) Since ^cUrḍī was one of Naṣīr al-Dīn's associates at the Marāgha observatory, one can probably accept this report as reliable, especially in that al-Shams correctly cites Mu^cīn al-Dīn b. Badrān as another of Ṭūsī's teachers.

expertise in these subjects;[24] it is less probable that Ṭūsī pursued religious studies with him since Kamāl al-Dīn was a Sunnī of the Shāfiᶜite school of law.[25] Ibn al-Fuwaṭī tells us that students came to study with him from east and west (*i.e.* of Islamdom); we may assume that Ṭūsī was one of them, most likely traveling to Mosul sometime in the 620's/1223–32 after completing his legal studies with Muᶜīn al-Dīn b. Badrān.[26]

Although there can be little doubt that both Naṣīr al-Dīn and his education were in many ways unique, we may yet draw some preliminary conclusions that can help us understand the initiation of a 13th c. individual into the "Ancient Sciences." An obvious point is that during this period the rift between those who studied the *awā'il* sciences and those who dealt mainly with the Islamic sciences was not as great as it had been in earlier centuries.[27] Farīd al-Dīn Dāmādh, Quṭb al-Dīn al-Miṣrī, Kamāl al-Dīn b. Yūnus, and Naṣīr al-Dīn himself are clear examples of this trend to do work in both the *manqūlāt* (the transmitted, *i.e.* traditional Islamic, sciences) and the *maᶜqūlāt* (the rational sciences).[28] It is also evident from the diverse religious affiliations of these individuals that we are here dealing with a tendency current among both Shīᶜites and Sunnīs.[29]

An important consequence of the closing of the gap between religious and nonreligious studies was the opportunity this afforded to bring the *awā'il*

---

[24] Ibn Khallikān, *Wafayāt*, 5: 311–312 (English trans., 3: 467–468); Ibn abī Uṣaybiᶜa, *ᶜUyūn*, 1: 306.

[25] The complexity of medieval Islam should always lead one to be tentative when making such assertions, however. It may not be irrelevant to note here that Kamāl al-Dīn is reported by Ibn Khallikān (*Wafayāt*, 5: 312; English trans., 3: 468) to have taught both the Torah and New Testament to Christian and Jewish students (*ahl al-dhimma*).

[26] Kamāl al-Dīn was associated with a number of schools in Mosul [Ibn al-Fuwaṭī, vol. V (in *Oriental College Magazine*, suppl. 1940), p. 293; Ibn Khallikān, *Wafayāt*, 5: 311, 313, 316 (English trans., 3: 467, 469, 472); Ibn abī Uṣaybiᶜa, *ᶜUyūn*, 1: 306] and would seem to have spent most of his time there except for several years of study at the Niẓāmiyya College in Baghdad in the early 570's/ca.1175 [Ibn Khallikān, *Wafayāt*, 5: 311 (English trans., 3: 467)].

[27] Cf. Sabra [1987a], pp. 236–238.

[28] For another example, see Makdisi's [1981] account of ᶜAbd al-Laṭīf al-Baghdādī (d. 629/1231), pp. 84–91.

[29] We should not conclude, however, that all the educated felt compelled to study philosophy or mathematics. Ibn Khallikān relates the story of a certain religious personage, Ibn al-Ṣalāḥ, who wished to study logic—in secret—with Kamāl al-Dīn b. Yūnus. After some time and little progress, Kamāl al-Dīn advised him to give up the study which was not going to do him any good on the one hand and might besmirch his reputation on the other. Kamāl al-Dīn himself was accused of suspect religious beliefs because of his assiduous devotion to the rational sciences. But it is important to note that Ibn Khallikān, himself a Shāfiᶜī jurist and sometime *Qāḍī al-Quḍāt* (Chief Justice) in Damascus, is apologetic for bringing up such matters; in general, he cannot restrain his admiration for Kamāl al-Dīn's erudition in *all* of the sciences.

sciences into the religious schools (*madrasa*; pl. *madāris*).[30] Though this probably did not mean that mathematics, the sciences, and philosophy became a part of the curriculum of the *madrasa*, whose purpose, above all, was the teaching of law, it did mean that someone like Kamāl al-Dīn b. Yūnus, who directed or taught at various schools in Mosul, had the opportunity to teach nonreligious subjects if he so chose.[31] Naṣīr al-Dīn's appropriation of *waqf* (religious endowment) funds at the Marāgha observatory, one of the few known instances of the use of such funds for a nonreligious purpose, may perhaps be better understood in the context of this changing relation between religious and nonreligious disciplines.[32]

A second point that should be stressed concerning Ṭūsī's education is that a student was willing to travel a considerable distance to study with the teacher of his choice. The trek of Naṣīr al-Dīn from Khurāsān to Mosul is an example of this; we have also seen that Quṭb al-Dīn al-Miṣrī traveled from the Maghrib to Egypt and from there on to Nīsābūr. Furthermore, one may conclude that, based on the example of Ṭūsī, a Shīᶜī would not hesitate to study with persons of other sects. Kamāl al-Dīn b. Yūnus and Quṭb al-Dīn al-Miṣrī, for example, were both Sunnīs of the Shāfiᶜī school of law. As we shall see in the concluding section below, this broad-minded attitude is an important part of Ṭūsī's mature philosophy.

3. The Ismāᶜīlī Stage

His studies completed, the young Naṣīr al-Dīn faced the unenviable task of finding an appropriate position. We know that sometime around 632/1235 Naṣīr al-Dīn found a patron in ᶜAbd al-Raḥīm b. abī Manṣūr Nāṣir al-Dīn Muḥtasham (died 655/1257), the last Ismāᶜīlī governor of Qūhistān, who ruled from the town of Qā'in during the reigns of the seventh and eighth Grand Masters of Alamūt, ᶜAlā' al-Dīn Muḥammad (ruled 618–53/1221–55) and his son Rukn

---

30 Makdisi in various places (*e.g.* [1981]), pp. 77–80) has emphasized that the foreign sciences were formally excluded from the *madrasa*; as he puts it, "neither the madrasa nor its cognate institutions harbored any but the religious sciences and their ancillary subjects." But even a superficial perusal of the situation, including Makdisi's own examples of *actual* curricula (pp. 81–91), is enough to convince one that the individualistic nature of the teaching in the *madrasa* allowed the "forbidden" subjects to be brought in, especially in the later Middle Ages. Makdisi's discussion takes this into account but without acknowledging the puzzle that he has foisted on his reader. For a rather more balanced approach and a critique of Makdisi, see Sabra [1987a], pp. 232–235.

31 For example, ᶜAlam al-Dīn Qayṣir b. abī al-Qāsim ᶜAbd al-Ghaniy Taᶜāsīf traveled specifically to study music with Kamāl al-Dīn, who was happy to be able to teach this rarely requested subject.

32 Sayili [1960], pp. 207–211.

al-Dīn Khurshāh (ruled 653–54/1255–56).[1] The Nizārī Ismāᶜīlīs, whose allegiance was to Nizār, the defeated son of the Fāṭimid[2] Imām al-Mustanṣir (died 487/1094), had in the late eleventh and early twelfth centuries established themselves in Syria and Iran under their redoubtable leader Ḥasan-i Ṣabbāḥ.[3] Their headquarters were located at the fortress of Alamūt in the Alburz mountains of Daylamān from which they ruled, in theory, the other Ismāᶜīlī strongholds in Iran and Syria, some of which were located in Qūhistān in eastern Iran.

What led Ṭūsī to enter the service of the Ismāᶜīlīs, a group greatly despised by the Sunnīs and more often than not shunned by the other Shīᶜa? Naṣīr al-Dīn gives us *one* answer in his *Risālah-i sayr wa-sulūk* (Epistle on the Journey and Conduct). Naṣīr al-Dīn in no uncertain terms tells us that he had become disillusioned with the exoteric Islam of his childhood and had "converted" to esoteric Ismāᶜīlism because he felt that one needed the divinely inspired Ismāᶜīlī Imām to make the Truth known.[4] But after the fall of the Ismāᶜīlī strongholds to the Mongols in 1256, Ṭūsī had a rather different story to tell. In the introduction to the *Īlkhānī Zīj*, for example, he states:

> At the time that [Hülegü] seized the dominions of the heretics [*i.e.* the Ismāᶜīlīs], I Naṣīr al-Dīn who am of Ṭūs and had fallen into the power of the heretics—me he brought forth from that place and ordered to observe the stars.[5]

Furthermore in the *Akhlāq-i Nāṣirī*, he tried to justify his earlier praise of the "heretics":

> ...to save both himself and his honour, he [*i.e.* Naṣīr al-Dīn speaking in the third person] completed the composition of an exordium in a style appropriate to the custom of that community for the eulogy and adulation of their lords and great ones. This is in accordance with the sense of the verse:
>
> And humour them while you remain in their house;
> And placate them while you are in their land.

---

[1] 632/1235 is the date for the completion of the *Risālah-i Muᶜīniyya*, which Ṭūsī dedicated to Muᶜīn al-Dīn, the son of Nāṣir al-Dīn Muḥtasham. The *Akhlāq-i Nāṣirī*, which he dedicated to Nāṣir al-Dīn himself, was probably completed about 633/1235–36 (see *Nasirean Ethics*, pp. 9, 178 and Riḍawī, *Aḥwāl*, p. 454).

[2] The Fāṭimids were an Ismāᶜīlī dynasty that ruled in parts of North Africa (and eventually Egypt) from 297/909 until 567/1171.

[3] The fundamental work on the Nizārī Ismāᶜīlīs is Marshall Hodgson [1955], *The Order of Assassins: The Struggle of the Early Nizārī Ismāᶜīlīs against the Islamic World.*

[4] See pp. 15–17 for a discussion of this "conversion."

[5] Translated by Arberry [1958], pp. 259–260.

> And also the well-attested tradition: "With whatsoever a man protects himself and his honour, it shall be recorded to him as a favour." While such a course is contrary to the belief, and divergent from the path, of the People of the Sharīᶜa and the Sunna, there was nothing else I could do.[6]

Lending some credence to Ṭūsī's obviously self-serving recantation is the historical reality that he, as many Persians, had to face during the period of the Mongol invasions. Naṣīr al-Dīn, after his sojourn in Iraq, would have needed a secure refuge if he wished to return to his homeland; but that refuge could hardly be Ṭūs or any other town in Khurāsān.[7] In Rabīᶜ I 617/May 1220 the initial contingent of the Mongol forces arrived at Nīsābūr; by the Spring of 618/1221, Ṭūs and its environs had been subjected to wholesale slaughter and destruction while in Nīsābūr itself virtually all the inhabitants who had remained in the city were killed.[8] The historian Juwaynī, who lived through the ensuing chaos, tells us that even by 658/1259–60, Khurāsān was still in ruin from years of turmoil.[9] A graphic example of this was Ṭūs where in 637/1239 only fifty houses remained in the once thriving city.[10] Naṣīr al-Dīn himself was, no doubt, away from Khurāsān during the years 617–18/1220–21; we have seen that he was studying in 619/1222 with Muᶜīn al-Dīn b. Badrān, probably in Baghdad.[11] But wherever he was at the time, returning to or remaining in Khurāsān was out of the question. One of his few options was to seek refuge with the Ismāᶜīlīs, who seemed to be the only group in Iran that could offer any security.[12] Ṭūsī himself confirms this in the revised introduction to his *Akhlāq-i Nāṣirī*:

> ...he [*i.e.* Naṣīr al-Dīn] had been compelled to leave his native land on account of the turmoil of the age, the hand of destiny having shackled him to residence in the territory of Qūhistān.[13]

But whether he felt shackled or not, Ṭūsī did spend many secure and productive years at the court of Nāṣir al-Dīn Muḥtasham. But sometime before 654/1256, Ṭūsī moved, or was relocated, from Qūhistān and ended up in Alamūt, which was the Ismāᶜīlī capital. The reason for this move is not entirely

---

[6] Ṭūsī, *Nāṣirean Ethics*, p. 24.

[7] This is obliquely, but clearly, referred to in Ṭūsī's *Sayr*, p. 41.

[8] Juwaynī, *History*, 1: 145–146, 171–172, 176–178.

[9] Ibid., 1: 96–97.

[10] Ibid., 1: 500–501.

[11] See pp. 6–7 and footnotes 14–15. The date 619/1222 is coincidentally that given by Riḍawī for the beginning of Nāṣir al-Dīn Muḥtasham's rule in Qūhistān (*Aḥwāl*, p. 9).

[12] Shūshtarī, *Majālis*, 2: 203; cf. Riḍawī, *Aḥwāl*, pp. 9–10. Shūshtarī seems to imply that Ṭūsī may not have been overly enthusiastic about going to Qūhistān since he had to be induced by an enticing offer (*laṭā'if al-ḥiyal*).

[13] Ṭūsī, *Nāṣirean Ethics*, p. 124.

clear and to a certain extent, its interpretation depends on how one is to understand Ṭūsī's relationship with his hosts. There is evidence that during this time Ṭūsī was not entirely content to remain with the Ismā^{c}īlīs; according to one account he communicated with Mu'ayyad al-Dīn Muḥammad ibn al-^{c}Alqamī, the Shī^{c}ite Imāmī (*i.e.* non-Ismā^{c}īlī) wazīr to al-Musta^{c}ṣim, the last ^{c}Abbāsid caliph (ruled 640–56/1242–58), apparently in hopes that his fellow Twelver would rescue him from Qūhistān with an invitation to the court in Baghdad.[14] But according to this version, Nāṣir al-Dīn Muḥtasham discovered his plans and placed him under arrest.[15] During his next trip to Alamūt, he deposited Ṭūsī with the Grand Master ^{c}Alā' al-Dīn Muḥammad for safekeeping.

Whether these latter events actually occurred in the way described or whether this account was a precautionary fiction invented by Ṭūsī or his friends, we do know that Naṣīr al-Dīn was in Alamūt at its fall to the Mongols in 654/1256. We do not know, however, when he arrived there from Qūhistān. A reasonable possibility is between 642/1244–45, the year of the assumption of the wazīrship by Ibn al-^{c}Alqamī and the date of composition of the *Asās al-iqtibās*, Ṭūsī's Persian work on logic written while still in the service of Nāṣir al-Dīn, and Ṣafar 644/June–July 1246, the date on which the *Ḥall mushkilāt "al-Ishārāt,"* a commentary on Ibn Sīnā's well-known *Al-Ishārāt wa-'l-tanbīhāt*, was completed. In this latter work, dedicated to Shihāb al-Dīn Muḥtasham who was probably in Alamūt at the time (thus providing us with some indication of Naṣīr al-Dīn's whereabouts), Ṭūsī complains bitterly about "the difficult circumstance" (*ḥāl ṣa^{c}b*) and the "disturbed state of mind" (*kudūrat bāl*) under which the work was written, which seems oddly out of character with the convert's fulsome prose in the *Sayr*. Since the latter can be dated between 644 and 654,[16] and it was almost certainly written while he was still in Qūhistān, 644/1246 becomes a likely candidate for a rather significant crisis that would eventually lead to his relocation to Alamūt.

However one takes his Ismā^{c}īlī stage, one must admit that some of his most creative work dates from this period. In addition to the influential work on ethics, the *Akhlāq-i Nāṣirī*, Ṭūsī presented the first of his new lunar and planetary models in his *Ḥall-i Mushkilāt-i Mu^{c}īniyya*, a work dedicated to Mu^{c}īn al-Dīn Shams, the son of Nāṣir al-Dīn Muḥtasham. These models were to appear many years later in the *Tadhkira*.[17] And if one accepts our assumption

---

[14] Thus the account in Shūshtarī, *Majālis*, 2: 203 and Khwāndmīr, *Ḥabīb al-siyar*, 3: 105–106.

[15] In the accounts of Khwāndmīr and of Waṣṣāf al-Ḥaḍra in *Tazjiyat al-amṣār*, Nāṣir al-Dīn Muḥtasham discovers this by way of Ibn al-^{c}Alqamī himself; see Riḍawī, *Aḥwāl*, pp. 11–12. But as Riḍawī points out, this is unlikely in view of the good relations between Ṭūsī and Ibn al-^{c}Alqamī in the aftermath of the fall of Baghdad in 656/1258 (p. 143).

[16] See Riḍawī, introduction to the *Sayr*, page *w*.

[17] See pp. 65–70 of this volume for details of the relationship between the *Tadhkira* and its earlier Persian version.

that Ṭūsī was relocated, for whatever reason, to Alamūt in or about 644/1246, then one may conclude that his scholarly activity continued unabated—indeed flourished—during this time when one may assume that he had access to the famous Alamūt library. One would then be able to date many of his recensions of Greek and early Islamic works—and perhaps even his decision to undertake such a formidable task—to the period of his Alamūt residence. Interestingly enough, he began with the most difficult first, completing his edition of the *Almagest* on 5 Shawwāl 644/13 February 1247 and of the *Elements* on 22 Shaᶜbān 646/10 December 1248. He continued to work on this project until it was concluded in Marāgha under the Mongols in Shaᶜbān 663/June 1265 with his edition of the *Spherics* of Menelaus.[18]

Naṣīr al-Dīn's relationship with the Ismāᶜīlīs, whatever it was, ended on the first of Dhū al-Qaᶜda 654/19 November 1256 with the fall of Alamūt to the Mongols. It is significant that he himself played some role in the final negotiations and in fact accompanied Rukn al-Dīn Khurshāh, who had replaced his assassinated father a year earlier, down the steep path from the mountaintop to Hülegü.[19] He may have even been instrumental in persuading Khurshāh to give up what by then must have appeared a hopeless cause.[20] This, of course, tends to put a somewhat different light on Ṭūsī's true position at Alamūt. On the other hand, it would not be unreasonable to assume that Ṭūsī's fidelity to Ismāᶜīlī dogma must have seemed far less important than the pressing need to use his skills during the sensitive negotiations.

4. The Mongol Period

Whatever the true explanation, Naṣīr al-Dīn was immediately enlisted into the service of Hülegü; he even accompanied him during the campaign against Baghdad and provided an eyewitness account of the end of the ᶜAbbāsid Caliphate.[1] When asked by Hülegü whether the attack on Baghdad would lead to a series of cataclysms as predicted by one of the Muslim astrologers in the Mongol entourage, Ṭūsī, "frightened and thinking he was being tested," answered in the negative.[2]

---

[18] For a convenient listing of Ṭūsī's recensions, see Krause [1936], pp. 499–505. Most of these, the so-called "Middle Books," which were to be studied between the *Elements* and the *Almagest*, have been published in Ṭūsī, *Majmūᶜ al-rasā'il* (2 vols.). Ṭūsī's *Elements* and *Almagest*, however, have yet to be edited, though the former has been printed several times. A work purporting to be his *Taḥrīr* of the *Elements* was published in 1594 in Rome, but Sabra [1969] has noted that this is a false attribution (p. 18).

[19] Rashīd al-Dīn, *Jāmiᶜ al-tawārīkh*, pp. 210–213.

[20] Hodgson [1955], pp. 267–268.

[1] See Boyle [1961], pp. 145–161; this contains a translation of the appendix of Naṣīr al-Dīn to the *Ta'rīkh-i jahān-gushā* of Juwaynī. See also Wickens [1962], pp. 23–35.

[2] Rashīd al-Dīn, *Jāmiᶜ al-tawārīkh*, pp. 260–263.

Naṣīr al-Dīn thereafter became a trusted advisor to Hülegü. Ṣafadī/Kutubī report that Hülegü was dependent on him to such an extent that he would not ride or travel without his sanction. Naṣīr al-Dīn was also made a *wazīr* and put in charge of *waqf* (religious endowments).[3]

But for our purposes, the major event that resulted from the relationship between Naṣīr al-Dīn and Hülegü was the building of the Marāgha observatory.[4] Apparently both Hülegü and his brother Möngke, who ruled the vast Mongol domains from China, were patrons of the sciences. The latter wished to construct an observatory in Beijing, but this was not accomplished in his lifetime. However an observatory was constructed—whether at the instigation of Hülegü or Naṣīr al-Dīn is not clear—in Ādharbayjān at Marāgha under the direction of Ṭūsī. It was begun shortly after the fall of Baghdad in Jumāda I 657/April–May 1259.

The significance of the Marāgha observatory has been well documented by Sayili. Naṣīr al-Dīn invited the most renowned scientists of the time, among whom we should mention Mu'ayyad al-Dīn al-ᶜUrḍī and Muḥyī al-Dīn al-Maghribī, to participate in the construction of the instruments as well as the actual observations. A fairly substantial number of students was also in attendance who benefited from an impressive library.[5] The observatory also had an international character due to the several Chinese astronomers on its staff. To pay for such a massive undertaking, Naṣīr al-Dīn, in his capacity as administrator of *waqf* funds, used the proceeds from these religious endowments to finance the observatory. This was a striking and significant departure from past practices; it would seem to indicate, as we have mentioned previously, a breaking down of the barrier between the Islamic and the pre-Islamic sciences. The use of *waqf* funds for the Marāgha observatory would also explain its relatively long life of some fifty years.

It is unfortunate that we know so little about the astronomical significance of the Marāgha observatory. The *Īlkhānī Zīj* itself contains surprisingly little that is new, but this seems to be more a function of the intended audience than of what actually occurred in Marāgha. It hardly seems likely that this work, in which Ṭūsī felt the necessity to inform his Mongol patrons that Muḥammad was from Mecca, was intended for the Muslim intellectual community.[6] Fortunately, other

---

[3] Ṣafadī, *Wāfī*, 1: 182; Kutubī, *Fawāt*, 3: 250.

[4] In what follows I merely summarize the account in Sayili [1960], pp. 187–223, whose full and excellent treatment need not be repeated here.

[5] The report by Ṣafadī/Kutubī of a library of 400,000 volumes seems to be an exaggeration. Recent excavations at the sight reveal a library structure that could not hold anything near that number [Vardjavand [1366 H. Sh.], p. 12 (French section), pp. 233–234 (Persian section)].

[6] Boyle [1963], pp. 251–252.

zījes that were based on the Marāgha observations exist and await scholarly examination.[7]

In 672/1274, Naṣīr al-Dīn left Marāgha with a group of his students for Baghdad. The reasons for this trip are unknown but it would seem that he intended to remain for awhile since his students had decided to accompany him. In the same year, Naṣīr al-Dīn died and was given a memorable funeral attended by a number of leading dignitaries.[8] It was during this brief final sojourn in Baghdad that Ṭūsī made his final revision of the *Tadhkira*, the version that is the basis for the edition contained herein.

5. Ṭūsī's Astronomy and Its Relation to His Hellenism and Religious Beliefs

Because Ṭūsī's religious beliefs ostensibly changed several times during his lifetime, it is well worth asking whether these changes had any effect upon his scientific, and in particular his astronomical, work. In other words, were Ṭūsī's Hellenistic cosmology and astronomy independent of his apparently distinct religious commitments that entailed, by his own account, different epistemological positions? An answer to this sort of question depends upon a much more thorough analysis of Ṭūsī's writings than is possible here; I would, though, like to offer a preliminary response since I believe that there are larger lessons to be learned concerning the attempt by Islamic astronomers to immunize the mathematical core of their discipline from the charge that it was inextricably tied to a suspect Aristotelian metaphysics.

Before discussing the implications of these "conversions" for his astronomy, we should briefly review the nature and justifications of his religious beliefs during the different stages of his life. For Ṭūsī's conversion to Ismāᶜīlism, we are in the fortunate position of having his spiritual autobiography written when he was in the service of Nāṣir al-Dīn Muḥtasham, the Ismāᶜīlī governor of Qūhistān. This work, which we have had occasion to mention previously, his *Risālah-i sayr wa-sulūk* (Epistle on the Journey and Conduct),[1] is a kind of intellectual and spiritual autobiography with a liberal dose of Ismāᶜīlī apologetics thrown in. In the introduction, the author states his name to be Muḥammad al-Ṭūsī, and several crucial details agree with what we know independently of his life; there

---

[7] One obvious example is Shams al-Munajjim al-Wābiknawī's *Al-Zīj al-Muḥaqqaq al-Sulṭānī ᶜalā uṣūl al-raṣad al-Īlkhānī* (the verified Sulṭānī *zīj* based upon the Īlkhānī observations); see Kennedy, *Survey*, p. 130 (no. 35). Cf. Sayili [1960], pp. 214–215. For recent discussions of this problem with particular reference to the work of al-Maghribī, one of the principals at Marāgha, see Saliba [1983], [1985], and [1986].

[8] Ṣafadī, *Wāfī*, 1: 183; Kutubī, *Fawāt*, 3: 252.

[1] Even the title of this controversial work is in dispute; Riḍawī states in his introduction to it (p. *h*) that it was not the title given it by the author but was made up at the time it was first printed.

can be little question of attributing the book to our Ṭūsī.[2] In the main body of the work itself, Ṭūsī tells of his growing disenchantment with exoteric theology (*kalām*), which he came increasingly to see as little more than apologetics; this transformation he seems to attribute in part to the subtle influences of his father, his uncle and his mathematics teacher Kamāl al-Dīn Muḥammad al-Ḥāsib.[3] He then became attracted to philosophy (*ḥikma*), which he felt gave an opportunity to the intellect for finding truth; but Ṭūsī came to believe that intellect was deficient in apprehending the "Giver of intellect" and thus philosophy could not give him answers to the ultimate questions. But philosophy, he tells us, did teach him the important lesson about potentiality and actuality. Taking the example of a body that is potentially in motion, he argues that it could never become actually in motion unless there is something else influencing it. Likewise a potential knower needed someone else, *i.e.* a teacher, to actualize that knowledge.[4] Eventually he chanced upon the *Fuṣūl-i muqaddas* (Sacred Articles) of the Ismā^c^īlī Lord of Alamūt Ḥasan ^c^alā dhikrihi al-salām (reigned 1162–66), and he at last found the religious teacher he was seeking, namely the divinely chosen Ismā^c^īlī Imām. He thus decided to join the court of Nāṣir al-Dīn Muḥtasham.[5]

After the fall of Alamūt, Ṭūsī, as we have seen, felt it necessary to repudiate his previous service to the "heretics." But Ṭūsī did more than protest his "innocence"; he also sought to disestablish his Ismā^c^īlism by returning to an essentially philosophical worldview. During this post-Ismā^c^īlī period he wrote a devastating attack upon Tāj al-Dīn al-Shahrastānī's refutation of Ibn Sīnā's philosophical position.[6] We should recall that it was this Tāj al-Dīn who had Ismā^c^īlī sympathies and had taught Naṣīr al-Dīn's great-uncle (also a Naṣīr al-Dīn). Ṭūsī in this work came down strongly in favor of (Hellenistic) philosophy, especially the Ibn Sīnā variety, and returned to his earlier notion that special pleading for a particular religious cause (and in Tāj al-Dīn's case this was Ismā^c^īlism) was incompatible with the search for truth. As he would state in his new introduction to the *Ethics*, philosophy (*ḥikmat*) "bears no relation to the agreement or disagreement of school or sect or denomination."[7]

---

[2] Although I at one time ([1982], 1: 12–17) agreed with Riḍawī that Ṭūsī was never an Ismā^c^īlī and thus was somewhat sceptical of the attribution of this work to Ṭūsī, and in any event very dubious about Ṭūsī's sincerity if he were indeed the author (cf. *Majmū^c^a*, pages *h-w* and Riḍawī, p. 591), I have now come to accept at least his authorship in large part because of the large number of statements about his life in the work that can be independently verified. Note that Poonawala [1977], p. 261 and Madelung [1985], pp. 88–89 accept the authenticity of the work and both see it as representing a sincerely held stage in his intellectual development.

[3] *Sayr*, p. 38.

[4] *Sayr*, p. 40.

[5] *Sayr*, p. 41.

[6] Tāj al-Dīn's book was called *Kitāb al-Muṣāra^c^a* (The Wrestling Match), while Ṭūsī's work was called *Maṣāri^c^ al-muṣāri^c^* (The Downfalls of the Wrestler). For a cogent discussion of these works see Madelung [1985], pp. 87–88 and Madelung [1976].

[7] Ṭūsī, *Nāṣirean Ethics*, p. 24; cf. Madelung [1985], pp. 87, 100.

One could engage in a long, but inevitably fruitless, debate about which Naṣīr al-Dīn was the true one. But leaving aside the question of the sincerity—or expediency—of his conversions, we should ask whether his association with the Ismāᶜīlīs and their doctrines had any significance for his scientific work. Was there any discernible change in it due to his new found faith? And more to the point, did the submission to a religious authority parallel any similar submission to scientific authority?[8] The answer must be negative for both; although Ṭūsī was not a scientific revolutionary—he remained committed throughout his career to a modified Ptolemaic worldview, for example—he was not hesitant to criticize and modify both his Greek and Islamic predecessors. And this attitude is clearly detectable during his Ismāᶜīlī period. As we shall see,[9] several of the original versions of the non-Ptolemaic models of the *Tadhkira* were first presented in his *Ḥall-i mushkilāt-i Muᶜīniyya*, which was written for Muᶜīn al-Dīn, the son of his patron Nāṣir al-Dīn Muḥtasham. And despite the ups and downs of a tumultuous life, Ṭūsī's interest in astronomy and the content of his work in it remained remarkably stable. But then we are faced with another question. Can we separate Ṭūsī's scientific work from his philosophical and religious views? Here I would go beyond a mere positive response to anticipate my argument in the following chapter on *hay'a*, namely that Islamic astronomers *themselves* were seeking—sometimes consciously, sometimes not—to free their physical theories from a metaphysical taint.[10] In short, I would argue that whatever the effects of Ṭūsī's conversion on his philosophical worldview, and thus on his "Hellenism" (taken here to mean a commitment to rational discourse in all matters), I believe that we may safely conclude that the effects on his astronomy were nonexistent.[11]

How influential was Ṭūsī's attitude? Was his attempt to place philosophy, and by implication the mathematical sciences, above religious disputes successful? As we shall see later, his position was quite influential in the East and we

---

[8] Cf. Sabra's [1984] attempt to draw a connection between Averroes's glorification of Aristotle and the theological literalism of Ibn Ḥazm (d. 1064). As he puts it, "it is difficult not to regard this attitude of Averroes's and the commentatorial style he adopted in most of his philosophical writings as a literalism in philosophy that paralleled the theological literalism of Ibn Ḥazm" (p. 144). Though Sabra does not dwell on the problem, he opens up the possibility that Averroes's change in attitude toward Ptolemaic astronomy—from embracing it in his earlier epitome (*talkhīṣ*) of Aristotle's *Metaphysics* to rejecting its fundamental basis in his large commentary (*tafsīr*) of the same work—had something to do with a parallel change in religious philosophy. Averroes's exaltation of Aristotle represents a rather extreme position in Islamic intellectual history. A much more sceptical approach toward authority is manifested in Ibn al-Haytham's *Shukūk*, pp. 3–4. (For a translation of this latter passage, see Sabra [1989], 2: 3.)

[9] See pp. 66, 69–70 of this volume.

[10] See pp. 38–41, 45–46.

[11] On the other hand there could be possible consequences for such things as his ethics; cf. Madelung [1985].

even have a very spirited defense of astronomy by the astronomer/theologian ᶜAlī al-Qūshjī (d. 879/1474) that explicitly attempts to raise it above religious objections.[12] But this philosophical position should be seen within the context of a broader intellectual trend in Eastern Islam, especially Iran, one that was not always well received further west in Syria and Egypt. To understand this we must examine how Ṭūsī's intellectual positions came to be played out on the political stage.

First one must acknowledge Ṭūsī's political genius,[13] which is nowhere more in evidence than in his ability not only to convince the Mongols, certainly not a gullible lot, that he had been an unwilling resident of Alamūt but more importantly in his skill in protecting the interests of the Shīᶜites in particular, and the Muslim community in general, from the caprice of the Mongol rulers. As we have seen, part of this involved writing new introductions to a number of works in which he had lavished praise on his former patrons,[14] and one could well conclude that this was the worst type of opportunism. One might even go so far as to claim that this is just an example of Ismāᶜīlī *taqiyya* (dissimulation); indeed, the author of *Rawḍat al-Taslīm*, which has been attributed to Ṭūsī, tells his audience that he must write "in a hurry, secretly, in a dark corner" during a period of *satr* (concealment) ordered by the Imām.[15] But this seems hardly in character with the reports in Ṣafadī/Kutubī that during his years of service under Hülegü, Muslims in general and, as one might expect, the Shīᶜa in particular benefited from him.[16] One of the Sunnīs he aided was none other than the historian Juwaynī, whom Naṣīr al-Dīn saved through elaborate astrological trickery from a death sentence imposed by Hülegü.[17] It is not beside the point to note here that it was this Juwaynī who, as he himself tells us, asked permission from Hülegü after the taking of Alamūt

---

[12] See Ragep [1982], 1: 141–145 for a translation of this passage as quoted by Tahānawī, *Dictionary*, 1: 48–49.

[13] Some would use a different characterization; see, for example, Reuben Levy [1923], p. 64: "The verdict of history on Naṣíru 'l Dín is a most unfavourable one. It might have been expected that the conduct of a man of his undoubted mental qualities would have been regulated by some standard higher than that of personal advantage. Yet he appears not only to have betrayed his Ismáᶜílí master to Húlágú, but to have been instrumental in bringing the last Caliph treacherously to his death at the hands of the Mongols."

[14] This included his ethical work, the *Akhlāq-i Nāṣirī*, and two astronomical works, the *Risālah-i Muᶜīniyya* and the *Ḥall-i mushkilāt-i Muᶜīniyya*.

[15] *Taṣawwurāt*, pp. 134, 126. Though Madelung [1985] and Poonawala [1977], p. 262 seem to accept that Ṭūsī is the author, there is, as far as I know, little substantial evidence. Even Ivanow, who more or less convinced himself that Ṭūsī was a lifelong Ismāᶜīlī, grudgingly acknowledged the insufficiency of the claims on Ṭūsī's authorship; nevertheless, he wished to "conventionally accept Ṭūsī's authorship...till...an incidental find may answer our quest decisively" (*Taṣawwurāt*, p. xxvi).

[16] Ṣafadī, *Wāfī*, 1: 182; Kutubī, *Fawāt*, 3: 250.

[17] See Ṣafadī, *Wāfī*, 1: 179–180. Ṣafadī is poking fun at the gullibility of the foreign conquerors, but it should be remembered that he is living safely in the following century.

> to examine the library, from which I extracted whatever I found in the way of copies of the Koran and [other] choice books...As for the remaining books, which related to their heresy and error and were neither founded on tradition nor supported by reason, *I burnt them all* [emphasis added].[18]

It is difficult to believe that Ṭūsī's *taqiyya* would extend to saving the life of someone who had almost single-handedly destroyed the literary heritage of his religious faction.

The reason this is an important point to keep in mind is that Ṭūsī's subsequent reputation in the East was very high among both Sunnīs and Shīᶜīs, a state of affairs that would be very hard to imagine if he had not acted to protect all Muslim interests during the Mongol hegemony. Indeed one need only refer to his theological work, the *Tajrīd al-ᶜaqā'id*, to note that it was commented on and used by both Sunnīs and Shīᶜīs. As with astronomy so also with *kalām* (dialectical theology) itself, Ṭūsī would seem to have been at least partially successful in transcending the religious disputes that had disturbed him so much as a youth.

But Ṭūsī did not fare so well further West in the Mamlūk areas of Syria and Egypt. There he was mercilessly denounced by such Ḥanbalī luminaries as Ibn Kathīr (ca. 700–74/1300–73)[19] and Ibn Qayyim al-Jawziyya (691–751/1292–1350). One may quote the latter to get the flavor of these invectives:

> We now reach the time of the defender (*naṣīr*) of idolatry, unbelief and heresy, the minister to the heretics, al-Naṣīr al-Ṭūsī, the minister to Hülegü. He relieved himself from following the Prophet and the members of his religion, putting them to the sword so as to satisfy his fellow heretics and save himself. He killed the Caliph, judges, jurists, and traditionists; he protected the philosophers, astrologers, natural philosophers, and magicians. He transferred the endowments of the religious schools, the mosques, and the hospices attached to them, making them his personal property...He also established schools for the heretics. He hoped to substitute the *Ishārāt* of the leader of the heretics Ibn Sīnā in place of the *Qur'ān*, but he could not accomplish that! He said: "The former is the *'Qur'ān'* of the elect while the latter is for the masses."...Finally he taught magic, for he was a sorcerer who worshipped idols.[20]

---

18 Juwaynī, *History*, 2: 719.

19 See his *Al-Bidāya wa-'l-nihāya*, 13: 267–268.

20 Ibn Qayyim, *Ighāthat al-lahfān min maṣāyid al-shayṭān*, 2: 267; this is quoted by Ziriklī, *Aᶜlām*, 7: 258. Later authors of the same persuasion as Ibn Qayyim have also used this passage in their works; see, *e.g.* Ibn al-ᶜImād al-Ḥanbalī, *Shadharāt al-dhahab*, 5: 339–340.

In wading through the hyperbole of this passage, one should keep in mind the real issue involved. Ṭūsī, insofar as he was a follower of Ibn Sīnā, did place philosophy beyond the pale of religious disputation or, as we have seen him put it in the *Ethics*, it "bears no relation to the agreement or disagreement of school or sect or denomination."[21] This was the crux of the battle between the philosophers and the traditionists of Islam, and Ṭūsī's power and influence were rightly seen by the latter as a direct threat.[22] They also rightly saw that Ṭūsī was seeking to carry on the work, if not the details, of Fakhr al-Dīn al-Rāzī in "naturalizing" the Hellenistic heritage in Islam.[23]

In this regard one might put into perspective the work on religious cosmology by the Egyptian al-Suyūṭī (849–911/1445–1505). As A. Heinen has noted, he hoped it would be *the* Islamic cosmology and he even went so far as to appropriate the word *hay'a*, the term for mathematical cosmology, into his title.[24] But in the East, Ṭūsī's legacy would have made the possibility of a religious cosmology *in competition* with (and potentially supplanting) a mathematical one far less likely. For Ṭūsī and his successors, science and religion could coexist within the intellectual scheme he was so instrumental in establishing.[25]

B. Works

A comprehensive evaluation of Ṭūsī's enormous corpus would be premature. Sheer size is only the beginning of the problem;[1] Naṣīr al-Dīn saw fit to try his hand at virtually all fields of both ancient and Islamic learning. Since neither my training, taste nor ability can match Ṭūsī's, my rather modest aim is simply to

---

[21] See p. 16, footnote 7.
[22] Cf. Madelung [1985], pp. 100–101.
[23] Cf. Sabra [1987a] and Peters [1968], pp. 193–202.
[24] Heinen [1982], p. vii.
[25] Here I would disagree strongly with Sabra's [1987a] implication that this coexistence in the later centuries of medieval Islam meant that science, and astronomy in particular, was afflicted with a kind of instrumentalism in the service of religion that "confines scientific research to very narrow, and essentially unprogressive areas" that did not generally include observational and theoretical astronomy (pp. 241–242). These "narrow, unprogressive" areas did include spherical astronomy, which was useful for such religious purposes as finding the *qibla* (direction of Mecca). While this might have been true in Mamlūk Egypt and Syria (cf. King [1983], pp. 537–539), where Ṭūsī was reviled, it was most emphatically not true in Persia and Central Asia where theoretical and observational astronomy flourished for many centuries after the Marāgha episode. My hope is that the following pages will be a modest contribution toward proving this contention.
[1] Riḍawī lists, apart from his poetry, some 190 works. But as a number of these are clearly misattributions and duplicates, Ivanow's estimate of something over 150 seems closer to the mark (*Taṣawwurāt*, p. xxvi).

give some sense of his writings and to provide a context within which to place his writings on *hay'a*.

Ṭūsī's work has often been characterized as being that of "revival" rather than of "origination."[2] But this seriously misrepresents his contribution to astronomy and, perhaps, to other fields as well. There is a kernel of truth to this "revivalist" tag, however. His role in post-13th century A.D. Islamic intellectual history cannot easily be underestimated when we consider that many of his works became the standard in a variety of disciplines up until modern times. His recensions of Euclid's *Elements*, Ptolemy's *Almagest* and the "Middle Books" of mathematics and astronomy virtually replaced earlier editions. His works in logic and *kalām* were also widely read and commented on, and it is hard to imagine Ibn Sīnā's *Al-Ishārāt wa-'l-tanbīhāt* being studied without Naṣīr al-Dīn's valuable commentary. His work on ethics, the *Akhlāq-i Nāṣirī*, was also of considerable importance, especially in the eastern domain of Islam. We should in addition mention here that among his works are books on *fiqh* and *ṣūfism*. In all these examples we have, for the most part, a consolidation of the endeavors of earlier generations of Islamic scholars. This occurred at an important point in Islamic history as there had been an unmistakable, but not easily understood, lull in intellectual activity in the century or so previous to that of Ṭūsī. Historically, then, Naṣīr al-Dīn easily falls into the role of consolidator rather than innovator.

But I would hesitate to leave the matter here. As is clear from astronomy, most of Ṭūsī's work can be viewed as simply a well-integrated account of what others had previously accomplished. Nevertheless his own contributions are not insubstantial and, when seen in the perspective of future developments, take on added importance.

I would like to venture a number of preliminary, working hypotheses about Ṭūsī's writings that, needless to say, may need to be revised as more of them are examined and edited. First, many of Ṭūsī's most creative and original work was composed while in the service of Nāṣir al-Dīn Muḥtasham in Qūhistān (ca. 630/1232 or 1233–ca. 643/1245 or 1246). Among these are the *Akhlāq-i Nāṣirī*, the *Risālah-i Mu^{c}īniyya* and its sequel the *Ḥall-i mushkilāt-i Mu^{c}īniyya* in which one finds Naṣīr al-Dīn's new models for the moon and planets, and the logical work *Asās al-iqtibās*. Furthermore, he worked on the commentary to Ibn Sīnā's *Al-Ishārāt wa-'l-tanbīhāt* during these years. As one might expect, all these works, with the notable exception of the latter, were written in Persian. Second, after going to Alamūt (ca. 643 or 644/1245, 1246 or 1247), Ṭūsī seems to have devoted himself in large measure to working on his recensions of Greek and early Islamic scientific work. Many of the editions of the so-called "Middle Books," in addition to his editions of the *Almagest* and *Elements*, date from this period. The next, and final, period of his life (654–72/1256–74), which coincides with his service under the Mongols, seems to have been a time of consolidating earlier gains rather than of making new departures. An important

---

[2] See, for example, Siddiqi [1963], pp. 566, 580.

preoccupation during this period was the Arabizing of earlier Persian works. The *Tadhkira*, completed in 659/1261, is basically an Arabic synthesis of the *Mu^{c}īniyya* and the *Ḥall-i mushkilāt-i Mu^{c}īniyya*, which were composed some 25 years previously. It is likely that we are dealing with a similar situation in the case of the Arabic treatise on logic, *Tajrīd al-manṭiq* (written 656/1258) and the earlier Persian work, the *Asās al-iqtibās* (written 642/1244 or 1245), though we should mention that the Arabic version is considerably abridged. It was also during this latter part of his life that Ṭūsī attempted to diminish the significance of his former relationship with the Ismā^{c}īlīs by claiming that he had only served them out of dire necessity. As we have noted, he even went so far as to rewrite the introductions of several works that had been composed for his Ismā^{c}īlī patrons.[3] A final point we should make is that longer and shorter versions of works on the same subject matter occur among Ṭūsī's writings. Whether the purpose of these different versions is pedagogical or otherwise is not immediately clear. One example we shall note later (pp. 66–67) is the *Zubdat al-idrāk fī al-hay'a*, which is an abridged version of the *Tadhkira*.

***

Inventories of Ṭūsī's works are given by Brockelmann (*GAL*, 1: 508–512 [= 670–676] and S1: 924–933) and Riḍawī (*Aḥwāl*, pp. 333–628), but each should be used with the standard caution. His Persian astronomical works are listed in Storey (*Persian Literature*, II.1, pp. 52–60); Matvievskaya/Rozenfeld catalogue his works in the exact sciences (*Mat. i astr.*, 2: 392–408).

---

[3] See p. 18.

C. Chronology of Ṭūsī's Life*

	597/1201	Naṣīr al-Dīn born in Ṭūs, N.E. Iran
ca.	603–10/1206–13	Early education with father, uncles, and Kamāl al-Dīn al-Ḥāsib
ca.	613/1216	Studies philosophy (*ḥikma*) and medicine in Nīsābūr
	618/1221	Chingiz Khān ravages Khurāsān
	619/1222	Ṭūsī studying with Shī^{c}ite jurist Mu^{c}īn al-Dīn in Iraq
ca.	625/1228	Studies astronomy and mathematics with Kamāl al-Dīn ibn Yūnus in Mosul
ca.	630/1233	Finds patron in Nāṣir al-Dīn Muḥtasham, Ismā^{c}īlī governor of Qūhistān
	632/1235	Dedicates *Risālah-i Mu^{c}īniyya* on astronomy to Mu^{c}īn al-Dīn, son of Nāṣir al-Dīn Muḥtasham
ca.	632–33/1235–36	Writes *Ḥall-i mushkilāt-i Mu^{c}īniyya*; first appearance of rectilinear Ṭūsī couple
ca.	633/1235–36	Dedicates *Akhlāq-i Nāṣirī* (*Nasirean Ethics*) to Nāṣir al-Dīn Muḥtasham
	642/1244–45	Completes *Asās al-iqtibās* on logic while still in Qūhistān
	644/1246	Probable date of completion of autobiography (*Sayr wa-sulūk*) as well as relocation to Ismā^{c}īlī fortress of Alamūt; completes *Ḥall mushkilāt "al-Ishārāt,"* a commentary on Ibn Sīnā's *Al-Ishārāt wa-'l-tanbīhāt*
	644/1247	Beginning of project to edit "Middle Books" of mathematics and astronomy as well as *Almagest* and Euclid's *Elements*; recension of *Almagest* completed
	646/1248	Recension of *Elements*
	654/1256	Alamūt falls to Mongols; Ṭūsī begins his service to Hülegü
	656/1258	Baghdad falls to Mongols; end of ^{c}Abbāsid caliphate
	657/1259	Construction of Marāgha observatory begins under Ṭūsī's direction
	659/1261	Marāgha version of *Tadhkira* appears
	663/1265	Recension of Menelaus's *Spherics*; "Middle Books" project completed; Hülegü dies
ca.	670/1271	*Zīj-i Īlkhānī* completed
	672/1274	Baghdad version of *Tadhkira* completed; Ṭūsī dies in Baghdad

* The first number is the *hijra* date; the second is the corresponding Christian date.

§2. The *Tadhkira*

The full title of the work is *Al-Tadhkira fī ᶜilm al-hay'a*, meaning "Memoir on the science of *hay'a*." Although *hay'a* is used to mean astronomy in a general sense, we shall find that it has a more specialized meaning as well that we shall explore to help us situate the *Tadhkira* within both an astronomical tradition and a literary genre. In addition to tradition, we shall also need to discuss its innovations, the rather remarkable history of the text, and its enormous influence.

A. Purpose of the *Tadhkira*

To understand a work, we would be well-advised to begin by asking what is its purpose. For the *Tadhkira*, we are fortunate that we have Ṭūsī's own words to guide us:

> The scientific exposition that we wish to undertake will be a summary account of [astronomy] presented in narrative form. The details are expounded and proofs of the validity of most of them are furnished in the *Almagest*. Indeed, ours would not be a complete science if taken in isolation from the *Almagest* for it is a report of what is established therein.

The idea of a summary seems innocent enough and, up to a point, it is. Anyone who has ventured beyond the introductory sections of the *Almagest* knows that a summary of the contents or a running commentary can be of great help (not to say indispensable). But the *Tadhkira* is neither a commentary nor a straightforward summary of the contents of the *Almagest*. It rather sets forth those contents within a textual structure meant to give a physical accounting of the Universe—in short, a cosmography. Why was it that someone in the 13th c. came to feel that the *Almagest* should be summarized from the point of view of physical bodies, bodies that made a cameo appearance there but can hardly be said to have been a main feature? And furthermore, why did this someone, namely our Naṣīr al-Dīn, adopt the textual structure that he did? To answer these questions, it will be convenient to distinguish the tradition within which the *Tadhkira* falls from the genre of astronomical work that came to be characterized by the *Tadhkira* itself. In what follows, I shall attempt to sketch an answer to these questions using a rather broad stroke indeed.[1]

B. The *Tadhkira*'s Ancient Forebears

To understand the tradition of the *Tadhkira*, we need to isolate three aspects of his work that Ṭūsī feels are crucial. First the notion that astronomy has a strong physical component is put forth explicitly in I.Intr. [2] when he states that "the subject of astronomy is the simple bodies." Second he makes clear how he views the relationship between his work and the *Almagest* in II.5 [10] with the following, revealing passage:

> These then are models and rules that should be known. We have only stated them here; their geometric proofs are given in the *Almagest*. Restricting oneself to circles is sufficient in the entirety of this science for whoever studies the proofs. However, one who attempts to understand the principles of the motions (*mabādi'*) must know the configuration (*hay'a*) of the bodies.

Here we can gain some understanding of what Ṭūsī meant in the introductory passage quoted above in which he indicated that the *Tadhkira* stood in a kind of dependent relationship to the *Almagest* inasmuch as the proofs of the statements being "narrated" in the former could presumably be found in the latter. But again, because Ptolemy in the *Almagest* did not go out of his way to discuss the bodies, Ṭūsī would surely be aware that he was doing something that, though arguably implicit there, was certainly not explicit.[2] The *Almagest* then, despite what Ṭūsī at times seems to be saying, is clearly not the only source for his work. Third Ṭūsī, though he never comes right out and says so, is clearly attempting to give a coherent and unified account of astronomy. Again we can contrast this with the *Almagest*, where Ptolemy treats each planetary system separately and is seemingly unconcerned there with such problems as how all

[1] Unfortunately the book in which "the details are expounded and proofs of the validity of most of them are furnished" has yet to be written.

[2] Several medieval Islamic commentators did not even think it was implicit. Ibn al-Akfānī (d. 1348), for example, claims that "the ancients were always constricting themselves to abstract circles for the configuration of the orbs, until Ibn al-Haytham openly declared their [the orbs'] corporeality...the later [scholars] have followed him with regard to this" [cited and translated by Langermann [1990], p. 13]. Ṭāshkubrīzāde (1495–1561), *Miftāḥ*, 1: 373 repeats this: "The ancients confined themselves (*iqtaṣarū*) in the configuration of the orbs to abstract circles."

the individual parts fit into a complete whole, how one system might affect another, and the actual (as opposed to relative) size and distance of the orbs. On the other hand, Ṭūsī provides just such an accounting.[3]

Can we find in antiquity works that could be considered earlier instances of a tradition of which the *Tadhkira* is a later member? I would venture to answer yes though I am thinking of an earlier tradition that should be considered in terms of providing inspiration rather than of establishing a specific genre. The first example I wish to consider occurs in Book λ, Chapter 8 of Aristotle's *Metaphysics*. In seeking to determine the actual number of eternal movers, Aristotle finds that he needs help from the "mathematicians" (τῶν μαθηματικῶν) so that "our thought may have some definite number to grasp." But though he is dependent on the astronomers, he is clearly not afraid to propose modifications in order to "explain the phenomena." In practice, this means that Aristotle accepts that the Eudoxan-Callippic mathematical theory of the celestial motions accounts for the phenomena of each individual planet. What he is seeking to rectify is the problem that results from Eudoxus having treated the orbs of each planet separately. When one fits the individual sets of homocentric spheres inside one another, the lower spheres will necessarily partake of the motion of the higher orbs thus upsetting the mathematical modeling for each planet. Hence Aristotle proposes "unrolling" spheres whose purpose is to undo the motions (with the exception of the daily motion) for a given planet so that one can in effect start over for the next one.

How should we understand this relationship between Aristotle and the astronomers? The traditional view would have it that Aristotle is acting as a physicist whereas Eudoxus and Callippus are astronomers whose purpose is to use geometrical (rather than physical) hypotheses to "save the phenomena."[4] Recent studies, however, have pointed out that the use of spheres and uniform rotation indicate physical commitments, and it would be a mistake to view Eudoxus and Callippus as exponents of some ancient version of instrumentalism.[5] How then should we view the relationship between Aristotle and Eudoxus? L. Wright has argued we should not regard Aristotle as someone who physicalized purely mathematical models, but rather as someone who took preexisting *physical* models and "showed how this model could be fitted into a

[3] For the individual parts, see II.2 (overview) and II.6–II.10 (details for each planet). For the dynamics of celestial motion, see II.4 [6–7]. And for the problem of planetary sizes and distances, Ṭūsī devotes the entirety of Book IV.

[4] The classic statement of this is by Duhem [1908], pp. 3–5 (Engl. trans. pp. 5–7). Heath [1913] states that "we have seen that that system [of Eudoxus and Callippus] was purely geometrical and theoretical; there was nothing mechanical about it...Aristotle, as we shall see, transformed the purely abstract and geometrical theory into a mechanical system of spheres" (p. 217). For other examples, see Wright [1973], p. 165. For a review of this viewpoint, see Lloyd [1978], pp. 202–204.

[5] This point was made forcefully by Wright [1973]; cf. Lloyd [1978], p. 219, who recants an earlier commitment to the conventional view.

unified planetary system which had consequences that the model as used by Eudoxus did not."[6] If this is the case, then we can point to the same three fundamental aspects that we have identified with Ṭūsī's *Tadhkira*: (1) Aristotle is providing a physical account of the All; (2) he is accepting as given the models of the astronomers, which he assumes to be at base physical; and (3) he is attempting to provide a coherent and unified account, which compels him to provide additional spheres. It is on this basis that I would claim kinship between the *Tadhkira* and this chapter of the *Metaphysics*.

A second example of an "inspirational" ancient predecessor of the *Tadhkira* is undoubtedly Ptolemy's *Planetary Hypotheses*, which is composed of two books.[7] Writing sometime after completing the *Almagest*, Ptolemy begins the first book by proclaiming his desire to set forth the results of the *Almagest* in a summary fashion (κεφαλαιωδῶς; Arabic trans. *jumal*)[8] so that they might be more easily conceived and also as an aid to those who wish to construct a mechanical model.[9] He then proceeds to catalogue the celestial circles, giving their positions, motions, and relative sizes (with the largest circle for each system of circles being 60). This part of Book I is followed by a discussion of the order of the planets, which, once determined, allows one to use the relative sizes of the circles to establish the absolute distances (in either Earth radii or stades) of the planets from the Earth. Computations of the sizes of the planets and stars followed by a passage on planetary visibility round out Book I. This part of the *Planetary Hypotheses* can then be understood as a summary of the main results of the *Almagest* presented without proofs for those who wish to gain a general picture of the heavens and their motions or who wish to construct actual physical models.

---

[6] Wright [1973], pp. 171–172. We need not follow him down the slippery slope of analogy with Kepler and Newton to accept his fundamental point.

[7] Only the first part of Book I is extant in Greek. Both books are available in an anonymous Arabic translation "corrected" by Thābit ibn Qurra. There is also a Hebrew translation from the Arabic. The Greek text of Book I (lacking the part on sizes and distances) with a French translation is in Halma [1820], pp. 41–56; the Greek text is also in Heiberg [1907], which in addition contains a German translation begun by L. Nix and completed by F. Buhl and P. Heegaard of the Arabic. Somehow the three translators missed the second part of Book I; this has been rectified by an English translation due to B. Goldstein [1967], who also included a commentary, a facsimile reproduction of one of the two extant Arabic manuscripts, namely British Museum MS Arab. 426 (hereafter referred to as BM 426) and variant readings of the other manuscript.

[8] Heiberg [1907], p. 70, line 12; BM 426, f. 81b, line 10 (Goldstein [1967], p. 13).

[9] Swerdlow/Neugebauer [1984] see this as basically an equatorium (p. 40).

Book II is a summary of a rather different sort. Ptolemy tells us that

> having pictured (*ja*ᶜ*alnā al-mithālāt*) their motions and the arrangements of their positions in a simple manner on great circles (*aflāk*), which they describe by their motions, it remains for us to describe the forms (*ashkāl*) of the bodies by which we may understand those circles (*aflāk*).[10]

Ptolemy then plunges into a rather involved discussion of the nature of the celestial bodies.[11] This is followed by a description of how the bodies are arranged for each planetary system.[12]

It is not difficult to cite numerous differences between the *Planetary Hypotheses* and the *Tadhkira* as well as other Arabic and Persian works of its type.[13] The two-book structure of the *Hypotheses*—the first a summary of the *Almagest* and the second an attempt to represent physically the orbs—is not followed in any Islamic astronomical text known to me. More importantly, the attempt to make the orbs into *manshūrāt* [πρίσματα or "sawed-off sections"] is rejected by Arabic astronomers out of hand.[14] Nevertheless, I would maintain that the *Planetary Hypotheses* is a prototype for the *Tadhkira*. The dual treatment of the motions and orbs in the two books of the *Hypotheses* have in the *Tadhkira* been combined to form a unified account for each body. The section on sizes and distances of the celestial bodies at the end of Book I of the *Hypotheses* has been placed in the final book (*bāb*) of the *Tadhkira*, which is meant to provide a unified view of the system taken as a whole. And most importantly, the purpose of each work is basically the same: (1) both seek to provide a summary of the *Almagest* and both assume the results and proofs of the *Almagest* as given; (2) both seek to provide a physical account of the celestial

---

[10] BM 426, f. 93a, lines 3–6 (Goldstein [1967], p. 36).

[11] Ptolemy's basic purpose in this part of Book II (BM 426, ff. 93a–95b; Goldstein [1967], pp. 36–41) would seem to be the defense of his proposal to do away with "complete spheres." He suggests using only those parts of the spheres where the planet is seen to move, leaving the so-called tambourines (*adfāf*) or sawed-off sections (*manshūrāt*) (BM 426, ff. 94b–95a). Ptolemy at no time, however, accepts a vacuum as some modern commentators would have us believe (*e.g.* Schramm [1963], pp. 19–20 and Neugebauer, *HAMA*, 2: 923). He instead suggests that an aether fills the gaps left after detaching the truncated orbs (BM 426, f. 95b, lines 8–13). Although the exact details of how the motion of the orb might occur in the aether remain obscure, Ptolemy's reason for adopting this somewhat radical position is explicit, namely that there is nothing useless in nature. As justification for this, he cites the fact that the orbs of Venus and Mercury fill almost exactly the space between the moon and sun (BM 426, f. 95a, lines 1–8).

[12] BM 426, ff. 96a–102b (Goldstein [1967], pp. 42–55).

[13] For an extensive discussion of this with regard to Ibn al-Haytham's *Hay'at al-*ᶜ*ālam*, see Langermann [1990], pp. 11–25.

[14] See, for example, the *Tadhkira*, II.11 [16].

bodies and their motions based on the mathematical models of the *Almagest*; and (3) both seek to give a unified account of the entire system, perhaps best exemplified by the sizes and distances section of each (I.1 of the *Planetary Hypotheses* and Book IV of the *Tadhkira*). It is these three fundamental aspects, which we have also identified in a somewhat different form in Aristotle's account of the Eudoxan-Callippan system, that I would maintain link the *Hypotheses* to the *Tadhkira*.

C. The Islamic Tradition of the *Tadhkira*

The existence of classical texts that correspond in some fashion with the Islamic tradition of which the *Tadhkira* is a member does not imply, of course, that those texts are in a direct historical line with, or are responsible for, that tradition. In particular, there is little evidence that any Islamic writer would have connected his work with Aristotle and Ptolemy in quite the way this 20th century writer has done above. We also are in the unfortunate position of not understanding very clearly how late Hellenistic astronomy was assimilated into the early Islamic science of the 8th and early 9th centuries. Though theoretical concerns seem to have been at a minimum during this period, there are at least hints that an interest in cosmography and the physics underlying astronomy may have been present at the earliest stages of Islamic science.[1] Ya^cqūb ibn Ṭāriq gave one of his works the suggestive title *Tarkīb al-aflāk* ("the arrangement of the orbs"), which was probably written in 161/777–78. Though only fragments have reached us, one of these, significantly for our purposes, is a table of planetary sizes and distances; although Bīrūnī thinks these are Indian, Pingree feels they are more in the spirit of *Planetary Hypotheses* I,2 though he is quick to point out that it is difficult to be definitive about a Greek influence.[2] Indeed, Ibn Hibintā (ca. 950) in his *Astrology* points to an Indian source for Ibn Ṭāriq's system of orbs.[3] Another work that is possibly from this period based on its attribution to the astrologer Māshā'allāh (762–ca. 815) is *De scientia motus orbis* (also known as *De elementis et orbibus coelestibus*).[4] This astronomical treatise in 27 chapters, which is extant only in Latin translation, has sections on the physical basis of astronomy, though the motivation here may be astrological.[5] Pingree has postulated a Syriac source that may be connected with the Hellenized pagan community of Ḥarrān; though this is far from proved, it does provide us with a

[1] Pingree [1973] provides a penetrating account of this period. For a taste of the muddled sources that must be used to reconstruct this period, see Hāshimī, *^cIlal*, who may simply be reflecting the eclectic nature of astronomy during the 9th century.

[2] Pingree [1968], pp. 105–109.

[3] Langermann [1990], p. 29.

[4] Pingree [1975] has summarized the contents, pp. 9–12. See also Sezgin, *GAS*, 6: 129.

[5] As Pingree [1975], p. 10 states: "Clearly this is Peripateticism for the astrologer."

possible conduit for the early interest in physical cosmography within Islam as well as a possible motivation having to do with the desire to put forth a unified system that would provide the rational basis for astrology.[6]

But were these early Islamic forays into physical cosmography sufficient to define a genre or to establish a *tradition* that can account for the *Tadhkira*? Based on the testimony of Ibn al-Haytham, a prominent, not to say crucial, figure in developing that tradition, the answer must be no. In the introduction to his *Al-Maqāla fī hay'at al-ᶜālam*, no doubt written at a fairly early stage in his career,[7] he identifies a group of mathematicians (*aṣḥāb al-taᶜlīm*) who have published works dealing with astronomical problems "by speaking generally" [*lit.*, "grossly"] (*bi-'l-jalīl min al-qawl*), but at the same time have "followed a procedure in apparent conformity with the detailed investigation of the science of astronomy" (*li-l-daqīq min al-naẓar fī ᶜilm al-hay'a*).[8]

> ...their goal is to collect that which has been stated in a detailed way and to present it as a summary (*jumal*) free of proofs...for the sole [purpose] of simplifying for those who would benefit from an acquaintance (*maᶜrifa*) with the configurations of those motions by acceptance rather than inquiry and by following the practitioners of the science rather than by reflection...[9]

One can see that Ibn al-Haytham is characterizing a relationship not unlike the one Ṭūsī describes between his *Tadhkira* and the *Almagest*, *i.e.* one of a general account that takes for granted an existing detailed study replete with mathematical proof. But the two other aspects we have identified as fundamental to the *Tadhkira*, a summary that takes the bodies as explicit starting points rather than simply implied and a presentation that is holistic rather than particular, are lacking in these earlier accounts according to Ibn al-Haytham. For he states that they are

> based upon the motions of imaginary points on the circumferences of intellected circles...resulting in what they [*i.e.* earlier writers] have expounded being limited to those circles and points alone since they have not made it their intention to clarify the way in which those various motions may be consummated while being assumed on the surface of solid spheres.[10]

---

[6] Pingree [1975], pp. 10–12.

[7] Cf. Sabra [1972], p. 197.

[8] Langermann [1990], p. 5 (Arabic numeration), lines 7 and 10–11. I have changed Langermann's translation (no. [1], p. 53) somewhat in order to emphasize what I take to be Ibn al-Haytham's rhetorical use of *jalīl* (general or gross) and *daqīq* (precise or detailed) to contrast two types of astronomical work.

[9] Langermann [1990], p. 5 (Arabic numeration), lines 19–23; cf. Langermann's translation and commentary, no. [3], pp. 53–54 and pp. 2–4.

[10] Langermann [1990], p. 5 (Arabic numeration), lines 12–18; cf. Langermann's translation, nos. [2–3], pp. 53–54.

Furthermore these summarizers have not provided a coherent account since neither have they clarified "the manner, given the various centers involved, that spheres can carry [other spheres]."[11]

It is important to recognize that Ibn al-Haytham is not saying that previous astronomical work has been "instrumentalist" in some Duhemian sense; in fact, he seems to go out of his way to indicate that previous work has assumed the existence of solid spheres. He will refrain, he tells us, from "criticism and rebuke" since the earlier summarizers "despite ignoring the things they neglected," a clear reference to the solid spheres, nevertheless "probably conceived of them and appreciated their importance" (*mutaṣawwirīn lahā wa-muqayyimīn* [text: *qayyimīn*] *bihā*).[12] Ibn al-Haytham even gives an explanation for this neglect. Since they were dependent upon works that used figures and proofs based on circles, their summaries likewise were "limited to those circles and points alone."[13] But what of those who made these detailed studies with proofs that ignored the solid bodies, Ptolemy being for Ibn al-Haytham the most eminent? Ibn al-Haytham excuses him by remarking that his "aim was to construct proofs for which circles and points sufficed";[14] there is no hint that he thinks that his great predecessor did not understand the importance of the solid orbs for astronomy.

Rather than seeing himself as a realist who must combat the instrumentalism of his predecessors, Ibn al-Haytham instead wishes to make explicit what is implicit in previous work,[15] namely to assume

> the imaginary circles and points that [Ptolemy] took to be detached (*akhadhahā mujarrada*)[16] to be on the surfaces of spheres moving on their own...since that is more true in terms of reality (*min ṣifat al-ḥāl*) and more clear for [providing] an understandable explanation.[17]

Here then is the first aspect that we have identified as crucial for the tradition we are tracing, namely providing a physical account that would explain the

---

[11] Langermann [1990], p. 5 (Arabic numeration), lines 18–19; cf. Langermann's translation, no. [3], p. 54.

[12] Langermann [1990], p. 6 (Arabic numeration), lines 1–3; cf. Langermann's translation, no. [5], p. 54.

[13] Langermann [1990], p. 5 (Arabic numeration), lines 13–17; cf. Langermann's translation, no. [2], p. 53.

[14] Langermann [1990], p. 6 (Arabic numeration), lines 14–15; cf. Langermann's translation, no. [8], p. 55.

[15] That the solid orbs were implicit in the work of ancient astronomers was not a unanimously held opinion; see footnote 2, p. 25.

[16] Here the context makes it clear that Ptolemy took these circles and points to be detached for the purposes of his proofs.

[17] Langermann [1990], p. 6 (Arabic numeration), lines 10–12; cf. Langermann's translation, no. [6], p. 54.

celestial motions. Ibn al-Haytham obliges us with clear statements of the other two aspects as well. He explicitly acknowledges the dependence of his work on the *Almagest* when he states that

> our statements concerning all the motions are according to the opinion and belief of Ptolemy; but all that we shall present is a summary by which one can understand the general picture of the form, the position and the motion but not the particulars of these.[18]

As for the problem of fitting all the parts together into a coherent whole, he addresses this as follows:

> ...it is possible to combine together all those bodies assumed for each one of the motions without there occurring any hindrance, resistance, or impediment; rather their motions in being combined are [still] continuous and permanent.[19]

Ibn al-Haytham's importance in establishing the *Tadhkira's* tradition should not be underestimated. Not only has he defined explicitly the crucial elements of that tradition, he has also placed himself in the historical position of being its originator. And insofar as his successors believed his story, and there is no reason to doubt that Ṭūsī, for one, did, this would more or less solve our problem of situating the *Tadhkira* within an Arabic tradition.

There remain, however, numerous problems with *our* privileging of Ibn al-Haytham's position. First there is the inescapable problem of the *Planetary Hypotheses* and the relation of the *Hay'at al-ᶜālam* to it. Since Ibn al-Haytham does not mention it in the *Hay'at al-ᶜālam*, we are in the rather uncomfortable position of claiming a work, *i.e.* the *Hypotheses*, as an important source for a tradition that is either unknown or ignored by a prime member of that tradition. I have little in the way of a solution of this puzzle except to offer the following facts. Ibn al-Haytham does know it by the time of writing his later *Al-Shukūk ᶜalā Baṭlamyūs* (Doubts about Ptolemy) and in fact presents a blistering attack on it.[20] Besides the fact that he considered the *Hypotheses* defective in not providing a physical body for each motion, a criticism applicable to his own ear-

---

[18] Langermann [1990], pp. 6–7 (Arabic numeration), lines 27–1; cf. Langermann's translation, no. [12], p. 55.

[19] Langermann [1990], p. 6 (Arabic numeration), lines 19–22; cf. Langermann's translation, no. [10], p. 55. Ṭūsī, as well as Ibn al-Haytham himself later in his life, would come to see this as an overly optimistic view. The point I wish to emphasize here, though, is simply that Ibn al-Haytham is explicitly acknowledging this as part of the work he is undertaking.

[20] *Shukūk*, pp. 42–64.

lier work, he also is disenchanted with the use of truncated orbs, the *manshūrāt* or πρίσματα.[21] Without attempting to be definitive, I would suppose that Ibn al-Haytham, if we assume that he knew the *Hypotheses* at the time of writing the *Hay'at al-ᶜālam*, may have regarded the use of anything other than spheres as so idiosyncratic as to disqualify it from the tradition he saw himself as establishing.[22] There is also the problem of how much we should accept Ibn al-Haytham's assertion that his predecessors had been derelict in attending to the physical side of astronomy.[23] Another problem I have not attempted to deal with here is that of his contemporaries, such as Ibn Sīnā and his student al-Jūzjānī, who may have been engaged in similar enterprises.[24]

Despite these difficulties in establishing Ibn al-Haytham as having single-handedly established physical cosmography in Islam, a position I would not at all wish to maintain, there remains the fact that he and several of his successors saw him in that light. The role of al-Khiraqī (d. 533/1138–39) seems to me particularly germane. He explicitly cites Ibn al-Haytham in both his *Muntahā al-idrāk* and *al-Tabṣira* as inspiration for his work. In particular, he states that he wishes to continue the work of Ibn al-Haytham in considering solid spheres as opposed to imaginary circles in astronomy.[25] And Ṭūsī specifically cites the *Muntahā* in his *Ḥall-i mushkilāt-i Muᶜīniyya*.[26] But of even more importance is the fact that it is this Khiraqī, working explicitly under the influence of Ibn al-Haytham, who establishes the basic structure, the genre, that Ṭūsī follows in both his *Muᶜīniyya* and *Tadhkira*.

D. The *Tadhkira* as Genre

Ibn al-Haytham's advocacy of an astronomical summary in which the celestial *bodies* played a fundamental role was quite influential. But his *Hay'at al-ᶜālam* acted more as a pioneering inspiration rather than as a prototype to be emulated. This latter role was assumed by Ṭūsī's Persian *Risālah-i Muᶜīniyya*

---

[21] *Shukūk*, pp. 47–58, especially p. 50.

[22] This is a widespread disenchantment; see Bīrūnī, *Qānūn*, 2: 635, lines 3–17 and Ṭūsī, *Tadhkira*, II.11 [16].

[23] For a discussion of this with particular reference to Battānī and Farghānī, see Langermann [1990], pp. 25–30. Whether Farghānī's *Elements of Astronomy* should be considered to be in line with Ibn al-Haytham's *Hay'a* is an interesting question deserving of serious study; I would merely point out here that Ibn al-Haytham certainly did not think so.

[24] See Saliba [1980].

[25] Shams al-Dīn abū Bakr Muḥammad b. Aḥmad al-Khiraqī, *Muntahā*, Tehran, Majlis-i Shūrā-i Millī MS (uncatalogued), ff. 2b–3a; idem, *Tabṣira*, Istanbul, Ayasofya(?) MS 3398, f. 2a. A German translation of the introductions to both works is due to Wiedemann and Kohl [1926–27]; reprinted in Wiedemann [1970], 2: 628–643.

[26] *Ḥall*, p. 16, line 1.

and by its later Arabic incarnation the *Tadhkira*; they defined by their structure and their approach to the problems of astronomy a certain category of work that would be the model for a distinct genre of astronomical writing.[1]

The development of this genre reflected a certain conceptualization of astronomy in Islam that seems to have evolved during the early centuries of Islamic science. Though a detailed examination would neither be appropriate nor possible here, even a brief discussion of this problem will be useful for situating the *Tadhkira* within the discipline of Islamic astronomy and for dispelling the notion that Ṭūsī's understanding of astronomy represents a new departure.

Encyclopedias and popular accounts of the sciences provide some of our primary sources for how astronomy was conceived. In the 10th century, *ᶜilm al-nujūm* (the science of the stars) was used by Fārābī, Abū ᶜAbd Allāh Muḥammad al-Khwārazmī, and the Ikhwān al-ṣafā' (Brethren of Purity) to designate astronomy in its widest sense, and Khwārazmī indicates that the term is equivalent to the Greek ἀστρονομία.[2] For all three, *ᶜilm al-nujūm* includes both mathematical astronomy[3] and astrology.[4] Khwārazmī and the Ikhwān al-ṣafā' use the term *ᶜilm al-hay'a* (the science of *hay'a*) to designate a certain branch of astronomy that Khwārazmī specifically identifies with "knowledge (*maᶜrifa*) of the arrangement (*tarkīb*) of the orbs, their configuration (*hay'a*), and the configuration of the Earth."[5]

With Avicenna (980–1037), however, we find something new. In his *Aqsām al-ᶜulūm al-ᶜaqliyya* (Classification of the Rational Sciences), *ᶜilm al-hay'a* has replaced *ᶜilm al-nujūm* as the general term for the discipline,[6] and astrology is no longer considered part of this reformulated astronomy.[7] This move is of more than passing significance. For what had once been considered a subdivision of astronomy has now been made to encompass the entire field. Thus whether one is composing an astronomical handbook (*zīj*), making instruments, finding the direction of Mecca, or doing anything astronomical, one would be engaged in this *ᶜilm al-hay'a*, which literally means "the science of configuration"; thus astronomy has implicitly come to designate, at least in this classification, a discipline having to do with bodies inasmuch as *ᶜilm al-hay'a* originally was used to indicate the study of the configuration of the Earth and orbs, and the arrangement of the latter. Without offering an explanation of why this might have occurred at this time, we can point to the related work of Avicenna's contemporary Ibn al-Haytham, who, as we have discussed at length above, was also intent on making the physical bodies central to astronomy.

---

[1] This distinct category was first pointed to by Livingston [1973].

[2] Fārābī, *Iḥṣā'*, pp. 84–86; Khwārazmī, *Mafātīḥ*, Bk. II, Ch. 6, esp. p. 215; Ikhwān al-ṣafā', *Rasā'il*, 1: 73.

[3] Fārābī calls this *ᶜilm al-nujūm al-taᶜlīmī*.

[4] For Khwārazmī, I base this on the fact that Ch. 6 of Bk. II, entitled "*Fī ᶜilm al-nujūm*," has a section devoted to astrology.

[5] Khwārazmī, *Mafātīḥ*, p. 215.

[6] Ibn Sīnā, *Aqsām*, p. 112.

[7] Astrology is here a subdivision of physics (*al-ḥikma al-ṭabīᶜiyya*); ibid., p. 110.

While Avicenna is making *hay'a* synonymous with astronomy, he also is, as we have seen, removing astrology from its purview. This may seem odd at first sight since one would assume that an emphasis on the physical aspects of astronomy would make astrology even more central to it. I believe the reason this did not happen was that ᶜ*ilm al-hay'a* was held to deal with the external aspect of the physical bodies and not their internal character, a point we shall return to below.

This new understanding of astronomy was not confined to Avicenna; one need merely look at late medieval (*i.e.* post-12th c.) classifiers of the sciences such as Ṭāshkubrīzāde (901–68/1495–1561) and Tahānawī (12th/ 18th c.) to see that this view of ᶜ*ilm al-hay'a* and all that it entailed had become commonplace.[8] Thus under this classification one finds, among other things, theoretical works, astronomical handbooks (*zīj*es), books on the making and use of instruments, treatises on observational astronomy, and tables of prayer times. In addition, certain topics in geography are accepted as being part of this astronomy.[9] But just as with Avicenna, ᶜ*ilm al-hay'a* among these later categorizers does not include works on astrology or the body of literature related to Aristotle's *De caelo* (*al-samā' wa-'l-*ᶜ*ālam*), both of which were usually considered to fall under the Peripatetic rubric of physics.

But while ᶜ*ilm al-hay'a* came to mean astronomy in its most general sense, the original meaning of the term was not completely lost; those works that were intended to give a general view of astronomy from the perspective of the configuration of the orbs would, of course, be *hay'a* works *par excellence*. But in order to distinguish this usage from the all-inclusive one, works of this genre were categorized under *hay'a basīṭa* (plain *hay'a*) since, as Ṭāshkubrīzāde tells us, they "lack (*yujarrada* ᶜ*an*) proofs; they are limited to conceptualizing (*taṣawwur*) and imagining (*takhayyul*) without certainty (*yaqīn*)."[10] It is this type of work that is exemplified both by *Al-Tadhkira fī* ᶜ*ilm al-hay'a* (Memoir on the science of *hay'a*) and by one of its more elementary, and consequently very widespread, offshoots, Jaghmīnī's *Al-Mulakhkhaṣ fī al-hay'a al-basīṭa* (Epitome of plain *hay'a*).[11]

---

[8] Ṭāshkubrīzāde, *Miftāḥ*, 1: 371; Tahānawī, *Dictionary*, 1: 41.

[9] The reason for this will be explored on p. 38.

[10] Ṭāshkubrīzāde, *Miftāḥ*, 1: 372.

[11] For Jaghmīnī's dependence on the *Tadhkira*, see commentary to III.1 [8]. There is a great deal of conflicting evidence concerning the date of this author, whose full name is Maḥmūd b. Muḥammad b. ᶜUmar al-Jaghmīnī. Suter [1900], p. 164 held that this Jaghmīnī was the same person who wrote a work on medicine called *Qānūnča* and who died in 745/1344–45; however a copy of the *Mulakhkhaṣ*, Istanbul MS Lâleli 2141, is purportedly dated 644/1246–47 (Krause [1936], p. 510), and King states that Ḥājjī Khalīfa, in a manuscript copy of the *Kashf al-ẓunūn*, gives the date of composition as 618/1221–22 (*EI*², 5: xvii). Cf. *GAL*, 1: 473 and S1: 865, King [1986], p. 150 and Suter and Vernet, *EI*², 2: 378. If Krause and King are correct, then Jaghmīnī would obviously not be dependent on the *Tadhkira*; however, at this point I find the evidence from III.1 [8] compelling. Clearly, though, further investigation of the manuscript evidence is needed to resolve the matter.

The three general aspects we have identified above as fundamental to the tradition of physical cosmography underlie, of course, *hay'a basīṭa*; it is at this level of generality that we can connect the *Tadhkira* with the *Planetary Hypotheses* and the *Hay'at al-ālam*. But the *Tadhkira* has certain other features—both stylistic and substantive—that distinguish it in varying degrees from these predecessors. Many, if not most, of these became codified in the large number of works modeled after the *Tadhkira* and it is in this sense that I believe we can refer to it as exemplifying a certain genre over and above its role of continuing the tradition of physical cosmography. In what follows, we shall list and discuss these features.

1. The Structure of the *Tadhkira*. As I have indicated above, we need to look at the *Tadhkira* not only as part of a tradition of cosmographical summaries but also as structured in a way that Ṭūsī felt would most effectively allow him to give a summary of astronomy from the point of view of the physical bodies. The structure comes from the four-part division of the work itself: (a) an introduction that gives the mathematical and physical principles used in the discipline; (b) a section divided into several chapters dealing with the configuration ("cosmography") of the celestial region (*hay'at al-samā'*); (c) another section also divided into several chapters dealing with topics related to the configuration ("geo-graphy") of the Earth (*hay'at al-arḍ*) in which Ṭūsī deals with the astronomically determined divisions of the Earth as well as with other astronomically determined phenomena that in general vary depending on the latitude of the observer; and (d) a section dealing with the size of the Earth and the celestial bodies, and the distances of those bodies from the Earth.

A four-part structure, as far as I have been able to determine, first appears in Khiraqī's *Muntahā al-idrāk*, which has an introduction and three books dealing with the arrangement (*tarkīb*) of the celestial bodies and their motions, the configuration (*hay'a*) of the Earth, and chronology. Ṭūsī himself has basically taken this structure as a model for both his *Risālāh-i Mu^{c}īniyya* and his *Tadhkira*, but he has made the sizes and distances chapter of the cosmographical section of the *Muntahā* into a separate part (Book IV) and has more or less dispensed with chronology.[12] It is this structure that came to define the genre of writing referred to as *hay'a basīṭa* (plain *hay'a*).

2. A *hay'a basīṭa* work should contain no geometrical proofs. Since a *hay'a basīṭa* work was an account meant to give a general overview of astronomy, it was generally held that it should be devoid of mathematical proof.

[12] He deals with it somewhat in III.10, but, perhaps with the *Muntahā* in mind, he insists that the details of chronology do not have a place in a work such as the *Tadhkira*; see III.10 [3]10–11.

Ṭūsī emphasizes this in I.Intr. [3], as we have seen earlier (p. 24), where he refers the reader to the *Almagest* for proofs of what is being presented "in narrative form." This is stated again in II.5 [10]: "These then are models and rules that should be known. We have only stated them here; their geometric proofs are given in the *Almagest*." And in IV.1 [4], he forgoes presenting one of Bīrūnī's methods for determining the size of the Earth since "it contains geometrical proofs." The one major exception occurs in II.11, where he gives several proofs related to his new models for planetary motion. But this exception is acknowledged by Ṭūsī in such a way that it proves the more general rule; for he states that he is presenting proofs "even though it was not our intention to provide geometrical proofs in this compendium" (II.11 [3]).

3. The *Tadhkira* as a summary for nonspecialists. In addition to not giving proofs, Ṭūsī also indicates his desire to avoid too much detail; for this he refers his reader to specialized accounts. That he could depend on these other works is a clear indication that Islamic astronomy was a highly developed field of study. I also think that this is an indication that Ṭūsī saw the *Tadhkira* as a work useful both for students of astronomy and for nonspecialists.[13]

As examples of this, we find in II.4 [12] that he cuts his discussion of the fixed stars short with the following remarks:

> Knowledge of the fixed stars and that which concerns them being a separate discipline, it is best that we confine ourselves to just what has been presented.

For those interested in numbers, he sends his readers to the "practical handbooks" (*kutub al-ᶜamal*) in II.14 [1]14–15 (for sector measurements) and in III.8 [8]4 (for the equation of time). When an example is needed, he usually will give the simplest one possible as in the case of the *qibla* determination in III.12 [4]; as we saw earlier, he avoided Bīrūnī's determination of the size of the Earth since it involved geometrical proofs.

---

[13] One difficulty with this is the presentation of non-Ptolemaic models in II.11, which would certainly seem to be written with the specialist in mind; evidently Ṭūsī himself recognizes this when he apologizes for going into detail "even though it was not our intention to provide geometrical proofs in this compendium" (II.11 [3]). Twenty-five years earlier this consideration had led him to delay publishing his newly discovered models in the *Risālah-i Muᶜīniyya*, the Persian prototype of the *Tadhkira*, since their "presentation in this compendium would not be appropriate"; instead he put them in his appendix to that work, the *Ḥall-i mushkilāt-i Muᶜīniyya*. (There are other cases in the *Muᶜīniyya* where Ṭūsī speaks of the inappropriateness of going into the details of complicated matters; see Section J, p. 68 of this volume.) Perhaps by the time of writing the *Tadhkira*, Ṭūsī felt that these "new" developments in the field had come to be an integral part of *hay'a* and were something the student or even general reader should be aware of despite their difficulty.

Finally we should note that Ṭūsī seems quite cognizant of disciplinary boundaries and will not hesitate curtailing his discussion when he feels a boundary is about to be traversed. As we have noted earlier with reference to Khiraqī's *Muntahā*, he dispenses with the historical aspects of chronology since he feels that it has nothing to do with astronomy.

> Every people has an epoch to which they refer the years of their history; understanding the details of that does not pertain to this science. (III.10 [3])

4. All simple bodies as the subject matter of astronomy. Ṭūsī is quite explicit on this point in that he states in I.Intr. [2]:

> The subject of astronomy is the simple bodies, both superior and inferior, with respect to their quantities, qualities, positions, and intrinsic motions.

Several points need to be noted. First it is the *simple* bodies that are the subject matter of astronomy, *i.e.* the four elements plus the celestial aether. Second the fact that the four elements are included means that a *hay'a* work would need to pay some attention to the sublunar world. In the *Tadhkira* we find that not only is the shape and situation of the Earth considered,[14] the cosmographical wrap-up given in II.2 includes a paragraph listing the various levels of the sublunar region. In addition, the entirety of Book III is devoted to "the configuration of the Earth" (*hay'at al-arḍ*), and the measurement of the Earth is taken up in IV.1.

The inclusion in a *hay'a* work of a distinct section on the configuration of the Earth was often taken to distinguish the Islamic and Greek astronomical tradition.[15] This seems something of an exaggeration since Ptolemy devoted Book II of the *Almagest* to many of the problems referred to in Book III of the *Tadhkira*. But the fact that some Islamic astronomers saw themselves as doing something new is significant and again highlights the importance they attached to the notion that astronomy should deal with *all* the simple bodies.

5. Hay'a should only deal with the external aspect of the bodies. Although Ṭūsī makes bodies the subject of astronomy, he is not thereby making astronomy a branch of physics. For one thing, it is clear that he follows Aristotle in taking it to be a "mixed science" inasmuch as he states that its principles come from metaphysics, geometry, and natural philosophy. Furthermore, he tells us that astronomy is interested in bodies in a particular way, namely "with respect to their quantities, qualities, positions, and intrinsic motions" (I.Intr. [2]).

---

[14] See II.1.

[15] See, for example, Qāḍīzāde al-Rūmī's (ca. 1364–ca. 1436) commentary to Jaghmīnī's *Mulakhkhaṣ*, f. 3a, lines 9–10.

The significance of such a restriction is underscored by Tahānawī who addresses this question in the introduction to his *Dictionary of Scientific Terms*. He notes that *ᶜilm al-hay'a* should be distinguished from works on *al-samā' wa-'l-ᶜālam, i.e.* the *De caelo* tradition of Aristotle, which was taken to be part of the physical or natural philosophy corpus. The reason was not because of subject matter but because of the different aspects of the subject matter studied by each discipline. Though both investigate the simple bodies, *ᶜilm al-hay'a* examines them with regard to their "quantities, qualities, positions, and motions." On the other hand, *al-samā' wa-'l-ᶜālam* studies them from the point of view of "their natures" (*ṭabā'iᶜihā*).[16] Thus it was for *ᶜilm al-hay'a* to examine the outward manifestations of simple bodies, whereas it was for *al-samā' wa-'l-ᶜālam* to investigate their essential nature.[17]

What this means in practice is illustrated in II.1, which deals with proofs for such things as the sphericity of the Earth and sky. In the final paragraph of the chapter, Ṭūsī tells us that the proofs presented

> are *inniyya,* which convey existence; those which convey the necessity of that existence are *limmiyyāt* proofs and are given in natural philosophy in the book *De caelo.*

These *innī* "proofs" are based upon observations; for example, the observations listed in II.1 [2] are the evidence Ṭūsī feels is sufficient to prove that [*inna*] the Earth is, generally speaking, a sphere. This sort of proof does not, however, indicate the physical or metaphysical reasons that would show us why [*limā*] the Earth must be spherical and no other shape. This would require that one investigate the *nature* of the bodies, but that, as we have seen, is the province of natural philosophy, not astronomy.

Although the notion of *innī* and *limmī* proofs ultimately derive from the fact/reasoned fact distinction made by Aristotle in the *Posterior Analytics* (Bk. II, Ch. 13), the manner being used here to differentiate two sciences that study the same subject matter has no real precedent in the Aristotelian corpus.[18] There are, however, other Greek works that one might point to as sources for the dichotomy made in Islamic astronomy. Simplicius (6th c. A.D.), in his commentary on Aristotle's *Physics*, quotes Geminus (1st c. A.D.)[19] to the effect that

> in many cases the astronomer and the physicist will propose to prove the same point, *e.g.* that the sun is of great size or that the Earth is spherical, but they will not proceed by the same road. The

[16] Tahānawī, *Dictionary*, 1: 47–48; actually Tahānawī is here quoting Bīrjandī's supercommentary on Qāḍīzāde's *Sharḥ al-Mulakhkhaṣ*. See also Nallino [1911], p. 32.

[17] This would also be the reason to exclude astrology from the purview of *hay'a*; cf. pp. 34–35.

[18] For an elaboration, see the commentary to II.1 [8].

[19] On dating Geminus, see *HAMA*, 2: 579–581.

> physicist will prove each fact by considerations of essence or substance, of force, of its being better that things should be as they are, or of coming into being and change; the astronomer will prove them by the properties of figures or magnitudes, or by the amount of movement and the time that is appropriate to it.[20]

Ptolemy in Book I of the *Almagest* would seem to exemplify this attitude since his proofs of the basic cosmological features (the sphericity of the Earth and universe, the Earth's centrality, and so forth) *generally* rely upon mathematics and observations. This, of course, would be perfectly in keeping with his stated preference for the potential certainty of mathematics over the "guesswork" of physics and metaphysics.[21]

But some Islamic astronomers felt he had not gone far enough. Bīrūnī, for example, chides Ptolemy for using arguments based on physics in *Almagest* I.3 to prove the sphericity of the heavens. Going outside one's disciplinary boundaries, he asserts, does not strengthen one's arguments but makes them merely persuasive rather than necessary.[22] Bīrūnī may be mildly upset with Ptolemy for some of his antics in the *Almagest*, but he is truly exasperated with what he sees as a blatant disregard for the confines of astronomy in the *Planetary Hypotheses*.

> Ptolemy in his book *al-Manshūrāt* [*i.e.* the *Planetary Hypotheses*] deviated from the path which he had followed in the *Almagest* [and took up] that which is related to opinions outside of this science, that is in the belief of people (*al-qawm*) that the celestial bodies have life (*ḥayāh*), perception (*shu*ᶜ*ūr*), sensation (*iḥsās*), and the choice (*ikhtiyār*) of the noblest (*al-afḍal*) motions...so that he even said that the paths of the planets traverse spheres that resemble anklets (*khalākhīl*) or bracelets (*aswira*) and are called *manshūrāt*. He then eliminated (*asqaṭa*) the rest of the sphere...he thus repudiated (*nabadha*) his physical and persuasive proofs (*istidlālātihi*) in the *Almagest* concerning the sphericity of the sky...and he did not explain what is on either side of the *manshūrāt*, whether it is of the genus of the aether thereby returning to what he rejected...or whether it is of a genus of what is below the aether...or whether it is a sixth genus...These are separate subjects of investigation that are [dealt with] in their own particular places.[23]

---

20 Heath [1913], p. 276; cf. Lloyd [1978], pp. 212–214.

21 *Alm.*, p. 36 (H6–H7).

22 Bīrūnī, *Qānūn*, 1: 27; see also 1: 49. Cf. Pines [1963], p. 199.

23 Bīrūnī, *Qānūn*, 2: 634–635.

Bīrūnī's attitude represents an extreme, but by no means unique, position among Islamic astronomers. Ṭūsī would certainly have agreed that Ptolemy had broken a cardinal rule of *hay'a* by discussing the essential nature of the heavenly bodies in an astronomical work. But he would not have agreed that astronomy could do without natural philosophy; in II.1 [6] he acknowledges that he must depend on the *physical* principle of the Earth's rectilinear inclination, rather than on observational evidence, to prove that the Earth does not rotate. This position did not go unchallenged, however; Ṭūsī's own student Quṭb al-Dīn al-Shīrāzī opposed him on this point and the debate continued for several centuries after Ṭūsī.[24]

That Islamic astronomers were insistent that *hay'a* should only deal with the external aspect of bodies, with some even holding that this meant that astronomy could be independent of physics and metaphysics, leads us into fascinating questions concerning the reasons this development took the form it did in Islam and the possible relationship of this with the emergence of non-Ptolemaic astronomy. Even to begin answering these questions, however, would take us very far afield; as Ṭūsī might have said, "it is best that we confine ourselves to just what has been presented."[25]

E. The Physical Principles Underlying the *Tadhkira*

Ṭūsī tells us in I.Intr. [2] that the principles (*mabādi'*) of astronomy come from metaphysics, geometry, and natural philosophy. Geometry is mainly dealt with in I.1,[1] while metaphysical principles of astronomy are generally ignored.[2] On the other hand, many of the principles of natural philosophy, or physics, are listed in I.2, while others are implied. Starting with Duhem, modern historians have tended to view these physical principles as somehow "metaphysical" in the modern sense; this is often reflected in the rather dismissive way in which they are referred to as "philosophical."[3] But this seriously misrepresents the historical situation of these principles and the manner in which they were understood by the practitioners of *hay'a*. For one thing, the physical principles of astronomy were conceived in a way that conformed to the notion that *hay'a* was about the external aspect of bodies. As we shall see later (p. 46), motion, within astronomy, was said to be "investigated [in terms of] its quantity and direction"; the ultimate origin of the motion and like questions was left to natural philoso-

---

[24] See the commentary to II.1 [6].

[25] For a preliminary viewpoint on these matters, see Ragep [1982], 1: 129–189.

[1] Other geometrical propositions are stated in IV.1 [1].

[2] I am not certain whether this is intentional.

[3] See, for example, Kennedy [1966], pp. 366–367; Neugebauer, *HAMA*, 1: 1 and 2: 572, 942; Hartner [1975], p. 9; and Goldstein [1980], p. 142. For different viewpoints, see Sabra [1984], especially note 3, pp. 145–146 and Ragep [1987], pp. 330–331. On Duhem's role in fostering this attitude, see Ragep [1990], especially p. 210.

phy. This is clearly analogous to the *innī/limmī* dichotomy discussed above. Thus Ṭūsī formulates his principles, in particular Principle 5, in a way that circumvents the "why" (*limā*) and concentrates on the "that" (*inna*).

But despite this disciplinary separation of astronomy from natural philosophy, Ṭūsī, as we have already noted, is not willing to dispense with natural philosophy.[4] As if to emphasize this, Ṭūsī gives his physical principles a certain prominence by placing them in a separate chapter at the beginning of his work (I.2). This is reinforced by a liberal sprinkling of references to these principles throughout the text in order to emphasize the inadequacies of certain of Ptolemy's models. Such a formalistic methodology is not nearly so manifest in earlier *hay'a* works, whose aims had not included the proposal of new models. Ibn al-Haytham's physical principles occur in an appendix to his *Hay'at al-ᶜālam* and this only in one manuscript.[5] In the *Tabṣira*, Khiraqī does provide a separate introduction on the geometrical principles but his physical principles are in a chapter on the ordering of the four elements.[6]

In the following, I shall present a listing of these principles, indicate their origins, which are mainly but not exclusively Aristotelian, and discuss their implications for Ṭūsī's astronomy.

1. A void is impossible (I.1 [1]4). This most fundamental of the principles requires that the universe be completely filled with bodies and thus provides the "raison d'etre" for *hay'a* itself. This, of course, is central to Aristotelian physics.[7]

2. The universe is finite. Although Ṭūsī does not specifically state this principle, it is obviously implied in II.2, which deals with the arrangement and order of the orbs; the ninth orb, responsible for the daily motion, is the "highest orb"

---

[4] As we have noted earlier (p. 41), this reluctance in the case of the basic cosmography was due to Ṭūsī's contention that the Earth's state of rest could not be proven from observations; he maintained that he needed to "borrow" a principle from natural philosophy. As for those physical principles themselves, he may have also felt that in order to be established they needed arguments that were neither mathematical nor observational. Needless to say, we can hardly be definitive without investigating Ṭūsī's other writings on the relationship between the various sciences. It is worth noting here that there were some Islamic astronomers who held that their science could be established without recourse to metaphysics or natural philosophy; for an example, see Ragep [1982], pp. 141–145.

[5] Langermann [1990], p. 7.

[6] This chapter, which is Bk. I, Ch. 1, is entitled *Fī bayān aqsām al-ajsām ᶜalā al-ijmāl* (A General Exposition on the Parts of the Bodies). In his *Muntahā*, Khiraqī places this material in his Bk. I, Ch. 2, which he calls *Fī sharḥ maᶜnā ism al-ᶜālam wa-taqsīm ajzā'ihi al-awwal* (An Explanation of the Meaning of the Term "World" and a Preliminary Division of Its Parts). In general, these chapters correspond to the last part of II.2 of the *Tadhkira*.

[7] Aristotle develops his views against the void in Book IV, Chapters 6–9 of the *Physics*.

(II.2 [4]) and all the other orbs and elemental levels are embedded within it. Aristotle sets forth his proofs for the impossibility of an infinite body in *De caelo*, Book I, Chapters 5–7. In particular, an infinite heavenly body could not complete a rotation in a finite time (*i.e.* every 24 hours).[8]

3. A body is either simple...or compound (I.2 [1]20–21). From (1) and (2), we have established that the universe is a finite plenum; therefore, it must be filled in some manner. Since the universe is finite in size and Aristotle precludes infinitely small bodies,[9] there must be a finite number of distinct bodies. Of these, some will be simple while others will be composed of simple bodies. One could probably take this to follow more or less automatically from Aristotle's often repeated position that the simple is prior to the compound.[10] The identity of these simple bodies is more problematic; in *De caelo*, Aristotle uses the reality of simple motion, which he takes to be straight-line and circular, to establish that there must be simple bodies that perform these simple motions.[11] It is fairly straightforward then to establish that the celestial region is made up of one of these simple bodies, the so-called *aether*, since the heavens as a whole perform a perfectly regular daily rotation. The identity and uniqueness of the four sublunar elements cannot be established from an appeal to simple motion alone, however, since there are only two simple, natural motions in the sublunar region (*i.e.* up and down). Ultimately Aristotle must depend on the different approach of *De Generatione et Corruptione* (Bk. II, Chs. 1–8) where he identifies four elementary tactile qualities—hot and cold, moist and dry—whose combination in the substratum accounts for the four elements; earth and water, the heavy elements, can then be characterized by simple downward motion while air and fire, the light elements, can be characterized by simple upward motion.[12]

***

With these three principles, Ṭūsī can begin the task, undertaken in Book II, of filling his universe with spherical celestial bodies and elements with the Earth at the center of the whole but not, as required by Aristotle, the center for each individual part.[13] He still needs a means for accounting for the motions of the simple bodies, which we should recall are the subject matter of astronomy,[14] and this is provided by two final principles.

---

[8] *De caelo*, Bk. I, Ch. 5, 272b30–273a4.

[9] Although a body may be potentially divided *ad infinitum*, an actual infinitely small body cannot exist (Aristotle, *Physics*, Bk. III, Chs. 5–7; Bk. VI, Ch. 2).

[10] Cf. Elders [1965], p. 86, where one may find references to the relevant passages.

[11] Aristotle, *De caelo*, Bk. I, Ch. 2, esp. 268b17–269a2.

[12] See I.2 [3].

[13] We shall return to this problem; see pp. 46–47 of this volume.

[14] See p. 38.

4. Every motion has a principle (I.2 [2]5). A moving body will, according to him, fall into one of the following four groups (I.2 [2]):

A. *Self-Moved*	B. *Moved by an External Agent*
1. monoform	3. accidental
2. nonuniform	4. by compulsion

In I.2 [2]5, Ṭūsī states that "every motion has a *mabda'*," which literally means "beginning," but since it translates Aristotle's ἀρχαί it is more appropriately rendered "principle [of motion]." In the case of a body moved by an external agent, the principle of motion is in the external agent, while in the case of a self-moved body the principle is internal or, in Ṭūsī's terms, "does not become positionally separate" from the moved body (I.2 [2]5–6).

Astronomy is primarily concerned with the monoform self-moved motion of simple bodies, which will be dealt with in Principle 5 below, and in the accidental motion resulting from one orb moving another. An example of the latter would be an epicycle embedded within a deferent that as a result is moved accidently as the deferent rotates.[15]

5. A simple body has a single nature and what issues forth from that nature does so monoformly (I.2 [1]20–21). Since this is the basis for holding that the celestial bodies move with uniform, circular motion, and that the elements move naturally with straight-line motion, we will need to examine with some care how Naṣīr al-Dīn conceives this principle in the name of which he will proffer models to replace those of Ptolemy.

Ṭūsī states the principle in a very general way, but he really only uses it to establish the principle of motion for the simple bodies. For the elements and the celestial bodies, Ṭūsī, as a corollary to Principle 5, states that each has a nature (*ṭab*ᶜ) as their particular "principle of motion."[16] This is meant to account for their self-moved motion that is "monoform" (ᶜ*alā nahj wāḥid*)[17]. In I.2 [3], he then identifies the three categories of "motion due to a nature" with the five elementary bodies: (1) earth and water, the heavy elements, move "by nature" toward the center; (2) air and fire, the light elements, move "by nature" away from the center; and (3) the celestial bodies move "by nature" with a circular motion about the center. The first two categories of natural motion are

[15] Ṭūsī considers another type of accidental motion, namely that in which "the mover is the place for [the mobile] naturally to be" (I.2 [2]11). The commentators tell us that an example of this is the "enclosing orb" of II.11 [4]. See the commentary to I.2 [2]10–11, II.4 [6–7] and II.11 [4].

[16] I.2 [2]7–9.

[17] Literally this means "in a single way." Because it includes the accelerated motion of falling bodies, "uniform" would not be an appropriate translation, and I have therefore resorted to the neologism "monoform." Cf. commentary to I.2 [1]21.

monoform in direction but not uniform in speed; celestial motion, however, is both monoform in its movement and uniform in speed. This uniformity has a precise definition: "every point on [a single, simple celestial body] produces at the center equal angles in equal times or cuts equal arcs from the circumference." This, of course, is the basic rule of celestial motion; because it arises as an immediate corollary from one of the physical principles, its violation by Ptolemy, as we shall see, is completely unacceptable to Ṭūsī.

In addition to establishing the simple motions and their character, this principle has other consequences. Because it results in an absolute dichotomy between straight-line and circular motion, Ṭūsī feels compelled in II.1 [6] to assert that the Earth's state of rest at the center of the Universe is the result of a physical principle, namely its "having a principle of rectilinear inclination" which causes it to be "precluded from moving naturally with a circular motion" (II.1 [6]13). His argument against the ability of observation to establish geostasis means that a geocentric Universe ultimately rests for him upon Principle 5.[18] Another consequence of this principle is that the celestial bodies do not undergo "any change of state except for their circular motion"; this is because they do not experience any rectilinear motion, which is the efficient cause of generation and corruption.[19] Thus the perfection of the heavens is not a metaphysical doctrine but rather the result of this physical principle.

The idea that simple bodies move with a natural motion is undeniably Aristotelian.[20] Naṣīr al-Dīn, however, has made a significant shift in the way he has presented this idea, a shift that, I believe, tells us a great deal about the way he understands the role of physical principles in astronomy. As we have seen, Ṭūsī divides self-moved motion into that which is due to what he calls "a nature" if it is monoform, "a soul" if it is not. Clearly, motion due to a nature is characteristic of simple bodies irrespective of the ultimate source of such motion since that of the elements is described as being natural (*ṭabī^{c}iyya*) while that of the orbs is voluntary (*irādiyya*).[21] Thus even though the motion of the orbs may be due to a soul, which would make it in some sense "unnatural," this is not relevant here since the motion of the orbs is regular. On the other hand, the principle of irregular motion, as is the case of vegetative and animal motion, is said to be a soul (*nafs*). In short, the principle of internal motion is dichotomized in terms of regularity and not in terms of soul.

Let us not underestimate what Ṭūsī has done; he has in essence disentangled the physical underpinnings of astronomy from the metaphysical by stating his

---

[18] This was not universally held by Ṭūsī's Islamic successors; see the commentary to II.1 [6].

[19] See the commentary to I.2 [4].

[20] Cf. *De caelo*, Bk. I, Ch. 2, 268b27–269a2 where Aristotle virtually states as an axiom that "by simple bodies I mean those which possess a principle of movement in their own nature...necessarily, then, movements also will be...simple in the case of the simple bodies."

[21] I.2 [2]8–9.

physical principles for the motion of the celestial bodies in a way that avoids any reference to the ultimate source of that motion. Tahānawī nicely summarizes the situation:

> In this science [*i.e. hay'a*], motion is investigated [in terms of] its quantity and direction. The inquiry into the origin (*aṣl*) of this motion and its attribution (*ithbāt*) to the orbs is part of Natural Philosophy (*al-ṭab*ᶜ*iyyāt* [*sic*]).[22]

Thus Ṭūsī *qua* astronomer can simply point to the observed regularity of celestial motions and, using Principle 5, attribute it to "a nature." The philosophical question of the ultimate source of the celestial motion is simply not of concern to him here. Uniform, circular motion of the celestial orbs is for him based on a physical principle that is self-evidently confirmed by the overwhelming weight of experience; the metaphysical explanation of that motion may interest him *qua* natural philosopher or metaphysician, but here in the *Tadhkira* he simply does not need it to develop his cosmography.[23]

F. Modeling

1. Ptolemaic Modeling

The physical principles enumerated by Ṭūsī provide the necessary basis for establishing a cosmography. They do not, however, correspond to the laws or assumptions of a deductive system; these principles are guidelines against which various models are judged, not laws from which they are deduced. This is most easily seen by simply noting that a Ptolemaic methodology could not lead unequivocally to a unique celestial configuration. Ptolemy himself had long before conceded the point when he remarked that the solar anomaly could be represented by either an epicyclic or eccentric model, which are mathematically, but not physically, equivalent;[1] this is echoed by Ṭūsī in II.6 [2]. A homocentric cosmology, with a single center for all the celestial bodies, could conceivably provide a unique set of orbs, and this is what would seem to be demanded by a strict adherence to Aristotelian physics; indeed, Ṭūsī himself states that celestial motion is "that which is about *the* center" (I.2 [3]14–15). Homocentrism, however, had not seriously been considered since at least—and more than likely

[22] Tahānawī, *Dictionary*, 1: 47.
[23] I hope to expand on these remarks in a forthcoming publication; for the present, see Ragep [1982], 1: 164–174 as well as notes 67–76, pp. 183–187.
[1] *Alm.*, III.4, p. 153 (H232).

before—the time of Apollonius (ca. 240–170 B.C.),[2] and its advocacy in 12th c. Spain was an exceedingly limited affair.[3]

That the Ptolemaic system could not provide a definitive cosmology has sometimes been taken to mean that the Ptolemaic orbs were not "real";[4] a further consequence that has been advocated is that Ptolemy himself held this position.[5] But the question of the reality, or lack thereof, of the orbs should be seen as both logically and historically distinct from what I consider a far more important question, namely how various astronomers viewed the relationship between the physical principles and the mathematical models. The antiquated and quaint notion that Ptolemy was an instrumentalist in the modern sense seems miraculously to be fading rapidly.[6] An alternative view, however, one that would take into account his various writings, actual procedures, and possibly changing positions, is still awaited. My objective here is simply to provide some rather preliminary and brief remarks in order to understand Ṭūsī's reservations and objections to his predecessor.

There can be little doubt that Ptolemy was committed to certain physical principles in astronomy despite his reservations about Physics as a discipline.[7] The evidence from the *Planetary Hypotheses* is incontrovertible, and even in the *Almagest* he relies on physical notions for one of his proofs for the sphericity of the World.[8] Furthermore Ptolemy commits himself in various places to the perfection of the heavens and to their uniform, circular motion.[9] Irregular motion is

---

[2] Neugebauer, *HAMA*, 1: 263. An explicit rejection of homocentrism was made by Sosigenes (2nd c. A.D.), the teacher of Alexander of Aphrodisias, as reported by Simplicius in his commentary on Aristotle's *De caelo*. He proposed that each body moved about its own center in the heavens while the heavens taken as a whole revolved about the center of the World, a position not dissimilar to what one finds in Ptolemy's *Planetary Hypotheses* as well as in a number of Arabic sources (see our commentary to II.4 [6–7]). Sosigenes gives the varying brightness of the planets and the changing size of the sun and moon as revealed by an annular solar eclipse as observational reasons for rejecting a homocentric cosmology. See Schramm [1963], pp. 56–57.

[3] See Sabra [1984], esp. p. 133. The influence of Biṭrūjī's homocentric cosmology in the Latin West should not blind us to its confined and episodic quality in the Islamic context.

[4] See, for example, Ibn Khaldūn, *The Muqaddimah*, Ch. VI, Section 21, 3: 134–135 (Quatremère's Arabic text, 3: 106); cf. Nallino [1911], p. 33, who also gives the Arabic text.

[5] See, for example, Duhem [1908], pp. 16–20 (English trans. [1969], pp. 16–18). Various Islamic writers would have agreed with Duhem; see p. 25, footnote 2 of this volume.

[6] See Lloyd [1978], pp. 215–217 and Ragep [1990]. In the latter, I pointed out that Duhem himself in his *Système du monde* partially recanted his earlier view and recognized that Ptolemy, by the time of writing the *Planetary Hypotheses*, took his models to be physical; in Duhem's inimitable style, he had become "a slave of the imagination."

[7] *Alm.*, p. 36; see also p. 40 of this volume.

[8] *Alm.*, p. 40 (H13–14).

[9] For example in the *Almagest*, p. 141 (H216–217).

to be explained by means of the "hypotheses" of eccentrics and epicycles.[10] Thus Ptolemaic "modeling" essentially comes down, in theory, to putting together an interlocked set of uniformly rotating concentric, eccentric, and epicyclic orbs. But Ptolemy's commitment to this scheme goes only so far. Whenever phenomenal reality intrudes, it is the physical principles that come to be modified. Ptolemy would seem not to have been oblivious to what he was doing; in his more chatty moments, such as in *Almagest* IX.2 and XIII.2, he evidently is attempting to justify some of his complicated models and his departures from standard procedure. My own tentative interpretation is that Ptolemy is a compromiser of a certain type; he is certainly willing to modify the physical principles when they cannot accommodate the phenomena but he is equally unwilling to abandon them. The result is a rather spectacular success as far as planetary prediction is concerned but a muddle for those interested in a coherent and consistent science. At least, this is the way many of his Islamic successors chose to view him.

2. *Ṭūsī's Criticism of Ptolemy's Models*

There are several types of objections one finds in Islamic astronomy to Ptolemy's models. First there are criticisms of deficiencies in their predictive ability. These begin at a surprisingly early period in the history of Islamic science; for example, Ptolemy's failure to discover the motion of the solar apogee led to a rather severe rebuke by the 9th c. author of *Fī sanat al-shams* (On the Solar Year).[11] Next there is the blanket objection to his multicentered cosmology that targets his use of epicycles and/or eccentrics; this type of criticism is associated with the Spanish Aristotelians of the 12th c.[12] Finally we have the criticisms that accept the modeling that uses eccentrics and epicycles but cannot abide any further compromises. Ṭūsī's criticisms fall into this last category.

It is difficult at present to date in a precise manner when this type of criticism began. Qabīṣī (d. 967) mentions a book of his called *Shukūk fī "al-Majisṭī"* (Doubts on the *Almagest*), but it is not clear to what these doubts refer.[13] We are on firmer ground by the time we reach the first half of the 11th c. Ibn Sīnā's student and biographer, Abū ᶜUbayd al-Jūzjānī, tells us that his teacher had informed him that he had spent some time trying to understand the equant as well

---

[10] On the term *hypothesis*, see Toomer, *Alm.*, p. 23–24. For this Ṭūsī uses the Arabic word *aṣl* (pl. *uṣūl*), which literally means source or basis. I have rendered it into English as model.

[11] Morelon [1987], p. 61; the author is not Thābit as has usually been thought but possibly, according to Morelon, one or more of the Banū Mūsā.

[12] For an overview and further bibliography, see Sabra [1984].

[13] Qabīṣī, *Risāla fī imtiḥān al-munajjimīn*, Damascus, Ẓāhiriyya MS 4871, f. 67a, lines 21–24.

as the components of latitude (*mayl, iltiwā'*, and *inḥirāf*) and eventually had reached an understanding of them. Reconstructing the details, however, has proven elusive, perhaps because, as he told his student Abū ᶜUbayd, he had informed no one of his result. Jūzjānī himself offered a solution to the equant problem that has recently been published.[14] As far as we know, Ibn al-Haytham provided the first comprehensive criticism of the *Almagest* and the *Planetary Hypotheses* (as well as Ptolemy's *Optics*) in his *Shukūk ᶜalā Baṭlamyūs* (Doubts Concerning Ptolemy).[15]

It is an open question whether Ṭūsī knew the *Shukūk* directly, but I am inclined to believe, despite what Naṣīr al-Dīn himself seems to say,[16] that the objections to Ptolemy's models were fairly well-known by his time.[17] After Ṭūsī, they were to become such a commonplace that they were referred to simply as the sixteen "difficulties" (*ishkālāt*).[18] Although Ṭūsī is not as systematic as his successors, he does manage to identify all sixteen and in the following table I list them, where he identifies them in the *Tadhkira*, and the location of his alternative solutions.

[14] Saliba [1980]. Jūzjānī's solution was little appreciated; see our commentary to II.11 [1].

[15] An edition is due to A. I. Sabra and N. Shehaby [1971]; a discussion of the text with an English translation of the objections to the prosneusis point of the Ptolemaic lunar model and to the equant is in Sabra [1978]. An English translation of the entire *Shukūk* has been done as part of a University of Chicago doctoral dissertation by D. Voss [1985].

[16] See the commentary to II.11 [1].

[17] Bīrūnī, for example, alludes to the difficulties raised by the moon's prosneusis point as well as the equant in the *Qānūn*, 2: 838.

[18] See, for example, Khafrī's commentary to the *Tadhkira*, II.11, ff. 189b–190a.

Table 1. The "Difficulties" of the *Almagest*.

Difficulty	Identified in *Tadhkira*	Ṭūsī's Solution
1. Irregular Motion of the Moon's Deferent	II.7 [25]	II.11 [1–8]
2. Irregular Motion of Mercury's Deferent	II.8 [19]	II.11 [11] admission of no solution
3–6. Irregular Motions of the Deferents of Venus, Mars, Jupiter, and Saturn	II.9 [15]	II.11 [10]
7–11. Motions on Small Circles of Epicyclic Apices of Upper and Lower Planets (Latitudinal Deviation)	II.10 [6] II.11 [15]	II.11 [19]
12–13. Motions on Small Circles of Endpoints of Mean Epicyclic Diameters of Lower Planets (Latitudinal Slant)	II.10 [6] II.11 [15]	II.11 [19]
14–15. Oscillation of Equators of Deferent Orbs of Lower Planets	II.10 [2]	II.11 [20]
16. Oscillation of Lunar Epicycle Due to Prosneusis Point	II.7 [25] II.11 [13]	II.11 [21]

For a detailed discussion of these objections and their solutions, the reader is referred to the commentary to II.11. For now we can note that these objections fall into two main categories: those that involve irregular motion and those that concern an incomplete rotation. Difficulties 1–6 are in the former category and arise from the motion of the deferent being about a point other than its own center. In the case of the upper and lower planets, this is due to Ptolemy's introduction of the equant as the point of reference for uniform motion, a point not coincident with the deferent center. For the moon, the center of the World, which is distinct from the moon's deferent center, has the role of reference point for uniform motion. Difficulties 14–16 violate the requirement that celestial motion be continuously circular. In all three cases there is an oscillation that clearly contravenes this. Difficulties 7–13 are "mixed" violations. In each case the endpoints of diameters perform circular motions without the epicycle as a whole doing so, which goes against the requirement that the entire celestial orb must be rotating uniformly. Furthermore since these motions are coordinated with the motions of the epicycle center on the deferent, they will be subject to the objections concerning irregular motion raised in Difficulties 2–6.

3. Ṭūsī's Models

Naṣīr al-Dīn dutifully presents his version of Ptolemy's planetary models in Book II, Chapters 6–10. He has a standard procedure that he follows: first, he gives the observations that need to be explained; he then lists the orbs and their motions that are meant to account for these phenomena; and finally he enumerates the anomalies that arise from these motions for the observer. Pursuant to his assessment in II.5 [10] that "one who attempts to understand the principles of the motions must know the configuration of the bodies," his orbs (*aflāk*) are solid bodies as defined in I.1 [15]. These are indicated in Figures T5, T6, T8, and T10. Although these models are avowedly Ptolemaic, they do not precisely correspond to those of Ptolemy in either the *Almagest* or the *Planetary Hypotheses*. In the case of the former, this is due to Ptolemy's use of circles rather than bodies to present his models. To give a simple example: an eccentric deferent in the *Almagest* is a circle on which the epicycle center moves whereas Ṭūsī represents it as a solid body bounded by two parallel spherical surfaces in which one embeds the solid epicycle sphere.[19] In Book II of the *Planetary Hypotheses*, Ptolemy does transform his circles into bodies, and the result is generally, though not exactly, that of the *Tadhkira*.[20]

Ṭūsī comes up with a Ptolemaic count of 22 solid orbs; he also gives a "circle" count that comes to 32. The difference results from the 2 additional circles for the upper and lower planets, the equant and inclined, that are not needed when only solid orbs are considered.[21] In Book II, Chapter 11, he presents his alternative models, which necessitate the addition of 45 extra orbs.

---

[19] See II.5 [10] and Fig. T4.

[20] In the absence of a careful study, or for that matter even an edition, of Book II of the *Hypotheses*, it is difficult at present to specify those differences in exhaustive detail. We can say, though, that Ptolemy himself has not eased our task since he at various places gives 41, 34, 29, and 22 as the count for his orbs. In part this results from his ambivalence about the need for 7 spheres to account for the daily motions of the planets; in part it comes about due to his initial presentation of the models in terms of spherical orbs and then his subsequent advocacy of truncated orbs (*manshūrāt*). Cf. Neugebauer, *HAMA*, 2: 922–926.

[21] See II.8 [21] and II.9 [17].

Table 2. Orbs and Circles.

Planet	Ptolemaic Models: Solid Orbs	Ptolemaic Models: Circles & Motions	Ṭūsī's Models: Additional Orbs
Sun	2	2	0
Moon	4	4	6 [3(long.)+3(prosneusis)]
Mercury	4	6	9 [0 (long.)+3 (deviation) +3(slant)+3(inclination)]
Venus	3	5	12 [3 (long.)+3 (deviation) +3(slant)+3(inclination)]
Mars	3	5	6 [3 (long.)+3 (deviation)]
Jupiter	3	5	6 [3 (long.)+3 (deviation)]
Saturn	3	5	6 [3 (long.)+3 (deviation)]
Totals	22	32	45

Ṭūsī therefore needs 67 orbs simply to accomplish what he admits is only a partial success.[22] This enormous loss in economy must be justifiable on some grounds, and it is clear that the benefit to Ṭūsī, as well as to his contemporaries and successors, came from the consistency of mathematical and physical principles resulting from these new models. Using the rectilinear version of his couple, Ṭūsī was able to use uniformly rotating orbs to solve the problem of the irregular motions of the planetary deferents with the exception of Mercury (Difficulties 1, 3–6) while his curvilinear version could bring about the oscillations of Difficulties 7–16, again with uniformly rotating orbs.[23]

There is another facet of Ṭūsī's achievement, namely his ability to isolate specific aspects of celestial motion, in particular rectilinear and curvilinear components. Thus with the rectilinear version of his couple he is able to bring the epicycle center linearly toward and away from the reference point for uniform motion (the equant for the upper and lower planets, the center of the World for the moon); this allows him to circumvent Ptolemy's need for a circumference, that of the deferent, to effect what is essentially a variation in rectilinear distance. For the curvilinear version, Ṭūsī can have an oscillation along an arc which can give him an inclination in either latitude or longitude; with Ptolemy, these inclinations, when he brings them about with a mechanism, are the result of motions on a small circle that result in both latitudinal and longitudinal displacements.[24] Historically it is interesting to note that Ibn al-Haytham's attempt

[22] This is because Mercury eludes him and because, as he notes, there are inadequacies associated with the curvilinear version of his couple; see II.11 [11], [19] and [21].

[23] For details, see the commentary to II.11 and Ragep [1987].

[24] See II.11 [14–15].

to give Ptolemy's small latitude circles a *hay'a*, or configuration, merely sought to reproduce those circles using a Eudoxan-style technique.[25] With Naṣīr al-Dīn, it is his additional circles, ironically, that helped break the stranglehold of circular motion.

G. Sources Named (and Unnamed) by Ṭūsī in the *Tadhkira*

Naṣīr al-Dīn was one of the best informed savants of the 13th century in both the ancient and religious sciences; there can be little doubt that he had a wide variety of sources to choose from in compiling the *Tadhkira*. But it is clear, both from his explicit statements and from his implicit choices, that he saw his primary source as the *Almagest*; even when more "modern" parameters are cited, Ṭūsī tends to use those from the *Almagest*.[1] There is little wonder then that "Ptolemy" and the "*Almagest*" are explicitly cited far more times (21) than any other author or text. (The next closest is Ibn al-Haytham at 2.)

In addition to explicitly cited sources, we can easily infer that Ṭūsī has depended on a number of classical and Islamic authors. For the mathematics introduction of I.1, Euclid is of course primary, but Ṭūsī has also used various works of Theodosius, Autolycus and possibly Hero. Archimedes is explicitly cited in IV.1 [1] for the measures of circles and spheres, and a proposition from Menelaus's *Spherics* is used in III.7 [2].[2] Euclid is referred to by name in IV.4 [2] for finding the volume of a sphere.

For physical principles, the Aristotelian corpus provides the fundamental basis, but there have been some important shifts in formulation and emphasis whose inspiration is by way of Ibn Sīnā.[3] But neither Aristotle nor any other philosopher or physicist is mentioned in this connection, and in fact the only philosophers or theologians referred to at all are Ibn Sīnā and Fakhr al-Dīn al-Rāzī, both with regard to the temperateness of the equatorial region in III.2 [2–4]. As for metaphysics, which is referred to as one of three disciplines from which *hay'a* derives its principles (the other two being natural philosophy and mathematics), it receives scant, if any, attention. There is one reference to divine providence in III.1 [6].[4]

Turning to practical astronomy, again the *Almagest* is fundamental, but my impression is that Bīrūnī's manifold and multifaceted works were an important

---

[25] See the commentary to II.11 [16].

[1] For example: he uses Ptolemy's eccentricity for the sun in IV.5 [1] rather than those of the moderns in calculating the nearest and farthest distances of the sun; in IV.3 [2] he gives the *Almagest* value for the moon's apparent diameter rather than the value from the *Planetary Hypotheses* or even his own stated values in II.13 [8]2–4.

[2] See the commentary for references.

[3] For these, see pp. 44–46 and the commentary to I.2.

[4] See our commentary for a brief discussion.

resource though he is explicitly referred to only once.[5] As for the early period of Islamic astronomy, Ṭūsī does mention Ma'mūn's astronomers in connection with the measurement of the Earth in IV.1 [2] and he knows of Ibrāhīm ibn Sinān's work on trepidation; in both cases, though, there is some question of how well-informed he actually is on these matters.[6]

I have discussed at some length above Ṭūsī's indebtedness to his predecessors in theoretical astronomy. Curiously he himself does not mention the major works I have identified as part of the *hay'a* tradition, namely Ptolemy's *Planetary Hypotheses*, Ibn al-Haytham's *Al-Maqāla fī hay'at al-ᶜālam* and Khiraqī's *Muntahā al-idrāk* and *al-Tabṣira*. (The *Muntahā*, though, is mentioned in Ṭūsī's *Ḥall-i mushkilāt-i Muᶜīniyya*.[7]) The lack of any explicit mention of the *Planetary Hypotheses* I find particularly surprising since a good deal of the material in Book IV on sizes and distances comes from there rather than the *Almagest*. But he uses expressions such as "they have stated" or "it seemed likely to them" rather than telling his readers that the material comes from the *Hypotheses*.[8]

As for the proposals to reform the Ptolemaic system, Ṭūsī leaves the reader with the strong impression that little, if anything, has been done by earlier astronomers. (He makes the point in both II.7 [25] and in II.11 [1] that none of his predecessors have ventured an explanation or solution of the irregular motion of the moon's deferent.) He does, though, mention an anonymous author in II.11 [12] who proposed a solution to the moon prosneusis problem. He also gives a fairly thorough explanation of Ibn al-Haytham's attempt to provide a *hay'a* (configuration) to resolve the difficulty of part of Ptolemy's latitude theory.

I do not wish to leave the impression that Naṣīr al-Dīn has willfully ignored his predecessors. Although he was quite well informed, there are real questions about what was generally available to someone in Persia during the middle (as opposed say to the end) of the 13th century. For example I am far from certain whether Ṭūsī knows the *Planetary Hypotheses* directly or even Ibn al-Haytham's *Shukūk ᶜalā Baṭlamyūs* (Doubts Concerning Ptolemy).[9] It is salutary to realize that Ṭūsī does not know the greatest of Ibn al-Haytham's works, namely his *Optics*.[10] We are still a long way from being able to write a true history of the influences on and the relationships between the principals of Islamic science.

[5] This in connection with measuring the Earth by the "dip of the horizon" method in IV.1 [4].

[6] See the commentary to II.4 [5] and IV.1 [2].

[7] See p. 33.

[8] For references, see the introductory remarks to the commentary of IV.5.

[9] See the commentary to II.11 [1].

[10] See the commentary to II.1 [1]21–1 and II.13 [1]16.

Explicit References in the Tadhkira *to Persons and Works*

1. Ptolemy: II.5 [6]; II.6 [1], [2], [3], and [4] (2 occurrences); II.11 [14], [16] and [18]; IV.2 [2] and [4]; IV.3 [1]; IV.5 [1]; IV.6 [1], [4] and [6].

2. Ibn al-Haytham: II.11 [16] (2 occurrences).

3. Abū ᶜAlī ibn Sīnā: III.2 [2] and [4].

4. Archimedes: IV.1 [1].

5. Abū al-Rayḥān al-Bīrūnī: IV.1 [4].

6. Euclid: IV.4 [2].

7. Fakhr al-Dīn al-Rāzī: III.2 [3].

8. Ma'mūn's scientists: IV.1 [2].

9. The *Almagest*: I.Intr. [3] (2 occurrences); II.5 [10]; II.10 [6]; II.11 [14].

10. Natural Philosophy Corpus (*al-ṭabīᶜiyyāt*): I.Intr. [4]; I.2 [title]; II.1 [8].

11. Geometry Corpus (*al-handasiyyāt*): I.Intr. [4]; I.1 [title].

12. *De caelo* (*al-samā' wa-'l-ᶜālam*): II.1 [8].

H. The Influence of the *Tadhkira*

The *Tadhkira* had an enormous influence on the subsequent history of astronomy—so much so that it would be foolhardy to pretend to do justice to it in a few pages. I propose here simply to sketch a program for dealing with that influence and to summarize what we know so far.

"Influence" is one of those tricky historical concepts whose meaning is usually assumed rather than delineated. Most of the discussion of the influence of the *Tadhkira* has focused on its non-Ptolemaic models in II.11, the part these played in the so-called "Marāgha school," and the significance of this "school" in the astronomy of the European Renaissance and, in particular, that of Copernicus. But besides displaying a Eurocentric bias, such a viewpoint has serious historical limitations. "Marāgha," for all its importance in the history of Islamic observational astronomy and teaching, was simply one *episode* in a very long and complex story of Islamic theoretical astronomy. As we shall see in Section J, Ṭūsī had developed his non-Ptolemaic models long before coming to

Marāgha and compiling his *Tadhkira*. ᶜUrḍī as well seems to have developed his models prior to his Marāgha residence,[1] and Shīrāzī's *Nihāya* and *Tuḥfa* were written after leaving Marāgha. And to call an Ibn al-Shāṭir, a Qūshjī, a Bīrjandī or any other late medieval Islamic astronomer writing in various regions of the Islamic world part of this "Marāgha school" substitutes shorthand for history. Later astronomers certainly acknowledged the importance of what had occurred in the 13th c., not only at Marāgha but elsewhere, but they would have seen this as part of a long historical process that, for some, had begun with Ibn al-Haytham, for others even earlier; in short, they considered themselves not part of some "school" but ongoing members of the *hay'a* tradition.

As we have discussed at length above, the *Tadhkira* was an important part of this tradition since it provided, first and foremost, a summary of astronomy from the point of view of the solid bodies. Its influence as a *hay'a basīṭa* work can be seen in numerous ways. The large number of extant manuscript copies stand in silent testimony to this. I suspect that from the 13th until the 18th centuries, the *Tadhkira* was the text of choice for beginning students of astronomy as well as educated laypersons who wanted an introduction providing more meat than Jaghmīnī's *Mulakhkhaṣ*.[2] It is, for example, quoted extensively by the encyclopaedists Ṭāshkubrīzāde (16th c.) and Tahānawī (18th c.) and is mentioned by Ṣafadī (14th c.) in one of his literary works.[3] In addition I believe that the *Tadhkira* was an important model for those texts promising a taste of *hay'a* without tears. In addition to Ṭūsī's own *Zubdat al-idrāk fī hay'at al-aflāk* and to a lesser degree his Persian *Zubdah-i hay'a*,[4] several elementary textbooks seem to derive both their form and contents from the *Tadhkira*. An instructive example is *Kitāb al-Nuzha al-ᶜAlā'iyya*, an otherwise inconsequential school text by a certain Tāj al-Dīn al-Tabrīzī. Though incomplete, the table of contents shows that this late simplified introduction to astronomy follows the basic form of the *Tadhkira*, an indication that its approach was influential at all levels of astronomical writing. An even more important example of this may be Jaghmīnī's *Al-Mulakhkhaṣ fī al-hay'a al-basīṭa* (Epitome of plain *hay'a*), which has been mentioned previously. There is some rather strong evidence that this extremely popular and simplified introduction to astronomy, which was the subject of numerous commentaries and supercommentaries, was dependent on the *Tadhkira*.[5]

---

[1] See Saliba [1979].

[2] Ṭāshkubrīzāde (901–68/1495–1561), for example, names the *Tadhkira* first as an epitome (*mukhtaṣar*) of *hay'a basīṭa* and then lists ᶜUrḍī's *Hay'a* and then Shīrāzī's *Tuḥfa* and *Nihāya* as more elaborated works of the genre. The *Mulakhkhaṣ* is then given as a "well-known" (or perhaps "much disseminated") epitome (*min al-mukhtaṣar al-mashhūr*) (*Miftāḥ*, 1: 372–373).

[3] Ṣafadī, *Al-Ghayth al-musjam fī sharḥ lāmiyyat al-ᶜajam*, 2: 257.

[4] The former seems never to have gained much popularity; the latter, in Persian and in Arabic translation, enjoyed somewhat greater success; see Section J.1, pp. 66–67.

[5] For the evidence, see p. 35, footnote 11 and the commentary to III.1 [8].

The *Tadhkira* was also influential on another level. It formed an important basis for more detailed and elaborated studies such as Shīrāzī's *Nihāya* and *Tuḥfa*, which themselves became the primary basis for future work in *hay'a*. It goes without saying that the many commentaries on the *Tadhkira*, listed in the next section, also continued, at times in very interesting and exciting ways, the development of *hay'a* in Islam. Ṭūsī's programmatic approach to the *ishkālāt* (difficulties) of astronomy in II.11, and his admission that several problems remained to be resolved such as Mercury and the latitude theory, provided a significant, indeed crucial, step—and challenge—to the further evolution of non-Ptolemaic modeling in Islam. One can hardly imagine any work after Ṭūsī dealing with the *ishkālāt* that did not mention the Ṭūsī couple, known in subsequent literature as the "model of the big and the small."

The influence of the *Tadhkira* was also felt in cultures beyond the borders of Islam. Jayasiṃha, who ruled in Rājasthāna from 1700 until 1743, clearly had an interest in the astronomy of the Yavanas (Muslims) that led to the acquisition, among other texts, of a copy of Nīsābūrī's commentary on the *Tadhkira*.[6] Of even greater interest, and somewhat surprising, is the Sanskrit translation by Nayanasukha and his assistant Muḥammad Ābida of II.11 of Bīrjandī's commentary on the *Tadhkira*, which is one of the more sophisticated and extensive discussions of non-Ptolemaic astronomy in Islam. But according to Pingree, the work had little lasting influence,[7] which is not surprising inasmuch as the *theoretical* aspects of Ptolemaic astronomy made few inroads into traditional Indian astronomy. Nevertheless this episode in Indian astronomy is of considerable historical interest, especially as it came when European influence was beginning to be felt.

Further west, the impact of the *Tadhkira* may also be detected. This was originally, but rather obliquely, suggested by Dreyer in a footnote in the course of his discussion of Ṭūsī's models that referred the reader to Book III, Chapter 4 of Copernicus's *De revolutionibus*, where the Ṭūsī couple is introduced.[8] But Dreyer suggests nothing beyond this curiosity; postulating a connection between late Islamic astronomy and Copernicus had to await another time and place. This came in the 1950s with the discovery by E. S. Kennedy of the models in Ibn al-Shāṭir's *Nihāyat al-sūl*, which were virtually identical with several of those used by Copernicus. In a series of articles culminating with Kennedy [1966], Kennedy and his collaborators laid out the circumstantial evidence linking late Islamic astronomy, including that of Ṭūsī, and Copernicus.[9] More substantive

[6] I depend here on the very informative article by Pingree [1987]. See also King [1980].

[7] Pingree [1987], p. 325. The Sanskrit translation has been studied and translated into English by T. Kusuba, who presented it as a Master's thesis to the History of Mathematics Department, Brown University.

[8] Dreyer [1906], p. 269. He knew of the *Tadhkira* from Carra de Vaux's [1893] translation .

[9] This research, as well as that of others, was exquisitely summarized by Swerdlow and Neugebauer [1984], 1: 41–48, where the reader may find extensive references to the relevant literature.

evidence came with the discovery by Neugebauer that MS Vatican Gr. 211, which was in Italy by 1475, has in it a short treatise dealing with planetary theory that contains diagrams of a Ṭūsī couple and lunar model; the treatise itself is a Greek translation by Gregory Chioniades of an Arabic original.[10] Furthermore Swerdlow [1972] has shown that the Ṭūsī couple was used by at least one other Renaissance astronomer, and Copernicus himself indicates that the device was hardly a novelty by his time.[11] As the evidence for contacts between late medieval Islamic and Renaissance astronomy has piled up, "the question," to quote Swerdlow and Neugebauer, "is not whether, but when, where, and in what form [Copernicus] learned of Marāgha theory."[12]

I would certainly concur in that judgment; but I think we should both refocus and reformulate the problem away from models and "Marāgha" and toward the *hay'a* tradition itself. Copernicus shared with his Islamic predecessors an approach to astronomy that emphasized the reintegration of physics into mathematical astronomy;[13] in that sense I would consider him as much part of the *hay'a* tradition as Ṭūsī—or for that matter Aristotle and Ptolemy. So the problem is not simply one of change in mathematical models but of the evolution of classical physics as well. In my commentary to II.1 [6], I present an interesting coincidence of views between Ṭūsī and Copernicus regarding the Earth's rotation. Both use the same arguments; each comes to a different conclusion. But as I point out, the debate involving Ṭūsī's position regarding the Earth's stasis went on in Islam until the 16th century—if not later—and involved major reformulations of Aristotelian physics. Did Copernicus know of these debates? I simply do not know. I would argue, though, that to foreclose the possibility is just as "biased" as to assert a connection without further proof. But whatever the outcome of the debate concerning Copernicus's predecessors and motivations, the role of the *Tadhkira* and its author in the history of astronomy would seem secure.

I. The Commentaries on the *Tadhkira*

The large number of commentaries written on the *Tadhkira* provide compelling evidence for its enormous influence. But they also provide a vast resource for studying the development and fate of *hay'a* in Islam. Unfortunately, our modern insistence on "creativity" and "originality" has led to a sharp downgrading of such "mere commentaries"; indeed the large number of them in astronomy and other fields has often been taken as a symptom of the declining

[10] Swerdlow and Neugebauer [1984], 1: 47–48 and Figs. 5–6, 2: 567–568.

[11] See our commentary to II.11 [2].

[12] Swerdlow and Neugebauer [1984], 1: 47.

[13] This was emphasized by Swerdlow [1976], who was responding to Rosen's [1975] ahistorical approach to the problem.

centuries of Islamic science. It is clear that those who have made these judgments have never read these works with any care or attention to detail. Even the more mediocre of them are written by competent scholars who are responding to a continuing interest in theoretical astronomy by both students and the educated public. The best of them are highly original works that provide new solutions to the *ishkālāt* (difficulties) of astronomy as well as very interesting passages concerning the status of astronomy, the relation of theory and observation, the role of physics in astronomy, and other theoretical concerns. The commentary style is simply that—a style within which the author can exhibit creativity and criticism to the extent of his abilities. In Bīrjandī's massive commentary of over 250 folios, the *Tadhkira* often becomes almost incidental as he discusses matters that Ṭūsī had either barely touched upon or ignored entirely. In short, these commentaries as a group represent an important part of the history of astronomy that need to be edited and studied if we are ever to go beyond our very fragmentary view of late medieval astronomy.

One also sees in many of these commentaries a very "modern" concern with textual criticism and historical reconstruction. This has greatly facilitated my own work, and I have not hesitated to use them as my basic "secondary" sources. My favorite quickly became that of Bīrjandī, who seems to have known almost everything there was to know about *hay'a*. My own commentary owes an enormous debt to his elaborations and insights. (I long since have forgiven him his arrogance and nit-picking.)

The following list contains those works that are self-declared commentaries or supercommentaries on one or more passages of the *Tadhkira*. They range in size from Kamāl al-Dīn al-Fārisī's 4-folio treatise, which is restricted to the *Tadhkira*'s discussion of retrograde and direct motion, all the way up to Shīrwānī's nearly 400-folio commentary on the entire work. It should be noted that virtually all the authors are Persians, but they worked and studied in places as far afield as Ḥamāh, Syria (no. 6) and Samarqand (no. 8). I have not listed those texts that are clearly derivative from the *Tadhkira* (which would include most subsequent works on *hay'a*) or that appropriate large parts of it (such as Quṭb al-Dīn al-Shīrāzī's *Nihāyat al-idrāk* and *Tuḥfa*) since they do not follow the standard format of a commentary or supercommentary. I should also mention that many copies of the *Tadhkira* and its commentaries contain extensive notes and glosses, which would be well worth studying to gain insight into the influence of the *Tadhkira* as well as the history of *hay'a*.

Unless otherwise indicated, all the works are in Arabic. (Note: As a general rule, I have throughout referred to the commentaries by the book, chapter, paragraph and line number of the *Tadhkira* established in my edition. In this way, I have avoided referring to specific copies of manuscripts that may be either unfoliated or inaccessible.)

(1) *Tibyān maqāṣid al-Tadhkira* (Exposition of the Intent of the *Tadhkira*) by Muḥammad b. ᶜAlī b. al-Ḥusayn al-Munajjim al-Ḥimādhī, composed some-

time between Jumādā I, 684/July–August 1285 and 4 Ramaḍān 710/ca. 25 January 1311; the former is the date of composition of *Al-Tuḥfa al-Shāhiyya* by Quṭb al-Dīn al-Shīrāzī, who claims in (2) that Ḥimādhī has substantially plagiarized his work, and the latter is the date of Shīrāzī's death. This commentary is only extant as an incorporated part (and therein substantially contracted) of (2).

Begins:

الحمد لله رب العالمين وصلوته على خير خلقه محمد وآله الطاهرين يقول أحوج خلق الله إليه محمد بن علي بن الحسين المنجم الحماذي لما كان كتاب التذكرة في علم الهيئة من مصنفات أفضل المتأخرين

(2) *Fa^c alta fa-lā talum* (*lit.*, You have Done It, So Do Not Condemn), by Maḥmūd b. Mas^c ūd Quṭb al-Dīn al-Shīrāzī, composed sometime between the boundary dates of (1). This work is a rather caustic reply to (1) that, among other things, attempts to defend the honor of the deceased Naṣīr al-Dīn, accuses Ḥimādhī of plagiarism with regard to the *Tuḥfa*, and generally directs sarcastic and at times vicious comments at that hapless astronomer.

I have used Tehran, Majlis-i Shūrā MS 3944 (= Arab League, Ba^c athat Īrān film no. 228); 232 ff.; copied 826/1422–23 from an autograph.

Cf. *GAL* I, p. 511 (= 675), SI, p. 931 (no. 40a); Matvievskaya/Rozenfeld, *Mat. i astr.*, 2: 432 (no. 387, A4).

Begins:

أما بعد حمد الله خالق الأفلاك ومديرها...فإن أحوج خلق الله إليه محمود بن مسعود بن المصلح الشيرازي ختم الله له بالحسنى يقول إني لأتفكّر في صناعة التصنيف وأجيل نظري على جمهور أصحاب التأليف

(3) *Tawḍīḥ al-Tadhkira* (Elucidation of the *Tadhkira*), by al-Ḥasan b. Muḥammad Niẓām al-Dīn al-Nīsābūrī, completed the first of Rabī^c I, 711/18 July 1311. Ḥājjī Khalīfa (under the entry for *al-Tadhkira*) remarks that this commentary is well-known and widely appreciated.

I have mainly used Najaf, Āyat Allāh al-Ḥakīm Library MS 649, 1 (= Arab League, uncatalogued *falak* film no. 315), 107 ff. I have also occasionally used London, British Library MS Add. 7472 for comparison.

Cf. *GAL* I, p. 511 (= 675), SI, p. 931 (no. 40b); Matvievskaya/Rozenfeld, *Mat. i astr.*, 2: 438–439 (no. 395, A3).

Begins:

الحمد لله الذي جعلنا من المتفكرين في خلق الأرض والسموات وشرفنا بالنظر في هيئة الأجرام المبدعات ، فهدانا التفكر في المصنوعات ، والتدبر في أمر المدبرات إلى وجود صانع قدير حكيم خبير

(4) Glosses (*Ḥāshiya*) on Nīsābūrī's *Tawḍīḥ al-Tadhkira*, by a certain Faṣīḥ al-Dīn. The only extant manuscript is Leiden or. MS 1010; 50 ff.

Cf. *GAL* I, p. 511 (= 675).

Begins:

أما على الإطلاق كالعدد للحساب قلت فيه بحث

(5) *Ḥāshiya* ᶜ*alā dhikr aṣl al-rujū*ᶜ *wa-'l-istiqāma fī al-Tadhkira* (Glosses on Retrograde and Direct Motion as Described in the *Tadhkira*), by Kamāl al-Dīn al-Ḥasan b. ᶜAlī b. al-Ḥusayn al-Fārisī (d. ca. 720/1320). Fārisī was a student of Shīrāzī and is particularly noted for his *Tanqīḥ al-Manāẓir*, an incisive commentary on Ibn al-Haytham's *Optics*.

The only copy of this short work known to me occurs at the end of a manuscript containing Nīsābūrī's commentary, namely Najaf, Āyat Allāh al-Ḥakīm Library MS 649, 2 (= Arab League, uncatalogued *falak* film no. 315), 4 ff.

Begins:

قال مولانا الأعظم الحبر الأعلم رئيس الحكماء سلطان المهندسين كمال الملة والدين الحسن بن علي بن الحسين الفارسي سقاه الله شآبيب رضوانه وكساه جلابيب غفرانه الحمد لوليه حاشية على ذكر أصل الرجوع والاستقامة في التذكرة الخط المستقيم المواجه في الغاية هو الذي ينصفه سهم شعاع البصر ويكون عمودا عليه

(6) *Takmīl al-Tadhkira* (Complement to the *Tadhkira*), by ᶜUmar b. Da'ūd b. Sulaymān al-Fārisī, completed during the last part of Ramaḍān 711/ January–February 1312 for Abū al-Fidā', the governor of Ḥamāh and author of a famous work on geography.

The unique manuscript is Cairo, Taymūr riyāḍa MS 128, 99 ff.; cf. King [1986], p. 152.

Begins:

قال المفتقر إلى رحمة ربه عمر بن داود بن الشيخ سليمان الفارسي... الحمد لله الذي فطر السموات بغير عمد إرشاداً للخلق الصمد

(7) *Bayān al-Tadhkira wa-tibyān al-tabṣira* (Explanation of the *Tadhkira* and Exposition of the Enlightenment [a pun on Khiraqī's *Tabṣira*]), by Jalāl al-Dīn Faḍl Allāh al-ᶜUbaydī (d. 751/1350), completed 24 Rabīᶜ I, 728/ca. 7 February 1328.

Two copies of this work are Ahmet III MS 3325, 2 (= Topkapı Saray MS 7058, 2), ff. 34b–131b and Ahmet III MS 3313 (= Topkapı Saray MS 7083). I have used the former.

Begins:

الحمد لله الذي خلق السماء متحركة على القطب والمحور والأرض ساكنة على شكل الكرة

(8) *Sharḥ al-Tadhkira* (Commentary on the *Tadhkira*), by al-Sayyid al-Sharīf ᶜAlī b. Muḥammad al-Jurjānī (740–816/1339–1413), completed Tuesday, in the middle of Dhū al-ḥijja, 811/(probably) 30 April 1409, in Shīrāz. Jurjānī is mainly noted today as a philosopher/theologian whose commentary on Ījī's *Mawāqif* became a standard work. He was for a time (1387–1405) part of Timur's entourage at Samarqand and has been described as having been a kind of court theologian there. If the evidence of extant MSS is any indication, this was an extremely popular work.

I have used Damascus, Ẓāhiriyya MS 3117, 160 ff.

Cf. *GAL* I, p. 511 (= 675), SI, p. 931 (no. 40c); Matvievskaya/Rozenfeld, *Mat. i astr.*, 2: 476 (no. 424, A1).

Begins:

تبارك الذى جعل في السماء بروجا متخالفة المراتب والآثار وزينها بثواقب الكواكب متفاوتة الأنوار والأنظار

(8_1) A work listed in *GAL* I, p. 511 (no. 40e) as a commentary by Mūsā Qāḍīzāde (Biblioteca Medicea Laurenziana or. MS 271 [= Palat. MS 311, not 313 as in *GAL*]) is actually a copy of Jurjānī's *Sharḥ* lacking the introduction and part of Book I, Chapter One.

(9) *Sharḥ al-Tadhkira* (Commentary on the *Tadhkira*), by Fatḥ Allāh b. ᶜAbd Allāh al-Shīrwānī (al-Rūmī al-Ḥanafī) (d. 891/1486), completed (according to the colophon of Ahmet III MS 3314) Wednesday evening, 3 Ramaḍān 879/Tuesday–Wednesday, 10–11 January 1475. Shīrwānī tells us that he wished to write a more detailed commentary than those of Nīsābūrī and Jurjānī; this mammoth work is certainly that. Among other things, it contains a long *tadhnīb* (appendix) on optics at the end of Book I of about 28 folios. The text is in Arabic.

Baghdādī (*Hadiyya*, 1: 815) claims that Shīrwānī was a resident of Qasṭamūnī (in Anatolia), studied with al-Sharīf al-Jurjānī, and died in 857/1453; the last two claims are untenable based on the date of composition given above as are Sarton's statements that this commentary was written in Turkish and completed in 1414 A.D. (*Introduction*, 2: 1007). I have taken Shīrwānī's death date from *GAL* SII, p. 290; repeated by Matvievskaya/Rozenfeld, *Mat. i astr.*, 2: 513. I do not know the basis of their information.

There do not seem to be many extant copies; I have used Ahmet III MS 3314 (= Topkapı Saray MS 7093), 370 ff.

Cf. Matvievskaya/Rozenfeld, *Mat. i astr.*, 2: 513 (no. 444v, A1).

Begins:

الحمد لله الذي هيّأ هيئة العالم بحكمته وشيّأ زمرتنا(؟) على التفكر فيها بشيئته

Cf. the anonymous commentary listed by Riḍawī, *Aḥwāl*, pp. 405–406 (no. 9).

(10) *Sharḥ al-Tadhkira* (Commentary on the *Tadhkira*), by ᶜAbd al-ᶜAlī b. Muḥammad b. Ḥusayn al-Bīrjandī (d. 932/1525–26), completed Rabīᶜ I, 913/July–August 1507. He was a student of Manṣūr al-Kāshī, the son of Muᶜīn al-Dīn al-Kāshī; both father and son may well have been on the staff of the Samarqand observatory (823–53/1420–49). Bīrjandī himself is noted for having written a commentary on Ulugh-beg's *Astronomical Tables* as well as a super-commentary on Qāḍīzāde's commentary on Jaghmīnī's *Mulakhkhaṣ* (Sayili [1960], p. 267).

The commentary on the *Tadhkira* makes it clear that he is well-informed not only about what took place at Samarqand, but also about earlier astronomical work. In addition to the commentaries on the *Tadhkira*, Bīrjandī refers to numerous other works in a variety of fields, all of which he is careful to cite. Curiously it contains both sophisticated criticism of previous work and rather extensive (some might say long-winded) explanations of elementary points; clearly Bīrjandī had colleagues as well as students in mind as an audience. The text is in Arabic and should not be confused with a Persian work by the same author entitled *Risālah-i hay'a* (also called *Sharḥ Mukhtaṣar al-hay'a*?); for the latter, see Storey, *Persian Literature*, II.1, p. 82 (no. 121, 3).

I have used Cambridge, Harvard College Library, Houghton MS Arabic 4285, 258 ff.

Cf. *GAL* SI, p. 931 (no. 40g) and Matvievskaya/Rozenfeld, *Mat. i astr.*, 2: 542 (no. 456, A9). (Note: Brockelmann apparently is confusing it with the Persian work noted above.) Book II, Chapter 11 was translated into Sanskrit; see p. 57 in this volume.

Begins:

الحمد لله الذي خلق السموات والأرض وجعل الظلمات والنور وبسط على بساط الساهرة بميامن[؟] قدرته الباهرة الظل والحرور

(11) *Al-Takmila fī sharḥ al-Tadhkira* (The Complement to the Commentary on the *Tadhkira*), by Shams al-Dīn Muḥammad b. Aḥmad al-Khafrī, completed

Monday, 4 Muḥarram 932/23 October 1525. The commentary to which this is a complement is Jurjānī's. *Khafr* is the name of a small village near Shīrāz (just east of Fīrūzābād) that I take to be the basis for the *nisba* of this writer. This reading is attested by Damascus, Ẓāhiriyya MS 6727 as well as by Riḍawī (*Aḥwāl*, p. 404). Other possibilities, which seem less plausible, are "Khiḍrī" and "Khafarī," which are given in *GAL*. This work appears to have been quite popular despite its late date. In addition to Jurjānī, Khafrī depends on Nīsābūrī's commentary as well as on Shīrāzī's *Tuḥfa*. Though written during the assumed precipitous decline of Islamic science, Khafrī shows real insight into and understanding of the major problems of *hay'a*.

I have mainly used Damascus, Ẓāhiriyya MS 6727, 323 ff.; Damascus, Ẓāhiriyya MS 6782, 297 ff. was used for comparison.

Cf. *GAL* SI, p. 931 (no. 40d) and Matvievskaya/Rozenfeld, *Mat. i astr.*, 2: 471 (no. 422a, A1).

Begins:

وتعاليت يا ذا العرش الأعلى وما أعظم شأنك وتباركت يا مبدع السموات العلي... فيقول الفقير إلى الله الغني محمد بن أحمد الخفري لما كان أجل العلوم بيانا وأوثقها تبيانا هو علم الهيئة

(12) *Taᶜlīqāt* (annotations) on the *Tadhkira*, by Ghiyāth al-Dīn Manṣūr. This is presumably Ibn Amīr Ṣadr al-Dīn Muḥammad al-Shīrāzī (d. ca. 950/1543–44), a prince who had been charged by the Ṣafavid King Shāh Ismāᶜīl I (ruled 1501–24) with restoring the Marāgha observatory, a task that was never brought to fruition (Sayili [1960], p. 288). Riḍawī (*Aḥwāl*, p. 406 [no. 10]) refers to a single manuscript of this work in the Madrasah-i ᶜālī-i Sipahsālār Library; it is not mentioned by any other source.

Begins (according to Riḍawī):

ان هذه تذكرة فمن شاء اتخذ إلى ربه سبيلا

(13) A commentary by Kamāl al-Dīn Ḥusayn b. Sharaf al-Dīn ᶜAbd al-Ḥaqq al-Ardabīlī (d. 950/1543–44). As far as I know, this work is not extant. See Riḍawī, *Aḥwāl*, p. 405 (no. 8), who cites Baghdādī, *Hadiyyat al-ᶜārifīn*, 1: 318.

(14) Another anonymous commentary is contained in Paris, Bibliothèque nationale MS ar. 6085. This incomplete manuscript lacks all of Book I and Book II, Chapters One through Seven. The illustrations are missing but otherwise the text seems complete. It was copied on Thursday, 18 Rajab 1091/15 August 1691.

Cf. *GAL* SI, p. 931 (no. 40f).

J. The Evolution of the Text of the *Tadhkira*

1. Ṭūsī's Other hay'a basīṭa *Works and Their Relationship to the* Tadhkira

The *Tadhkira* was in large measure modeled after one of Ṭūsī's Persian works, the *Risālah-i Muᶜīniyya*, written during the early period of his residence at the Ismāᶜīlī stronghold in Qūhistān. Ṭūsī subsequently published an appendix (*dhayl*) to it, usually called *Ḥall-i mushkilāt-i Muᶜīniyya*, whose contents, which included the first appearance of the rectilinear Ṭūsī couple, were also incorporated into the *Tadhkira*. The *Muᶜīniyya* thus forms an essential part of the background to the *Tadhkira*. Of the many astronomical and astrological writings attributed to Naṣīr al-Dīn, the only other work that I would classify within the genre of *hay'a basīṭa* (plain or simplified *hay'a*) is his *Zubdat al-idrāk fī hay'at al-aflāk*. This small treatise seems to be an abridgment of his larger *hay'a* productions, but further study will be needed to establish a more precise relationship.

Brief accounts of these works follow:

(a) *Risālah-i Muᶜīniyya* [or, mistakenly, *al-mufīd* and *al-mughniya*] *dar hay'a* (the *Muᶜīniyya* treatise on *hay'a*), a treatise written in Persian and completed Thursday, 2 Rajab 632/22 March 1235.[1] This work was written in Qūhistān for Muᶜīn al-Dīn Abū al-Shams, the son of Nāṣir al-Dīn Muḥtasham, the Ismāᶜīlī governor of Qūhistān and patron of Naṣīr al-Dīn. There are two different versions of the introduction, one in which Ṭūsī lavishes praise upon his Ismāᶜīlī patrons and the other in which he leaves out any direct mention of them altogether. Clearly, the second was written after the fall of Alamūt in 654/1256.

The text itself has the characteristic four-part division of a *hay'a basīṭa* work and in fact quite closely, but not exactly, anticipates the *Tadhkira* in both structure and content: Both have an introductory section consisting of two chapters dealing with geometry and natural philosophy; both have a section on the configuration of the celestial bodies (*hay'at-i ajrām-i ᶜalawī*) divided into fourteen chapters; both have a section on the configuration of the Earth (*hay'at-i zamīn*) divided into twelve chapters; and both have a section on distances and sizes (*abᶜād wa-ajrām*), the *Muᶜīniyya* having six chapters whereas the *Tadhkira* has seven.

A facsimile reproduction of this work is due to Muḥammad Taqī Dānish-Pizhūh, who provides both versions of the introduction in his intro-

[1] This is the date given in the colophon of Tehran, Dānishgāh (Kitābkhānah-i markazī), Āqā Mishkāt MS 1014 (1), which is transcribed by Dānish-Pizhūh on p. *z* of his introductory remarks to the facsimile reproduction of another manuscript of the *Muᶜīniyya*. (There is an indication that the former copy may be an autograph.) The date of composition is also confirmed by Riḍawī, *Aḥwāl*, p. 388, based on another manuscript source.

ductory remarks.[2] Kennedy [1984] gives a table of contents and reports that the work has been described by Usmanov [1978].

For further bibliographical details, see *GAL* I, p. 511 (no. 40), SI, p. 931; Krause [1936], pp. 494–495 (no. 2); Storey, *Persian Literature*, II.1, p. 56 (no. 7); Riḍawī, *Aḥwāl*, pp. 384–388 (no. 30)); King [1986], p. 152; and Matvievskaya/Rozenfeld, *Mat. i astr.*, 2: 404 (A10).

(b) *Ḥall-i mushkilāt-i Mu^cīniyya = Sharḥ-i Mu^cīniyya = Dhayl-i Mu^cīniyya* ("A Solution to the Problems," "A Commentary," or "An Appendix" to the *Risālah-i Mu^cīniyya*), also in Persian, and meant to be an appendix or completion of the *Mu^cīniyya*. This short work was also written for Mu^cīn al-Dīn Abū al-Shams, who asked that certain obscure points from the *Mu^cīniyya* be clarified. Once again we have the curious situation of two prefaces, one extolling the virtues of Ṭūsī's Ismā^cīlī patrons and the other neglecting any mention of them altogether.

There are nine chapters with a rather wide range of content.[3] Dānish-Pizhūh has also brought forth a facsimile reproduction of a manuscript of this work that begins with the second preface;[4] in his own introduction, he supplies the other preface. Kennedy [1984] gives a table of contents. An unpublished edition has been completed by Wheeler M. Thackston; a translation, also unpublished, is due to him and the present writer.

For further bibliographical details, see Krause [1936], p. 495 (no. 3); Storey, *Persian Literature*, pp. 56–57 (no.7); Riḍawī, *Aḥwāl*, pp. 388–390 (no. 31); and Matvievskaya/Rozenfeld, *Mat. i astr.*, 2: 403–404 (A18 & A18a).

(c) *Zubdat al-idrāk fī hay'at al-aflāk* (The Essential Understanding of the Configuration of the Orbs), a short work in Arabic of unknown date.[5] It consists of an introduction and two chapters (one on the *hay'a* of the celestial region, the other on the *hay'a* of the Earth), and a concluding section on sizes and distances. Ṭūsī states in the introduction that his intention is to epitomize the available

[2] Tehran: Intishārāt Dānishgāh Tahrān (no. 300 in the series), 1335 H. Sh.

[3] For example, Chapter Nine deals with the rather mundane topic of using the Indian circle for finding directions whereas Chapter Five deals with the difficult subject of Ibn al-Haytham's latitude theory. In Chapter Three, he presents the Ṭūsī couple for the first time and his alternative models for the deferents of the moon and planets.

[4] Tehran: Intishārāt Dānishgāh Tahrān (no. 304 in the series), 1335 H. Sh.

[5] It should not be confused with the Persian work *Zubdah-i hay'a* (Essentials of *hay'a*), which was Arabized on several occasions; the Princeton translation has the title *Al-Zubda fī al-hay'a*. Though this work overlaps the *Zubdat al-idrāk* in numerous ways, it does not have the characteristic 4-part division of a *hay'a basīṭa* work but is instead divided into 30 chapters. It is thus comparable to Farghānī's *Elements of Astronomy*. For further information on this work, see *GAL* SI, p. 931 (no. 44a); Krause [1936], p. 497 (no. 13); Storey, *Persian Literature*, p. 60 (no. 15); Riḍawī, *Aḥwāl*, pp. 390–391 (no. 32)); and Mach, *Catalogue* (Princeton), p. 421.

works on *hay'a*. Ḥājjī Khalīfa compares it, sizewise, to *Al-Mulakhkhaṣ* by Jaghmīnī.

A copy of the work, which I have examined, is Istanbul, Topkapı Saray, Ahmet III MS 3430 (5), ff. 59b–92b (=Arab League *falak*, no. 123). Another copy is Paris, B. N. MS ar. 2511 (1). Further bibliographical details may be found in *GAL* I, p. 511 (no. 44) and SI, p. 931; Krause [1936], p. 497 (no. 14); Riḍawī, *Aḥwāl*, p. 391 (no. 33)); and Matvievskaya/Rozenfeld, *Mat. i astr.*, 2: 403–404 (A11).

***

An obvious question that needs to be addressed is the relationship of these works to the *Tadhkira*. The *Zubdat al-idrāk* is clearly meant to be an even more abridged and simplified work than the *Tadhkira*. For the most part Ṭūsī here eschews criticisms of Ptolemy; the one exception I have been able to find is a very brief reference to Ibn al-Haytham's alternative models for latitude.[6] But without a date of composition, I am unable to say whether it was made with the *Tadhkira* in hand. The audience is also not entirely self-evident; unlike what one finds in the introduction to the *Zubdah-i hay'a*, Ṭūsī does not say here that the work was meant for his students. Perhaps it was intended, as implied by Ḥājjī Khalīfa, for the lay audience that was captured by Jaghmīnī's *Mulakhkhaṣ*; if so, it was a dismal failure, to judge from the small number of extant manuscripts and the apparent lack of any commentaries on it. The Persian *Zubdah-i hay'a* and its Arabic translations were clearly more successful in that regard.

We are on firmer ground when we come to the question of the relationship of the *Tadhkira* to the *Mu*ᶜ*īniyya* and its *Ḥall*. As stated above, the *Mu*ᶜ*īniyya* itself is a straightforward exposition of Ptolemaic astronomy in the *hay'a* mode, such as one finds in Khiraqī's *Muntahā* or *Tabṣira*. However the young Naṣīr al-Dīn, who was 34 at the time, had clearly been thinking about the "difficulties" of Ptolemaic astronomy and their solutions when he was writing the *Mu*ᶜ*īniyya*, which was completed in 632/1235. For in the midst of his exposition of the moon (Section II, Chapter 5), he remarks that there is a doubt (*shakk*) concerning the motion of the moon's epicycle center on the deferent. In a passage rather reminiscent, but in important ways different, from II.7 [25] of the *Tadhkira*, he concludes:

> Then one of two things must follow: either the invariability of distance and closeness of the epicycle center from the center or the variability in speed and slowness in the motion of the center. And

[6] *Zubdat al-idrāk*, Istanbul, Topkapı Saray, Ahmet III MS 3430 (5), f. 76a–b. It is interesting that Ṭūsī states that neither Ptolemy's nor Ibn al-Haytham's proposals are "free from defect (*khalal*)," which is reminiscent of what we find in the *Mu*ᶜ*īniyya* (see p. 68 of this volume).

> these two are prohibited. This is a serious doubt (*shakk-i ᶜaẓīm*) concerning this account, which no one among the practitioners of the science has ventured anything or, if they have put it forth, it has not reached us. There is an elegant way (*wajh-i laṭīf*) of solving (*ḥall*) this doubt whose presentation in this compendium would not be appropriate. If in the future the blessed disposition of the prince of Iran, may God multiply his eminence, commands the pleasure of investigating this question, in that chapter (*bāb*) an account will be given, God willing.[7]

In the next chapter (Section II, Chapter 6), which deals with the upper planets and Venus,[8] Ṭūsī again points to the problem of the irregular motion of the deferent, this time due to the equant, and claims a solution:

> The doubt which occurs with regard to the moon occurs in exactly the same way for the motion of the epicycle center on the deferent equator with the lack of uniformity about its center and the uniformity about another center different from it...the solution of this doubt, which no one from the practitioners of the science has ventured anything, is among the puzzles (*asrār*) of the science of *hay'a*. God willing, in the future an explanation (*bayān*) of that will be made.[9]

Naṣīr al-Dīn also has doubts about Ptolemy's latitude theory, which he expresses in Section II, Chapter 8. Here, though, the work of one of his predecessors, namely Ibn al-Haytham, has reached him, and he gives a sketch of his theory.[10] But Ṭūsī is not content with what he finds:

> Yet even with this establishing [of orbs?], this variation (*ikhtilāf*) has not become ordered (*manẓūm*). Concerning that, several other defects come to the fore; but this [work] is not the place to expound on that.[11]

---

[7] *Muᶜīniyya*, p. 31, lines 7–12.

[8] This corresponds to II.9 of the *Tadhkira*.

[9] *Muᶜīniyya*, p. 37, lines 6–11. Ṭūsī points to the same problem for Mercury on p. 42, lines 7–9.

[10] *Muᶜīniyya*, p. 44, line 20 – p. 45, line 2. For a discussion of Ibn al-Haytham's model, see our commentary to II.11 [16].

[11] *Muᶜīniyya*, p. 45, lines 2–3. Other references to Ibn al-Haytham's treatise occur on p. 46, lines 8–9, where Ṭūsī repeats that this is not the place to go into details, and on p. 47, lines 11–14, where he gives the Ibn al-Haytham's additional orbs needed for the latitude of Venus and Mercury.

To summarize, we have the following situation in March 1235: (a) Ṭūsī has published his *Muᶜīniyya* as a relatively short summary of astronomy in the *hay'a* mode and dedicated it to the son of his patron; (b) he claims to have a solution to the irregular motion of the deferents of the moon and planets; (c) he has been studying Ibn al-Haytham's treatise on latitude and though he presents its main results he believes there remain problems that need to be resolved; (d) he promises to present his results in a future work.

From this, it is reasonable to draw the following conclusions. Naṣīr al-Dīn had been inspired at a fairly early period of his career with the rectilinear version of his couple, which he can use to resolve the irregular deferent motions in longitude (Difficulties 1–6; see p. 50).[12] He has also studied Ibn al-Haytham's treatise on latitude and accepts that it can give a *hay'a* (configuration) to the small circles of Ptolemy's latitude theory. But he still feels that Ibn al-Haytham has not resolved the problem completely and refers to unspecified "defects" that remain. These presumably would be Objections 2 and 3 that he refers to in II.11 [15].[13] Unlike the case with the irregular motion on the deferents, he does not claim a solution for these problems. As far as I can tell, he also does not mention the difficulty of the moon's prosneusis nor does he make the critical connection between that problem and the difficulties of Ptolemy's latitude theory.[14] In other words, though he points to Difficulties 7–15 (see pp. 50–51), he does not yet have the curvilinear version of his couple with which to resolve them. He still depends on Ibn al-Haytham despite his reservations.

The confirmation of this reconstruction comes in the *Ḥall*. There can be little if any question that it is the work that Naṣīr al-Dīn tantalizingly dangled before his patron in the *Muᶜīniyya*. And since there is nothing in the *Ḥall* that is not forecast in the *Muᶜīniyya*, there is little to recommend putting its date of composition more than a few months, perhaps a year at most, after that of the *Muᶜīniyya*.[15] It could then be viewed, as it was, simply as an appendage to the

[12] At this stage in his investigations, Ṭūsī seems to have assumed that Mercury could be dealt with in the same manner as the moon and other planets; see *Muᶜīniyya*, p. 42, lines 7–9. In the *Ḥall* Ṭūsī does not mention Mercury explicitly but it seems to be included when he states that "the difficulty encountered in the other planets may also be resolved [with the couple] if the equant takes the place of the inclined orb and the deferent that of the eccentric" (*Ḥall*, p. 12, lines 3–4). In the *Tadhkira*, he realizes that Mercury is a special case defying an easy solution; see II.11 [11].

[13] These have to do with the irregular motion of the epicycle apex on the small circle and the disruption of the position in longitude. See also pp. 50–51.

[14] See the commentary to II.11 [13].

[15] The recent attempt by Saliba [1987a] to whisk the *Ḥall* "closer to the date of the *Tadhkira* (1260/1261) [*sic*], which it resembles much more closely" does not take into account crucial historical and intellectual developments of the 13th c. For Saliba's dating would force us to assume that Naṣīr al-Dīn was still dedicating books to the Ismāᶜīlīs after they had been completely decimated by the Mongols and after he had joined the Mongol entourage. For other reasons to reject Saliba's dating, see the commentary to II.11 [18].

earlier work.[16] Of the nine chapters of the *Ḥall*, seven are of a fairly pedestrian nature. The two that are of greatest interest are Chapter Three, rather cumbrously entitled "Concerning the Solution of the Doubt Occurring with Regard to the Motion of the Center of the Lunar Epicycle on the Circumference of the Deferent, and the Uniformity of That Motion About the Center of the World," and Chapter Five, entitled "On the Configuration of the Epicycles of the Wandering Planets According to the Doctrine of Abū ᶜAlī ibn al-Haytham." Chapter Three presents the rectilinear version of the couple, while Chapter Five simply presents Ibn al-Haytham's theory. There is no hint of the curvilinear version of the couple that he would use in the *Tadhkira* for resolving Difficulties 7–16 (on latitude and the moon's prosneusis); he is basically at the same place he was when he completed the *Muᶜīniyya*.

Between 632/1235 and 644/1247, the date of the completion of his recension of the *Almagest*, Naṣīr al-Dīn has developed the idea, if not the final form, of the curvilinear version of the couple. He has also made the crucial connection between the prosneusis problem and the latitude problem.[17] The stage for the final synthesis in the *Tadhkira* is now set.

2. The Marāgha (α) Version of the Tadhkira

The process of Arabizing Ṭūsī's Persian works went on both in his lifetime and afterwards; in the former category we may note his Arabic treatise on logic, *Tajrīd al-manṭiq* (written 656/1258), which seems to be an abridged version of his earlier Persian work, the *Asās al-iqtibās* (written 642/1244–45). In the latter category, there are several Arabic translations of the *Zubdah-i hay'a*. One translator of the treatise, a certain ᶜImād al-Dīn ᶜAlī al-Qāshī (?), tells us in his introduction that the work, though "of great usefulness," was not "of general benefit" since it was in the "Persian language whose understanding is denied to the intelligent Arab."[18]

One of the motivations for composing the *Tadhkira* was, no doubt, likewise to have his Persian *Muᶜīniyya* reach a wider audience. He would furthermore have wished in the latter part of his life to synthesize the *Ḥall* as well as his more recent discoveries mentioned in passing in the *Taḥrīr al-Majisṭī* (Recen-

---

16 This is clear in the way the *Muᶜīniyya* is referred to in the *Ḥall*, namely simply by chapter and section. For example, Chapter Three begins: "In the 5th Chapter of Section II dealing with the arrangement of the orbs of the moon, this problem occurs." Sometimes Ṭūsī speaks as if the reader has the *Muᶜīniyya* in hand as when in the sentence immediately following the above he states: "as already mentioned [*i.e.* in the *Muᶜīniyya*], the same problem also occurs in the orbs of the other planets..." The intimate connection between the *Muᶜīniyya* and the *Ḥall* may be the reason that it apparently lacks a date of composition in the extant manuscripts.

17 See the commentary to II.11 [18] for details.

18 Princeton, Mach MS 4884 (4066), f. 2b.

sion of the *Almagest*) into a more coherent whole. The immediate occasion for the *Tadhkira* was provided by his friend ᶜIzz al-Dīn al-Zinjānī (d. 660/1261–62), who requested a work in Arabic on *hay'a*.[19] It is this Zinjānī who is evidently being referred to in the preface as "one of our dear friends" and to whom the work is being presented as a "memento" (*tadhkira*).[20]

This first version of the *Tadhkira*, which was completed in Marāgha, may be confidently dated as the beginning part (*awā'il*) of Dhū al-qaᶜda 659/September or October 1261, based on the following note that occurs in the margin of the last page of a copy of Faḍl Allāh al-ᶜUbaydī's commentary on the *Tadhkira* (Istanbul, Topkapı Saray, Ahmet III MS 3325, f. 131b):

> وفرغ المصنف من تسويد المتن أوائل ذي القعدة سنة تسع وخمسين وستمائة ببلدة مراغه
>
> The author completed drafting (*taswīd*) the text (*matn*) during the beginning part (*awā'il*) of Dhū al-qaᶜda in the year 659 in the town of Marāgha.[21]

It is rather curious that none of the twenty or so manuscripts of the *Tadhkira* that I have examined gives the date of composition, a point to which we shall return.

3. The Baghdad (β) Version of the Tadhkira

The completion of a *muswadda* ("draft," *i.e.* the original manuscript) did not imply an unalterable text; corrections could be made and revisions carried out either by the author alone or in the course of having it read to him by a student or copyist. There is ample evidence to indicate that this is precisely what happened to the *Tadhkira* between the time of its original completion in 659/1261 and Ṭūsī's death in 672/1274. In the following, I shall first present the evidence, both textual and secondary, for the existence of the final revision of the *Tadhkira*, the Baghdad (β) version, which was probably compiled sometime during the latter part of 672 H. (January–June 1274). I shall then discuss the nature of these revisions and the evidence that some of them were made before 672 H.

---

[19] This information reliably comes from Ibn al-Fuwaṭī, (*Talkhīṣ*, Vol. IV, 1: 234–235), who was the librarian at the Marāgha observatory.

[20] Cf. commentary to I.Pref. [2]5.

[21] Virtually the same statement occurs in a copy of Jurjānī's commentary (Istanbul, Köprülü MS 927 (2), f. 218a). Riḍawī, *Aḥwāl*, p. 400 gives this same information but without indicating his source.

The most clear-cut evidence for a later, revised text comes from a number of the commentators in their discussion of II.4 [2]23–[3]16, which has to do with the possible motion of the ecliptic equator with respect to the equinoctial. One version of the text has four possibilities for this motion, another has eight.[22] Jurjānī, whose commentary is dated 811/1409, states that the newer version with eight possibilities was made an extended period (*mudda madīda*) after the first. Shīrwānī, writing some 65 years later in 879/1474, adds that the new version was completed more than 20 years after the first and that this was done in Baghdad. Bīrjandī in 913/1507 also states that the change was made in Baghdad. Khafrī, writing in 932/1525, gives the 20-year period of separation.

Certain problems immediately arise. The first, which is not serious, concerns the 20-year period between the two versions. Obviously the greatest length of time possible between the Marāgha version and this Baghdad revision would be 13 years. Could some version earlier than Marāgha, or perhaps the *Mu*ᶜ*īniyya*, be intended by the earlier one? There is absolutely no evidence for the first possibility and I think that it can be safely dismissed. The second is also untenable since the *Mu*ᶜ*īniyya* has neither the four nor eight possibilities.[23] Clearly the 20 years is simply a minor exaggeration.

Another more serious problem is whether we are justified in attributing the revisions to Ṭūsī himself since they were seemingly done in the last year of his life. For we know from the historical sources that he left Marāgha for Baghdad in 672 H. (July 1273–June 1274), just a few months before his death.[24] We also know that he was accompanied by a number of his students, and it might be supposed that one of them revised the text after his death. However, there is considerable evidence that Ṭūsī himself was responsible for several of the revisions and some of these can be dated prior to the relocation to Baghdad. The best evidence for this comes from MS M, which is based upon Shīrāzī's working copy of the *Tadhkira*.[25] Now we know that Shīrāzī left Marāgha, whether or not estranged from Ṭūsī being unclear, in the period[26] between 667/1268 and 672/1274; it is then reasonable to conclude that the revisions recorded in the text of MS M were effected before Shīrāzī's departure. And since Shīrāzī, according

[22] Suter [1902], p. 175, depending on Nallino, who in turn depended on Jurjānī, originally pointed out that there were two versions of the *Tadhkira*. Note that Nallino mistakenly thought that the incomplete copy of Jurjānī's *Sharḥ* contained in Biblioteca Medicea Laurenziana or. MS 271 was by Mūsā Qāḍīzāde.

[23] *Mu*ᶜ*īniyya*, Section II, Ch. 2, pp. 15–16. There is a brief mention of the possibility of the motion of the ecliptic but little in the way of the extensive treatment in the *Tadhkira*.

[24] See p. 15. Ṣafadī indicates that he was only in Baghdad a few months before his death, which occurred on 18 Dhū al-Ḥijja 672 H./25 June 1274 A.D. I have inferred from this that the trip to Baghdad, and hence the Baghdad version of the *Tadhkira*, should be dated sometime between January and June 1274.

[25] Istanbul, Süleymaniye Kütüphanesi, Lâleli MS 2116. For the colophon that establishes this, see p. 78.

[26] Walbridge [1983], pp. 17–18. A date closer to 667/1268 seems the most likely.

to the note at the end of MS M, read the entire text to Ṭūsī (*qara'ahā ilā ākhirihā ᶜalā muṣannifihā*), all the revisions to the Marāgha version were authorized ones.

An important example of a revision done before Shīrāzī's departure occurs in I.1 [2]19–20, where Ṭūsī gives his definition of a plane surface.[27] MSS DGT[28] contain what Jurjānī refers to as the older and incorrect version. MS M contains a definition that Jurjānī tells us was read (and presumably approved) by Ṭūsī; from what we have established above this would have been done while Shīrāzī was in Marāgha. Now there is a second definition contained in MSS FL and in the margin of MS T, which Jurjānī gives as an alternative to the definition of MS M. For reasons that I shall come to presently, I take this latter definition to be of a later date than MS M and more than likely to be after Shīrāzī's departure from Marāgha.

Very strong evidence that revisions continued to be made after Shīrāzī left Marāgha come from Shīrāzī himself. In his *Faᶜalta fa-lā talum*, which is a supercommentary on an early commentary of the *Tadhkira*, Shīrāzī informs us that Ṭūsī's addition to II.2 [3]22–23 of the *Tadhkira* occurred *after* he left the service (*khidma*) of Naṣīr al-Dīn.[29] MS M gives the addition but a note, perhaps due to Shīrāzī, advises the reader to ignore it.[30] MSS GFL, however, incorporate the revision into their texts without further comment whereas MSS DT do not have it at all.

We have now established that certain revisions were made at various times between 659/1261 and 672/1274, but we still need some means for identifying the original Marāgha text and the various stages in the revision process. The manner in which certain critical revisions are handled in MSS DT provide the key we need for establishing the original text. In the major revision to II.4 [2]23–[3]16 mentioned above, both manuscripts give the original version in the body of the text and provide the revision in the margin.[31] In virtually all the cases in which we can identify an older and newer version, this is the pattern that emerges. For example, in the text of MS D we find what Jurjānī, Khafrī, and Bīrjandī identify as the original version of I.1 [16]8–9, while in the margin of MS D there occurs what they identify as the revision; the copyist has also added the following remark: "The author, may God have mercy on him, changed to this text." Khafrī also specifically indicates that the change is due to the author. For II.5 [8]6–8, MS D specifically attributes the variant given in the margin to "the new, emended version" (*al-iṣlāḥ al-jadīd*); the variant occurs in the text of

---

[27] For this and all subsequent examples, the reader is referred to the commentary where a detailed discussion occurs.

[28] For a listing of manuscripts, see pp. 76–81.

[29] See our commentary to II.2 [3]21–23 for additional details.

[30] It is interesting that Shīrāzī made sure to be informed of continuing changes to the *Tadhkira* even after leaving Marāgha.

[31] MS T places "*kh aṣaḥḥ*" after the variant indicating that it is from a corrected manuscript.

MSS FL and in the margin of MS T, where the copyist indicates it comes from another version or manuscript (*kh*).

Now MS D has another interesting feature; in the colophon it claims that it is from an autograph. The conclusion then seems inescapable that the *texts* of MSS DT represent to a fairly high degree the original Marāgha version. In the margins, the copyists have placed the revisions.[32] My very strong suspicion is that this is precisely the way revisions were handled during much of the period between 659/1261 and 672/1274. Ṭūsī would approve an original version and then have the copyist place the revisions he had made up to that point in the margins;[33] or he might do the job himself as is indicated by a note at the end of a copy of the *Tadhkira* preserved in Tehran, Sipahsālār MS 4727:

> I voweled and collated [this copy] with a copy that was read to its author, may God continue to protect him, and I transcribed its marginal notes that were in his noble handwriting and what he added to it; I indicated some of them in haste according to [my] effort and ability.[34]

At various points an approved text with the revisions incorporated into a fair copy of the text might be produced; this is what Shīrāzī's copy, exemplified by MS M, represents.[35] (This could also help us understand why no copies of the *Tadhkira* I have examined give the date of composition; it would not have made sense to date these later fair copies with the original 659/1261 year of composition.) After Ṭūsī went to Baghdad, there seems to have been a final version produced and this is represented by MSS FL, which incorporate virtually all the revisions previously made into a new fair copy of the text; it cannot be ascertained whether this occurred before or after his death but because the major revisions can be verified as being due to Ṭūsī himself (either from the evidence of MSS DT or from the commentators), there is no reason to doubt that the other minor changes one finds in MSS FL are also due to him.

What are the nature of these changes? Compared with the changes from the *Muʿīniyya* and the *Ḥall* to the *Tadhkira*, they are fairly insignificant. A large number simply correct grammatical mistakes or stylistic infelicities, such as one finds in I.1 [14]1. In the following list of revisions, I generally ignore these but

[32] The copyist of MS T often indicates that the revisions are corrections with the abbreviation *kh aṣaḥḥ*.

[33] There is also evidence that Ṭūsī provided glosses for certain words and phrases, which he did not intend to be incorporated into the text. The one example I know is a *ḥāshiya* (gloss) in the margin of MS M, f. 48b (II.13 [3]2) for the word *makth*: "What is intended by *makth* is the amount (*masāfa*) occurring on the moon from its being cut; this has been copied from the handwriting of the author, may God be pleased with him."

[34] Mudarrisī, *Sar-gudhasht*, p. 114.

[35] But even in such a case, additional corrections might be put in the margins by a diligent copyist as we have seen.

list the others in somewhat arbitrary categories. Starred citations (*) indicate corrections that have been discussed in the commentary.

(a) Simplifications: I.Intr. [1]10; II.11 [6]2.

(b) Corrections of Mistakes: *I.1 [2]19–20; *I.1 [16]8–9; *I.1 [17]12–13; II.7 [1]1; *II.9 [1]9; II.10 [2]14–15; II.10 [2]17; *II.11 [5]22; II.12 [5]12; II.12 [6]18; II.14 [2]21; III.2 [1]9; III.8 [7]1; *III.10 [1]13–14; *IV.1 [2]7; *IV.1 [3]18–19; IV.3 [1]24; *IV.4 [2]17–18; IV.6 [5]11; *IV.6 [5]17.

(c) Amplifications: I.1 [13]19; II.1 [7]14; *II.2 [3]22–23; *II.4 [2]2–[3]16; II.8 [14]10–11; *II.9 [6]12; II.10 [3]10; II.10 [3]11–12; II.11 [2]5; *II.11 [5]14–15; II.11 [5]19–20.

(d) Clarifications: I.1 [14]21; I.1 [14]22–23; II.3 [10]20; II.4 [2]23; *II.5 [8]6–8; *II.5 [9]3_2; II.7 [9]8; II.7 [14]10; II.7 [16]2–4; II.8 [1]15; II.8 [1]20; *II.8 [15]5; II.8 [16]8; II.9 [1]14; *II.9 [8]21; II.9 [14]11; *II.11 [4]2–4; *II.11 [4]9–10; II.11 [10]9–10; II.12 [1]16; II.12 [3]8; II.13 [9]15; III.3 [1]15; III.3 [2]5; III.3 [3]1; III.9 [1]11; IV.1 [3]12; IV.1 [3]17; IV.5 [3]19.

In addition the original of Figure T25 in III.9 [1] was revised in the Baghdad version.

In general, the revisions are more numerous in the first two books than the last two. Few, if any, can be called sensational; in particular, the revisions made in the chapter that presents Ṭūsī's new models (II.11) are fairly minor. Some of the changes are surprisingly for the worse. This is especially true of the one in II.5 [8]6–8, which in the Baghdad version gives an incorrect proportion for finding retrograde motion. Other examples are I.1 [13]18 and II.11 [5]19–20. In this regard, it is interesting that a revision might be marked for deletion; this seems to be the case for II.9 [6]12, where a phrase that was added in the margins of MSS MT, and into the text of MS G, is marked for deletion in MS F and is missing entirely in MS L, these latter two manuscripts being my best witnesses to the Baghdad version. Finally we should note that though the vast majority of the revisions are due to Ṭūsī, there are a few suspicious ones that might be due to Shīrāzī or perhaps another student or colleague. Three that I strongly suspect are due to Shīrāzī, and that occur in MS M, are variants to II.9 [1]9, II.11 [9]1–2_2 and IV.3 [5]19; both Jurjānī and Bīrjandī suggest that this is so for II.9 [1]9.

K. List of Manuscripts

In the following list, nos. 1–6 are the principal manuscripts that have been used to establish the edition (see Section M.2). The other five manuscripts that have sigla were used at an earlier stage of editing, but their variants added little to the final edition; consequently I decided to leave them out of the final apparatus. On occasion, though, they are referred to in the apparatus and commentary. In addition to the eleven manuscripts with sigla, I have also examined, with varying degrees of thoroughness, nos. 7, 10, 11, 12, 13, 16, 29, 31, 32, 33, 35.

A convenient listing of manuscript collections and their catalogues is in Sezgin, *GAS* 6.

Sigla	Description of Manuscript

D ض 1. Istanbul, Feyzullah MS 1330, 1; ff. 1b–83a; ca. 14–16 lines/page; numerous annotations, especially in the beginning; copy completed on 17 Ṣafar 757/ca. 20 February 1356 from an autograph by Aḥmad b. Maḥmūd b. Muḥammad al-Qazwīnī.

Colophon:

نقلت هذا الكتاب من خط مصنفه محمد بن محمد بن الحسن الطوسي رحمه الله في السابع عشر من الصفر ختم بالخير والظفر[؟] لسنة سبع وخمسين وسبعمائة وأنا أحوج خلق الله تعالى إلى غفرانه أحمد بن محمود بن محمد القزويني

I copied this book from the handwriting of its author Muḥammad b. Muḥammad b. al-Ḥasan al-Ṭūsī, may God have mercy on him, successfully completing it on the 17th of Ṣafar of the year 757. And I am Aḥmad b. Maḥmūd b. Muḥammad al-Qazwīnī, of God's creation the most in need of His forgiveness.

F ف 2. Vatican, ar. MS 319, 1; ff. 1b–64a; 17 lines/page; has occasional Latin annotations [15th–16th c. ?]; copied Friday, 5 Muḥarram 683/24 March 1284 by Maḥmūd b. Muḥammad b. al-Qāḍī Taqī al-Dīn in Baghdad at the Niẓāmiyya College.

Sigla	Description of Manuscript

Colophon:

فرغ من تحريره العبد المذنب المفتقر إلى رحمة ربه الغفور
محمود بن محمد بن القاضي تقي الدين يوم الجمعة خامس
محرم سنة ثلاث وثمانين وستمائة في مدرسة نظام الملك بمدينة
السلام حماها الله تعالى عن نوائب الدهر وحدثانه حامداً لله
تعالى ومصلياً على نبيه والحمد لله رب العالمين

The writing of this was completed by the sinner in need of the mercy of his forgiving Lord Maḥmūd b. Muḥammad son of the Judge Taqī al-Dīn on Friday, the 5th of Muḥarram of the year 683 at the College of Niẓām al-Mulk in the City of Peace, may God Almighty protect it from the misfortunes and afflictions of time, praising God Almighty and praying for His Prophet; Praise be to God, Lord of the Worlds.

G غ 3. St. Petersburg (a.k.a. Leningrad), Oriental Institute MS A 437; ff. 1b–43b; 17 lines/page; incomplete, folios disordered; copy completed 3 Ramaḍān 673/ca. 2 March 1275.

Colophon:

تمت بعون الله وحسن توفيقه في الثالث من الرمضان سنة
ثلاث وسبعين وستمئة

Completed with the aid of God and the goodness of His granting success on the 3rd of Ramaḍān of the year 673.

L ل 4. Leiden, University Library MS or. 905 (= 1093); ff. 1b–95a; 15 lines/page; very readable pointed script with numerous vowel markings; no colophon but a marginal note on f. 95a indicates that it was compared with a "reliable copy."

M م 5. Istanbul, Lâleli MS 2116; ff. 1b–83b; 17 lines/page; copy completed on 8 Ramaḍān 681/ca. 10 December 1282.

Sigla — Description of Manuscript

Colophon (in margin of f. 83b):

قوبل بالنسخة التي بخط مولانا المعظم علامة العالم أفضل
المتقدمين والمتأخرين زبدة خلق الله في العالمين قطب الملة
والدين أسبغ الله ظلال جلاله وكان يحدَث عنها وقراها إلى
آخرها على مصنفها محمد بن محمد الطوسي رضي الله عنه
في ثامن[؟] رمضان المبارك[؟] في[؟] إحدى[؟] ⟨و⟩ثمانين[؟؟]
⟨و⟩ستمايه

This has been collated with the copy in the hand of our great master, the most erudite in the World, the most excellent of the ancients and the moderns, the cream of God's worldly creation, the pole (*quṭb*) of the community and creed, may God widen the shadows of his grandeur, and he used to narrate from it and he read it in its entirety to its author Muḥammad ibn Muḥammad al-Ṭūsī, may God be pleased with him, on the 8th of blessed Ramaḍān in 681 [Krause [1936] reads 684].

T ط — 6. Istanbul, Ahmet III MS 3453, 19 (= Topkapı Saray MS 7005, 19); ff. 261b–281a; 27 lines/page; an important manuscript containing, in addition to the *Tadhkira*, Ṭūsī's recensions of the "Middle Books"; copied 12 Rabī^{c} I, 677/ca. 3 August 1278 by ^{c}Abd al-Kāfī b. ^{c}Abd al-Majīd b. ^{c}Abd Allāh al-Tabrīzī in Baghdad.

Colophon:

وأكتب المصنف رحمة الله إليه هذا سوّده أحوج خلق الله
محمد بن محمد الحسن الطوسي فرغ نسخ هذا الكتاب عبد
الله الفقير إليه عبد الكافي بن عبد المجيد بن عبد الله
التبريزي في الثاني عشر من شهر ربيع الأول سنة سبع
وسبعين وستمائة في مدينة بغداد دامت محميه عن الآفات
حامدا لله تعالى ومصليا على أشرف خلقه محمد وآله الطيبين
الطاهرين

Sigla	Description of Manuscript

The author, God's mercy upon him, dictated this: The most needy of God's creation, Muḥammad b. Muḥammad b. al-Ḥasan al-Ṭūsī, drafted this. The copying of this book was completed by the needful servant of God ^{c}Abd al-Kāfī b. ^{c}Abd al-Majīd b. ^{c}Abd Allāh al-Tabrīzī on the 12th of the month of Rabī^{c} I in the year 677 in the city of Baghdad, may its protection from harm continue, praising God Almighty and praying for Muḥammad, the most noble of His creation, and for his most excellent and virtuous family.

7. Aleppo, Aḥmadiyya (*waqf*) Library MS 1284.

8. Aligarh, Subḥānullāh Oriental Library MS 121, 3; 44 ff.; Sha^{c}bān 1300/June–July 1883.

9. Baghdad, MS 2958 [? see Matvievskaya/Rozenfeld, *Mat. i Astr.*, 2: 403 (no. 368, A9)].

10. Cairo, al-Azhar MS 18079, 1; ff. 1–94; copied 1290/1873–74 by Muḥammad Amīn Riḍā.

11. Cairo, National Library MS K 3957, 3; ff. 155b–253a; n. d. (ca. 900/1500) [see King [1986], p. 151].

12. Cairo, National Library, Ṭal^{c}at hay'a MS 38, 1; ff. 1a–62a; copied 1076/1665–66 [see King [1986], p. 151].

13. Cairo, National Library, Taymūr majāmī^{c} MS 181, 1; pp. (*sic*) 1–120; copied ca. 1114/1702–03 [see King [1986], p. 151].

14. Diyarbakır (Turkey), MS 2213 A, 8; ff. 82a–113a; copied in 727/1326–27 (see Şeşen, 3: 33).

15. Edirne (Turkey), Selimiye MS 1244, 3; ff. 197b–250b; copied in 879/1474–75 (see Şeşen, 3: 33).

16. Heidelberg, University Library MS or. A 144; 88 ff.; n. d. (ca. 1650 A. D.).

Sigla	Description of Manuscript
S س	17. Istanbul, Ahmet III MS 3317 (= Topkapı Saray MS 7081); ff. 1b–60a; copy owned by Sultan Bāyazīd II.
H ح	18. Istanbul, Ahmet III MS 3333, 1 (= Topkapı Saray MS 7082, 1); ff. 1b–32a; copied in 728/1327–28 by Muḥammad b. Muḥammad al-Samarqandī.
	19. Istanbul, Ali Emiri Arabi MS 2735; 150 ff.; n. d. (ca. 9th–10th c./ 15th–16th c.) [see Krause [1936], p. 494].
	20. Istanbul, Aşir Hafid MS 203, 2; 90 ff.; copied 797/1394–95 [see Krause [1936], p. 494].
	21. Istanbul, Carullah MS 1457, 2; 26 ff.; copied ca. 712/1312–13 in Tabrīz; incomplete [see Krause [1936], p. 494].
	22. Istanbul, Fatih MS 3388; 77 ff.; copied 955/1548–49 in Istanbul [see Krause [1936], p. 494].
	23. Istanbul, Fatih MS 3389; 56 ff.; n. d. (ca. 7th–8th c./13th–14th c.) [see Krause [1936], p. 494].
	24. Istanbul, Feyzullah MS 1331; 66 ff.; copied 749/1348–49 in Aleppo [see Krause [1936], p. 494].
	25. Istanbul, Lâleli MS 2115; 88 ff.; n. d. (ca. 9th–10th/15th–16th c.) [see Krause [1936], p. 494].
	26. Istanbul, Ragıp Paşa MS 919, 2; ff. 71b–134a; copied 877/ 1472–73 [see Krause [1936], p. 494].
N ن	27. Leiden, University Library MS 188, 4 (= 1092); ff. 38b–111a; copied in 785/1383–84.
C ص	28. Leipzig, University Library MS 261 (= K. 203); ff. 1b–40a; copied 14 Shawwāl 790/ca. 16 October 1388.

Sigla	Description of Manuscript
	29. London, British Library MS 1339, 1 (= Add. 23,394); copied in 1107/1695–96 by Muḥammad Riḍā b. ᶜAzīz Allāh al-Tūnī.
	30. Los Angeles, UCLA Library Arabic MS 1117 [courtesy of Aḥmad Harīdī, Cairo].
	31. Mashhad, Āstān Quds Riḍawī MS 8568.
	32. Najaf, Āyat Allāh al-Ḥakīm MS 1099 (= Arab League uncatalogued *falak* film no. 321); 91 ff.; copied 1324/1906–07.
	33. Paris, Bibliothèque nationale MS ar. 2330, 8; ff. 83b–86b (contains only part of Book I).
B ب	34. Paris, Bibliothèque nationale MS ar. 2509 (= Suppl. ar. 962); ff. 2b–82a; used by Carra de Vaux [1893] for his translation of II.11, but the text is a jumble of the two versions and the figures are rather inaccurate; copied Monday (end of day) 2 Shaᶜbān 791/26 July 1389.
	35. Princeton, MS 4881 (Mach); 56 ff.; copied Muḥarram 771/August–September 1369.
	36. Tashkent, Oriental Institute MS 8990, 1 [see Matvievskaya/Rozenfeld, *Mat. i Astr.*, 2: 403 (no. 368, A9)].
	37. Tehran, Dānishkadah-i Ilāhiyyāt Library MS 275 G; 45 ff.; n. d.
	38. Tehran, Kitābkhāna Najm Ābādī; copied in 671/1272–73 [see Riḍawī, *Aḥwāl,* p. 400].
	39. Tehran, Sipahsālār MS 4727; 19 ff.; collated with a copy that had been read to Ṭūsī and that had marginal notes in his hand, which were incorporated into this manuscript; copy completed 10 Rabīᶜ II, 761/ca. 29 Feb. 1360 [see Mudarrisī, *Sar-gudhasht*, p. 114].

L. Concordance of Manuscripts

page of edition	D	F	G	L	M	T
91.	1b,1	1b,1	1b,1	1b,1	1b,1	261b,1
93.	1b,14	1b,13	1b,11	2a,1	2a,1	261b,7
95.	2b,1	2a,8	2a,3	2b,3	2b,1	261b,14
97.	3a,6	2b,8	2a,16	3a,10	3a,7	261b,23
99.	3b,9	3a,6	2b,11	4a,1	3b,10	262a,4
101.	4a,10	3b,2	3a,3	4b,4	4a,9	262a,10
103.	5a,6	4a,5	3b,2	5b,1	5a,2	262a,21
105.	5b,11	4b,3	3b,14	6a,7	5b,6	262b,2
107.	6b,9	5a,6	4a,12	7a,6	6b,1	262b,12
109.	7b,3	5b,7	4b,9	8a,1	7a,11	262b,22
111.	8a,9	6a,7	5a,5	8b,10	7b,16	263a,3
113.	9a,2	6b,8	5b,3	9b,7	8b,9	263a,14
115.	9b,6	7a,6	5b,15	10a,13	9a,15	263a,22
117.	10a,14	7b,9	6a,13	11a,11	10a,9	263b,6
119.	11a,4	8a,9	6b,10	12a,5	10b,17	263b,15
121.	11b,9	8b,9	7a,6	12b,14	11b,8	263b,25
123.	12a,12	9a,7	7b,2	13b,6	12a,14	264a,6
125.	12b,9	9b,4	7b,10	14a,12	12b,11	264a,11
127.	13b,2	10a,7	8a,7	15a,9	13b,4	264a,20
129.	14a,10	10b,10	8b,5	16a,6	14a,15	264b,2
131.	15a,3	11a,13	9a,3	17a,4	15a,9	264b,13
133.	15b,12	11b,14	9a,15	18a,2	16a,3	264b,23
135.	16a,8	12a,9	9b,7	18b,8	16b,7	265a,5
137.	16b,7	12b,7	33a,3	19a,14	17a,9	265a,13
139.	17b,6	13b,3	33a,17	20a,13	18a,3	265a,23
141.	18a,5	13b,15	33b,7	20b,14	18b,2	265b,2
143.	19a,10	14b,7	34a,9	22a,2	19b,10	265b,17
145.	19b,6	15a,1	34b,3	22b,1	20a,7	265b,23
147.	20b,7	15b,9	34b,16	23b,3	21a,7	266a,7
149.	21a,9	16a,5	35a,8	24a,6	21b,8	266a,16
151.	22a,4	16b,16	35b,5	25a,9	22b,5	266a,27
153.	22b,10	17a,16	35b,16	26a,2	23a,13	266b,9
155.	23b,4	17b,17	36a,12	26b,15	24a,6	266b,20
157.	24a,8	18a,16	36b,6	27b,7	24b,12	266b,27
159.	24b,13	18b,16	36b,17	28b,1	25b,4^{m}	267a,10
161.	25b,7	19b,2	37a,13	29a,14	26a,13	267a,21
163.	26a,3	20a,1	37b,2	30a,1	27a,1	267a,25
165.	26b,6	20a,16	37b,12	30b,8	27b,3	267b,6
167.	27a,12	20b,16	38a,8	31a,15	28a,9	267b,15

page of edition	D	F	G	L	M	T
169.	28a,6	21a,17	38b,3	32a,12	29a,2	268a,5
171.	28b,13	22a,1	38b,14	33a,10	29b,9	268a,15
173.	29b,9	22b,3	39a,9	34a,7	30b,4	268a,25
175.	30b,2	23a,5	39b,4	35a,3	31a,15	268b,9
177.	30b,5	23b,1	39b,6	35a,6	31b,2	268b,11
179.	31a,1	23b,7	39b,12	35b,9	32a,1	268b,16
181.	31b,5	24a,7	40a,9	36a,15	32b,9	268b,24
183.	32a,10	24b,7	40b,3	37a,7	33a,16	269a,9
185.	33a,5	25a,9	40b,16	38a,3	34a,8	269a,20
187.	33b,7	25b,6	41a,8	38b,7	34b,11	269a,27
189.	34a,2	26a,4	41a,15	39a,5	35a,1	269b,6
191.	34b,9	26b,5	41b,10	39b,15	36a,2	269b,18
193.	35b,4	27a,7	42a,6	40b,12	36b,17	270a,1
195.	36b,2	27b,12	42b,4	41b,13	37b,7	270a,13
197.	37a,8	28a,12	42b,16	42b,6	38a,11	270a,22
199.	37b,1	28b,1	12a,1	43a,1	38b,1	270b,1
201.	38a,1	29a,4	12a,15	43b,13	39a,7	270b,9
203.	38b,4	29b,6	12b,9	44b,9	40a,3	270b,17
205.	39a,7	30a,5	13a,3	45b,4	40b,9	270b,26
207.	39b,2	30b,2	13b,1	46a,3	41a,2	271a,6
209.	39b,7	30b,8	13b,5	46a,11	41a,10	271a,18
211.	40a,14	31a,11	14a,1	47a,12	42a,3	271b,3
213.	41a,8	31b,14	14a,16	48a,11	42b,12	271b,13
215.	42a,2	32b,1	14b,13	49a,10	43b,6	271b,23
217.	42b,13	33a,6	15a,11	50a,8	44a,17	272a,6
219.	43a,8	33a,15	15a,16	50b,6	44b,11	272a,10
221.	43b,3	33b,5	15b,5	51a,7	45a,4	272a,14
223.	44b,4	34a,15	16a,5	52a,9	46a,3	272a,27
225.	45a,8	34b,14	16a,17	53a,3	46b,8	272b,9
227.	45b,2	35a,7	16b,8	53b,1	47a,2	272b,12
229.	46a,5	35b,10	17a,4	54a,11	47b,9	272b,21
231.	46b,10	36a,13	17a,17	55a,3	48a,15	273a,2
233.	46b,14	36b,2	17b,2	55a,7	48b,2	273a,5
235.	48a,1	37b,4	18a,4	56a,11	49b,9	273a,21
237.	48b,9	38a,7	18b,1	57a,7	50b,1	273b,4
239.	49a,3	38a,13	18b,4	57b,1	50b,10	273b,6
241.	49b,7	38b,16	19a,3	58a,10	51a,16	273b,18
243.	50a,11	39a,15	19a,15	59a,3	52a,6	273b,26
245.	50b,15	39b,13	19b,8	59b,8	52b,12	274a,6
247.	51b,1	40a,9	20a,1	60a,11	53a,15	274a,13
249.	52a,11	40b,15	20a,16	61a,9	54a,11	274a,23
251.	53a,6	41b,2	20b,14	62a,6	55a,7	274b,6

page of edition	D	F	G	L	M	T
253.	53b,16	42a,7	21a,11	63a,4	56a,2	274b,16
255.	54b,9	42b,10	21b,7	64a,2	56b,10	274b,24
257.	55a,12	43a,11	23a,2	64b,11	57b,1	275a,6
259.	56a,5	43b,13	23a,15	65b,6	58a,10	275a,15
261.	56b,12	44a,13	23b,11	66a,14	58b,17	275a,24
263.	57a,8	44b,5	23b,17	66b,12	59a,11	275b,1
265.	57b,12	45a,4	24a,12	67b,6	59b,16	275b,9
267.	58b,3	45b,5	24b,7	68b,1	60b,5	275b,18
269.	59a,11	46a,7	25a,3	69a,15	61a,13	275b,27
271.	60a,6	46b,10	25a,16	70a,13	62a,7	276a,10
273.	60b,3	47a,2	25b,5	70b,10	62b,3	276a,16
275.	61b,2	47b,4	25b,13	71a,14	63a,4	276a,22
277.	62a,1	48a,6	26a,5	72a,1	63b,7^{m}	276b,3
279.	62b,4	48b,6	26a,12	72a,10	63b,8	276b,8
281.	63b,5	49a,15	26b,9	73a,7	64b,1	276b,20
283.	64a,12	49b,16	27a,4	74a,4	65a,7	277a,1
285.	65a,6	50a,16	27a,16	74b,13	65b,13	277a,9
287.	66a,4	51a,4	27b,14	75b,12	66b,8	277a,19
289.	66b,11	51b,5	28a,10	76b,5	67a,15	277a,27
291.	67b,7	52a,10	28b,7	77b,3	68a,12	277b,11
293.	68a,6	52b,4	28b,16	78a,3	68b,12	277b,19
295.	68b,5	53a,3	29a,4	78b,2	69a,6	277b,23
297.	69a,1	53a,14	29a,16	79a,4	69b,5	278a,4
299.	69a,5	53b,1	29b,2	79a,12	69b,13	278a,7
301.	69b,13	54a,10	29b,13	80a,3	70a,16	278a,15
303.	70b,3	54b,11	30a,8	80b,13	71a,6	278a,24
305.	71a,8	55a,12	30b,4	81b,10	71b,14	278b,6
307.	71b,15	55b,14	30b,16	82b,6	72b,6	278b,16
309.	72b,3	56a,11	31a,9	83a,11	73a,9	278b,24
311.	73a,4	56b,7	31b,1	83b,15	73b,10	279a,3
313.	73b,7	57a,5	31b,11	84b,7	74a,14	279a,11
315.	74a,15	57b,6	32a,7	85b,1	75a,5	279a,20
317.	75a,5	58a,6	32b,2	86a,9	75b,12	279b,1
319.	75b,4	58b,1	32b,10	86b,8	76a,9	279b,7
321.	76a,13	59a,4	22a,8	87b,7	76b,18	279b,19
323.	76b,3	59a,8	22a,11	88a,1	77a,1	279b,21
325.	77b,3	59b,15	22b,10	89a,1	78a,2	280a,6
327.	78a,7	60a,13	–	89b,6	78b,3	280a,13
329.	78b,6	60b,7	–	90a,9	79a,2	280a,19
331.	79a,9	61a,5	–	90b,14	79b,6	280a,27
333.	80a,2	61b,6	–	91b,8	80a,15	280b,9
335.	80b,11	62a,8	–	92b,3	81a,6	280b,19

page of edition	D	F	G	L	M	T
337.	81a,14	62b,7	43a	93a,10	81b,11	280b,27
339.	82a,5	63a,9	43a,4	94a,6	82b,4	281a,10
341.	82b,10	63b,7	43a,13	94b,12	83a,9	281a,17

Notes: (1) In MS G, pp. 327–335 are missing. The text ends on p. 325, line 14 and resumes on p. 337, line 20. (2) ^{m} = margin.

M. Editorial Procedures

1. Previous Work on the Tadhkira

The French orientalist Carra de Vaux seems to have been the first modern European to have become interested in the *Tadhkira.* He made a translation into French of Book II, Chapter 11 (*i.e.* the part in which Naṣīr al-Dīn presents his non-Ptolemaic models), which Paul Tannery saw fit to include as Appendix VI of his *Recherches sur l'histoire de l'astronomie ancienne* published in 1893. Except for a brief discussion in Dreyer [1906], which followed Carra de Vaux's translation, interest in the *Tadhkira* waned until Roberts [1957], in which the similarity between the models of the 14th c. Damascene astronomer Ibn al-Shāṭir and Copernicus was first noted. (This was based on a discovery by E. S. Kennedy and the recognition of the connection to Copernicus by O. Neugebauer.) Kennedy [1966] established the historical continuity of the work of Ibn al-Shāṭir with that of Ṭūsī and his immediate successors, dubbing the whole lot, rather ahistorically, the "Marāgha School." Hartner as well dealt with various aspects of Ṭūsī's work, in particular the lunar theory [1969] and the connection to Copernicus ([1973] and [1975]). Except for Livingston [1973], who discussed the *Tadhkira* as a genre of astronomical writing, the main interest in it has centered on the non-Ptolemaic modeling.

I have seen little need of providing a systematic criticism of my predecessors; obviously I owe them an enormous debt. But this has not meant that I have avoided indicating those places where my own understanding and interpretation have differed from theirs.

2. Establishment of the Edition

The very large numbers of extant manuscripts of the *Tadhkira*, either standing as independent texts or incorporated into a commentary, considerably complicates the problem of establishing *the* or even *a* text. I had originally conceived of the task as one of collating as many manuscripts as possible and producing a

"Platonic text" that assumed Naṣīr al-Dīn was flawless to a fault. But after producing shoebox after shoebox of variants, and an endless assortment of stemmata that bore an increasing resemblance to a New York subway map, I concluded that I, and I hope my readers, would be better off if I followed the classicists a bit less and the evidence of the manuscripts and the commentators a bit more. That evidence, which I have discussed in detail in Section J, leads inexorably to the conclusion that there was an original version of the *Tadhkira* produced in Marāgha in 659/1261 and that there was an emended version (which the copyist of MS D refers to as *al-iṣlāḥ al-jadīd*) finalized in Baghdad in the last few months of Ṭūsī's life—or perhaps shortly after his death—in 672/1274. I therefore decided to present as the main edited text this final revision (or more precisely as close an approximation to it as I could manage). I refer to this throughout as the Baghdad (β) version, which would have been Ṭūsī's final word on the subject. Revisions are marked off by slashes and the readings from the original Marāgha (α) version are presented at the foot of the page as are those from an intermediate stage (MS M).

No manuscript can be said to establish alone the Marāgha or Baghdad version; of the manuscripts I have examined, the ones that had most completely incorporated into their texts the revisions that had been made from the time of the completion of the Marāgha version in 659/1261 until Ṭūsī's death in 672/1274 were MS F (Vatican ar. MS 319, 1) and MS L (Leiden, University Library MS or. 905). I do not know the date or provenance of MS L, but MS F was completed just 10 years after Ṭūsī's death in 683/1284 and, by a happy coincidence, at the Niẓāmiyya College in Baghdad itself. MS F has thus served as my main source, but MS L has been used to corroborate it. Furthermore the marginal variants in MSS DMT have also been used to confirm that what one finds in MSS FL were indeed Ṭūsī's revisions. The main text of MS D (Istanbul, Feyzullah 1330, 1), which was copied from an autograph, is in my estimation very close to the original Marāgha version; revisions have usually been scrupulously put into the margins. Another manuscript that is very close to MS D, and in which the revisions are marginalized, is MS T (Istanbul, Ahmet III MS 3453, 19). As I discuss in Section J.3, MS M (Istanbul, Lâleli MS 2116) is a special case; it has been used to establish the revisions to the Marāgha version that were made before Shīrāzī left Marāgha (and Ṭūsī), probably in 667/1268.

Although I have examined some twenty manuscripts of the *Tadhkira* (excluding commentaries) and compared at least eleven of them in detail, I decided to restrict the final apparatus for the most part to the variants from only six manuscripts.[1] In addition to the five mentioned above (MSS DFLMT), I also included MS G (St. Petersburg (a.k.a. Leningrad), Oriental Institute, MS A 437) mainly because of its early date (673/1275). But in addition to being incomplete and hopelessly corrupt due to an incompetent copyist, it incorporates material

[1] Occasionally there are references in the apparatus or the commentary to one of the other manuscripts.

from the new version without indication and sometimes puts it in the wrong place, which would seem to show that the copyist found some of Ṭūsī's marginal revisions but did not know quite where to put them (see, for example, II.10 [3]). What is interesting is that it has very late revisions, such as II.2 [3]22–23, which was added after Shīrāzī's departure from Marāgha. On the other hand, it has a variant on II.4 [2]23–[3]16 that I have not found elsewhere, which would indicate a very early revision since it retained but modified the original four-possibility version before it was changed to eight possibilities.[2] MS G is indicative of a highly contaminated text that brings together original texts and revisions in a haphazard and confusing way; in fact, many, if not most, of the manuscripts I have looked at do this. It is only because of the care taken by the commentators and by a few copyists that "purer" textual versions can be presented here.

Let me turn to some specific points. Though MS F is fairly close to my Baghdad version, there are some differences. In the main this is due to my reluctance to assume that a variant in it represents one of Ṭūsī's revisions unless there are other witnesses or corroborating evidence. An important example where MS F was not used is IV.1 [3]18–19, where I could not confirm that its recalculated value of 3,756,231½ was due to Ṭūsī. On the other hand, MS F does not contain all the revisions that can be ascertained elsewhere; examples are I.1 [14]22–23 and IV.6 [5]11. But I would maintain that it is a better witness to the final revised version than MS L, which lacks the revisions of II.8 [14]10–11, II.9 [1]9, II.13 [9]15, II.13 [9]19–20, III.8 [7]1, III.10 [1]13–14, III.12 [2]15, and IV.5 [3]19. As for my best witnesses to the original (α) version, namely MSS DT, they occasionally seem to place revisions into the text rather than the margins. For MS T, this is the case for II.9 [8]21; for MS D, examples are II.10 [3]10 and II.10 [3]11–12.

I have not seen it as my job to correct Ṭūsī's mistakes; where it would seem that these were not corrected in the Baghdad revision, they have been retained. In addition to the particularly egregious case of II.5 [8]6–8, where the revision itself is nonsensical, examples are II.1 [6]11, II.11 [10]11&12, II.13 [6]12, and IV.3 [5]19.[3] In general, such "mistakes" are indicated by an exclamation mark. On the other hand, I have standardized the spelling of numbers as well as such words as *hāhunā* and have used the modern rules for *hamzas*. The spellings in the manuscripts are indicated in the text apparatus. Furthermore I have attempted to have subjects and verbs agree despite the distance between them; even though the grammarians disagree on this, as do the copyists, such a course seemed best for a scientific treatise where the minimization of ambiguity is desirable.

As for the text figures, I have generally followed MSS FL and have given variations in the figure apparatus. Labels that are absent in MSS FL but present

[2] See the commentary for more details.

[3] The last is particularly significant; see our commentary for details.

in the other manuscripts are put in parentheses. For Figure T25, where the divergence between the Marāgha and Baghdad versions was especially great, I placed the Baghdad figure in the text and the Marāgha version in the commentary (Figure C34).

3. The Translation

I have attempted to be as literal as possible, but I have also been concerned to make the translation sound like reasonable English; the *Tadhkira* is written with a precise, structured, scientific Arabic that developed over many centuries, and I have felt strongly that the English should reflect this. For the most part, I have followed the English vocabulary that has evolved during this century to translate scientific Arabic (which is fairly consistent with that of the classicists and Latin medievalists), but I have not hesitated to coin new renditions when this seemed appropriate (for example, dirigent for *mudīr* and monoform for *ᶜalā nahj wāḥid*). A complete listing of technical terms and their translations is given in the glossary.

Ṭūsī normally writes his numbers and fractions in words; an exception occurs in II.6 [4], where he gives the sun's eccentricity alphanumerically. He is not consistent in the way he gives fractions, employing sexagesimals, sums of unit fractions, and non-unit fractions. I have retained this inconsistency in the translation, but, for the sake of clarity, generally rendered his numbers numerically.

The translation itself follows the text of the Baghdad (β) version; with the exception of untranslatable grammatical variants, the variant readings from the Marāgha (α) version and from MS M have been indicated at the bottom of the page. Line numbers in the margin follow the edited text and not the actual lines of the translation.

4. The Commentary

Neugebauer's *History of Ancient Mathematical Astronomy*, Pedersen's *Survey of the Almagest*, and Toomer's translation of the *Almagest* have made it unnecessary to explain the Ptolemaic system in detail in a work such as this. I have therefore not belabored Ṭūsī's "summary" of the *Almagest* in the commentary but rather concentrated on those points that are rather more obscure. In general these involve matters in which Islamic astronomers had modified or recast the science they had inherited from antiquity. But I have also felt it necessary to elaborate on points that were of interest, or that may have been unclear, to Ṭūsī's successors; on these matters I have relied heavily upon Shīrāzī and the commentators.

Each comment is generally numbered in the following sequence: Roman numeral of the book; number of the chapter; bracketed number of the paragraph; and line number. The line number invariably refers to the edited text. This is usually followed by an identifying transliterated phrase followed by the English translation in parentheses.

Part II

Edition and Translation

In the Name of God the Beneficent, the Merciful

[1] Praise be to God Who brings forth good and Who inspires truth. And may His blessings be upon Muḥammad, who was sent with the Final Message, upon his family, the most excellent of families, and upon his Companions, the most excellent of Companions.

[2] We wish to present a summary of astronomy as a memento for one of our dear friends, and we ask God to grant us success for its completion; for He is the One who grants success and to Him is the Final Return. Let us now set forth what we have in mind in chapters that are contained within four books.

BOOK I

Concerning That Which Must Be Presented by Way of Introduction

[1] / Every science has: [a] a subject, which is investigated in that discipline; / [b] principles, which are either self-evident or else obscure, in which case they are proved in another science and are taken for granted in this science; and [c] problems, which are proved in this science.

[2] The subject of astronomy is the simple bodies, both superior and inferior, with respect to their quantities, qualities, positions, and intrinsic motions. Those of its principles that need proof are demonstrated in three sciences: metaphysics, geometry, and natural philosophy. Its problems aim at gaining knowledge of these bodies in and of themselves, of their shapes, of the manner of their arrangement and motions, of the amounts of their motions and distances, and of the reasons for changes in position.

/10/ …] β, M = Every science has: [a] a subject whose essential attributes are investigated in it] α.

بسم الله الرحمن الرحيم

[١] الحمد لله مفيض الخير وملهم الصواب ، وصلواته على محمد المبعوث بفصل الخطاب وعلى آله خير آل وأصحابه خير أصحاب .

[٢] نريد أن نورد جملاً من علم الهيئة تذكرة لبعض الأحباب ونسأل الله أن يوفق لإتمامه ، إنه الموفق وإليه المآب . فلنورد ما قصدناه في فصول تشتمل عليها أربعة أبواب .

الباب الأول
فيما يجب تقديمه

[١] لكل علم موضوع يُبحث / في ذلك العلم عنه / ، ومبادئ إما بينة بنفسها وإما خفية تُبيَن في علم آخر وتُستعمل في ذلك العلم على أنها مسلمة ، ومسائل تُبيَن في ذلك العلم .

[٢] وموضوع الهيئة الأجرام البسيطة العلوية والسفلية من حيث كمياتها وكيفياتها وأوضاعها وحركاتها اللازمة لها . ومبادئها المحتاجة إلى البيان تتبين في علوم ثلاثة : ما بعد الطبيعة، والهندسة ، والطبيعيات . ومسائلها معرفة تلك الأجرام بأعيانها وأشكالها ، وكيفية نضدها وحركاتها ، ومقادير الحركات والأبعاد، وعلل اختلاف الأوضاع .

/10/ ...] M , β = فيه عن أعراضه الذاتية] α .

[3] The scientific exposition that we wish to undertake will be a summary account of this presented in narrative form. The details are expounded and proofs of the validity of most of them are furnished in the *Almagest*. Indeed, ours would not be a complete science if taken in isolation from the *Almagest* for it is a report of what is established therein.

[4] In this [study], it is necessary that there be a familiarization with certain definitions and rules, which will be presented by way of introduction. For their elucidation, one is referred to the above-mentioned sciences. Despite the fact that the explanations of these [definitions and rules] occur in various places, they will be grouped into two divisions: one of them pertaining to the geometry [corpus] and the other pertaining to the natural philosophy [corpus]. Let us then present an account of them in two chapters.

CHAPTER ONE

An Account of What Needs to Be Known That Pertains to the Geometry [Corpus]

[1] Among those things having position, *i.e.* that can be indicated by the senses, are: [a] a **point**, which is that without part; [b] a **line**, which is that with length only and which ends in a point; [c] a **surface**, which is that with length and width only and which ends in a line; and [d] a **solid**, which is that with length, width, and depth and which ends in a surface. These "ends" are called **boundaries**.

[2] A **straight line** is the one on which all the given points are facing one another. / A **plane surface** is the one on which the assumption of straight lines in all directions is possible. /

/19–20/ …] β = a plane surface is the one of which it is possible to produce straight lines in all directions] M = a plane surface is the one whereby the lines assumed on it in all directions are straight] α.

[٣] والفن الذي نريد أن نشرع فيه تقرير جمل من ذلك تورد على سبيل الحكاية . وتُبيّن تفاصيلها وتُقام البراهين على صحة أكثرها في المجسطي ، فهو ليس بعلم تام إذا أفرز عن المجسطي لأنه حكاية / ما / عما ثبت فيه .

[٤] ولا بد / فيه / من تعرّف حدود وأحكام تورد على سبيل التصدير ، ويحال بيانها على العلوم المذكورة . وهي ، على اختلاف مواضع بياناتها ، تنقسم إلى قسمين : أحدهما ما يتعلق بالهندسيات والآخر ما يتعلق بالطبيعيات . فلنقدم ذكرها في فصلين .

الفصل الأول
في ذكر ما يحتاج إلى معرفته مما يتعلق بالهندسيات

[١] من الأشياء التي لها وضع أي التي يمكن أن يشار إليها بالحس : النقطة وهي ما لا جزء له ، والخط وهو ما له طول فقط وينتهي بالنقطة ، والسطح وهو ما له طول وعرض لا غير وينتهي بالخط ، والجسم وهو ما له طول وعرض وعمق وينتهي بالسطح . وتسمى النهايات حدوداً .

[٢] والمستقيم من الخطوط هو الذي يتحاذى جميع النقط التي تفرض عليه . / والمستوي من السطوح هو الذي يكون فرض الخطوط المستقيمة عليه في جميع الجهات ممكناً / .

/4/ ما] β , M = α– . /5/ فيه] β , M = α– . /19–20/ ...] β = والمستوي من السطوح هو الذي يمكن أن تخرج منه الخطوط المستقيمة في جميع الجهات؛ M = والمستوي من السطوح هو الذي تكون الخطوط المفروضة عليه في جميع الجهات مستقيمة] α .

[3] An **angle** is either: [a] a surface bounded by two lines meeting at a point in such a way that they do not form a single line; or [b] a solid bounded by surfaces meeting at a point in such a way that any two of its surfaces join at a line without forming a single surface.

[4] The point at which two lines join or intersect is a **common part** for them; and, similarly, a line for surfaces and a surface for solids.

[5] If a straight line stands erect on another straight line and there result two equal angles on its two sides, then these are **right angles**, and each one of the two lines is **perpendicular** to the other. An angle which is smaller than a right angle is **acute**, and that which is larger is **obtuse**.

[6] A straight line that stands erect on a plane surface in such a way that it along with any given line lying in the [plane] and meeting it bound a right angle is **perpendicular** to that surface. When one plane surface stands erect on [another] plane surface in such a way that any two perpendiculars that are produced in the [planes] from any given point on their common part bound a right angle, then the two [planes] **intersect at right angles**.

[7] Straight lines occurring in a plane that do not meet even if extended without limit in both directions are **parallel**. Similarly planes that do not meet even if extended without limit in all directions are also [parallel]. Non-straight [lines] and non-plane [surfaces] may also be called parallel if distances between them are completely invariable.

[8] A **circle** is a plane surface bounded by a curved line in whose interior is a point such that all the straight lines extending from it to the [curved line] are equal. That line is its **circumference**, and the point is its **center**. The extended lines are **radii**. A line extending in both directions from the [point] to the circumference is a **diameter** of the circle, which it bisects.

[٣] والزاوية سطح أحاط به خطان ملتقيان عند نقطة من غير أن يتحدا خطاً واحداً أو جسم أحاط به سطوح ملتقية عند نقطة يتصل كل سطحين منها عند خط من غير أن يتحدا سطحاً واحداً .

[٤] والنقطة التي يتصل أو يتقاطع عليها خطان فصل مشترك لهما ، وكذلك الخط للسطوح والسطح للأجسام .

[٥] وإذا قام خط مستقيم على خط مستقيم وحدثت عن جنبتيه زاويتان متساويتان فهما قائمتان وكل من الخطين عمود على صاحبه . والزاوية التي هي أصغر من قائمة حادة والتي هي أعظم منفرجة .

[٦] والخط المستقيم القائم على سطح مستوٍ بحيث يحيط مع كل خط يفرض فيه ملاقياً له بقائمة عمودٌ على السطح . وإذا قام سطح مستوٍ على سطح مستوٍ بحيث يحيط كل عمودين يخرجان فيهما من أي نقطة تفرض على فصلهما المشترك بقائمة فهما يتقاطعان على قوائم .

[٧] والخطوط المستقيمة الكائنة في سطح مستوٍ التي لا تتلاقى وإن أخرجت في الجهتين إلى غير / نهاية / هي المتوازية ، وكذلك السطوح المستوية التي لا تتلاقى وإن أخرجت في جميع الجهات إلى غير / نهاية / . وقد يقال في غير المستقيمة والمستوية منهما متوازية إذا لم تختلف الأبعاد بينهما أصلاً .

[٨] الدائرة سطح مستوٍ يحيط به خط مستدير في داخله نقطة يكون جميع الخطوط المستقيمة الخارجة منها إليه متساوية ، وذلك الخط محيطها وتلك النقطة مركزها والخطوط الخارجة أنصاف أقطارها . والخارج منها إلى المحيط في الجهتين قطر لها وهو يُنصّف الدائرة .

/17/ ...] M , β = النهاية] α . /19/ ...] M , β = النهاية] α .

[9] Every straight line that divides the circle into any two parts whatever is a **chord**, and what is separated off from the circumference is an **arc**. Half a chord is the **sine** of half the arc. The perpendicular extending from the midpoint of an arc to the midpoint of the chord is the **versed sine** of half that arc.

[10] A **sphere** is a solid bounded by a curved surface in whose interior is a point such that all straight lines extending from it to the [curved surface] are equal. That surface is its **circumference**, and the point is its **center**. The extended lines are **radii**. A line extending in both directions from the [point] to the circumference is a **diameter** of [the sphere].

[11] Every plane that divides a sphere into two segments produces a circle in it, which is the common part between them. If it bisects it, then this is the largest circle that can occur in that sphere and it will pass through its center; thus their two centers coincide.

[12] If a sphere rotates about itself, every point marked on it describes by its motion in / each / complete rotation a circle, which is the point's **circuit**. The exceptions to this are two points, which are the **poles** of the sphere. The diameter that connects them does not move either; it is the **axis**.

[13] The great circle that is equidistant from the two poles is the sphere's **equator**. All the circuits are parallel to one another as well as being parallel to the equator. The axis is perpendicular to each. Any two circuits on the two sides of the equator and equidistant from it are equal. Every great circle / or small one in the sphere / has an axis and two poles as is the case with the equator.

[14] Two given great / circles / on a sphere bisect one another at two points, and their [common] part is a straight line / that passes through the center. / The greatest distance between the two circles is the same as the distance between

/13/ each] β = –α, –M. /19/ or small one in the sphere] β = –α, –M. /21/ circles] β = –α, –M. /22–23/ that passes through the center] β = –α, –M.

[٩] وكل خط مستقيم يقطعها بقطعتين كيف اتفق فهو وتر وما يُفرَز من المحيط قوس . ونصف الوتر لنصف القوس جيب . والعمود الخارج من منتصف القوس إلى منتصف الوتر سهم لنصف القوس .

[١٠] الكرة جسم يحيط به سطح مستدير في داخله نقطة يكون كل الخطوط المستقيمة الخارجة منها إليه متساوية ، وذلك السطح محيطها وتلك النقطة مركزها والخطوط أنصاف أقطارها . والخارج منها إلى المحيط في الجهتين قطر لها .

[١١] وكل سطح مستوٍ يقطع الكرة إلى قطعتين يُحدِث دائرة فيها هي الفصل المشترك بينهما . فإن نصّفها فهي أعظم دائرة تقع في تلك الكرة وتمرّ بمركزها فيتّحد مركزاهما .

[١٢] وإذا دارت الكرة على نفسها / فعلت / كل نقطة ترسم عليها بحركتها في / كل / دورة تامة دائرة هي مدارها ، إلا نقطتين هما قطبا الكرة . والقطر الواصل بينهما أيضاً لا يتحرك وهو المحور .

[١٣] والدائرة العظيمة المتساوية البعد عن القطبين منطقتها . وتكون المدارات جميعاً متوازية وموازية للمنطقة ، والمحور عموداً على الكل . وكل مدارين عن جنبتي المنطقة / متساويتي / البعد عنها متساويان . ولكل دائرة عظمى / أو صغرى في الكرة / محور وقطبان كما للمنطقة .

[١٤] وإذا فرضت على كرة / دائرتان / عظيمتان فهما تتناصفان على نقطتين ويكون فصلهما خطاً مستقيماً / ماراً بالمركز / ، ويكون أعظم الأبعاد بين الدائرتين كالبعد بين

/12/ ...] M , β = فعل] α . /13/ ...] β = M– , α–. /18/ ...] M , β = متساوبي] α.
/19/ ...] β = M– , α– . /21/ ...] β = M– , α– . /22–23/ ...] β = M– , α– .

their two poles. If they intersect at right angles, each of the [circles] passes through the poles of the other; the converse also holds.

[15] An **orb** is a spherical solid bounded by two parallel surfaces having the same center. The outer [surface] is called **convex** and the other **concave**. Sometimes the concave is not taken into account, as in the case of epicycles.

[16] A **circular cylinder** is a solid bounded by two equal, parallel circles, which are its **bases**, and by a circular surface that connects their circumferences. / The line that joins the two centers is an **axis** of the [cylinder]; if it is perpendicular to the planes of the two circles, the **cylinder** is a **right** one. /

[17] A **circular cone** is a circular solid that rises from a circle, which is its **base**, to a point, which is its **apex**. The line that joins the point and the center of the base is / the [cone's] **axis**; if it is perpendicular to its base, the **cone** is a **right** one. /

[18] If a cylinder or a cone is cut by a plane passing through the axis, it produces a quadrilateral in the cylinder and a triangle in the cone. If the plane is parallel to the base, then in both cases it will produce a circle.

CHAPTER TWO

An Account of What Needs to Be Accepted from Natural Philosophy in This Science

[1] A body is either **simple**, in which case it has a single nature and what issues forth from that nature does so monoformly, or else **compound**, in which case it is composed of

---

/8–9/ …] β = The line that joins the two centers is perpendicular to the planes of the two circles and it is the axis of the cylinder] α, M. /12–13/ …] β = perpendicular to its base and it is its axis] α, M.

قطبيهما . فإن / تقاطعتا / على قوائم مرّ كل منهما بقطبي الأخرى وبالعكس .

[١٥] الفلك جسم كري يحيط به سطحان متوازيان / مركزاهما / واحد . ويسمى الخارج منهما محدّباً والآخر مقعّراً ، وربما لا يعتبر المقعّر كما في التداوير .

[١٦] الأسطوانة المستديرة جسم تحيط به دائرتان متساويتان ومتوازيتان هما قاعدتاها وسطح مستدير واصل بين محيطيهما . / ويكون الخط الواصل بين المركزين سهماً لها ، فإن كان عموداً على سطحي الدائرتين كانت الأسطوانة قائمة / .

[١٧] والمخروط المستدير جسم مستدير يرتفع من دائرة هي قاعدته إلى نقطة هي رأسه . والخط الواصل بين النقطة ومركز القاعدة / يكون سهمه ، فإن كان عموداً على قاعدته كان المخروط قائماً / .

[١٨] وإذا فُصل الأسطوانة والمخروط بسطح يمرّ بالسهم أحدث في الأسطوانة ذا أربعة أضلاع وفي المخروط مثلّثاً . فإن كان السطح موازياً للقاعدة أحدث فيهما دائرة .

الفصل الثاني
في ذكر ما يحتاج في هذا العلم
إلى تسليمه من الطبيعيات

[١] الجسم إما بسيط وهو الذي له طبيعة واحدة يصدر عنها ما يصدر على نهج واحد ، وإما مركّب وهو الذي يتركب من

/1/ ...] M , β = تقاطعا] α . /3/ ...] β = مركزهما] M , α . /8–9/ ...] β = ويكون الخط الواصل بين المركزين عموداً على سطحي الدائرتين وهو سهم الأسطوانة] M , α . /12–13/ ...] β = يكون عموداً على قاعدته وهو سهمه] M , α.

simple bodies and may turn out to be a species distinct from them. A simple body is either celestial or elemental. The **celestial bodies** are the orbs and the luminous bodies whose proper place is the orbs. The **elemental bodies** are the four well-known elements. The compound bodies are those things composed of them, namely minerals, plants, and animals. Their **proper places** are those of the elements. A void is impossible.

[2] Every motion has a **principle**. The mobile is said to be **self-moved** if its principle of motion does not become positionally separated from it. If it does separate from it, the movement is ascribed to the mobile, but the action of moving the mobile is attributed to that in which its principle of motion occurs. If the motion of a self-moved mobile is monoform, its principle of motion is called a **nature** whether the motion is natural and elemental or voluntary and celestial. Otherwise its [principle of motion] is called a **soul** whether the [motion] be vegetative or animal. When a mobile is moved by something other than itself, its motion is **accidental** if it is part of the mover or if the mover is the place for it naturally to be; otherwise its motion is by **compulsion**.

[3] Motion due to a nature is divided into: [a] that which is toward the center; its principle is heaviness and it is characteristic of the two heavy elements; [b] that which is away from the center; its principle is lightness and it is characteristic of the two light elements. These two [motions] are **displacing** and **rectilinear**. And [c] that which is about the center; this motion is **in place** and **circular**, and it is characteristic of the celestial bodies. The [latter motion] is divided into **simple**, which arises from a single, simple body with every given point on it producing at the center equal angles in equal times or cutting equal arcs from the circumference, and into **composed**, which arises from a combination of more than one simple body. Every motion whose angles or arcs vary in equal times is composed; however, the converse does not hold.

[4] Nothing having the principle of circular motion can undergo any rectilinear motion at all, and conversely, except by compulsion. Thus the celestial bodies neither tear nor mend, grow nor diminish, expand nor contract; neither does their motion intensify nor weaken. They do not reverse direction, turn, stop, depart from their confines, nor undergo any change of state except for their circular motion, which is uniform at all times.

بسائط /وقد يصير/ نوعاً غيرها. والبسيط إما فلكي وإما عنصري: والفلكي هو الأفلاك والأجرام النيّرة التي مكانها الأفلاك، والعنصري هو العناصر الأربعة المشهورة. والمركّب ما يتركب منها من المعادن والنبات والحيوانات، وأمكنتها أمكنة العناصر. والخلاء محال.

[٢] ولكل حركة مبدأ. والمتحرك إن لم يفارقه مبدؤه بالوضع قيل إنه يتحرك بنفسه؛ وإن فارقه نُسب التحرك إليه والتحريك إلى ما فيه مبدؤه. والمتحرك بنفسه إن كانت حركته على نهج واحد سمي المبدأ طبعاً سواء كانت الحركة طبيعية عنصرية أو إرادية فلكية. وإن لم يكن كذلك سمي نفساً سواء كانت نباتية أو حيوانية. والمتحرك بغيره إن كان كجزء من المحرك أو كان المحرك مكاناً له بالطبع فالحركة عرضية، وإلا فقسرية.

[٣] والحركة بالطبع تنقسم إلى ما إلى المركز ومبدؤه الثقل وتختص بالعنصرين الثقيلين؛ وإلى ما من المركز ومبدؤه الخفة وتختص بالعنصرين الخفيفين، وهما أينيّتان مستقيمتان؛ وإلى ما على المركز وهي وضعية مستديرة وتختص بالفلكيات، وتنقسم إلى بسيطة تصدر عن جرم واحد بسيط كل نقطة تفرض عليه تفعل عند المركز في أزمنة متساوية زوايا متساوية أو تقطع من المحيط قسياً متساوية، وإلى مركبة تصدر عن جملة بسائط فوق واحدة. وكل حركة تختلف زواياها أو قسيها في الأزمنة المتساوية مركبة، ولا ينعكس.

[٤] وكل ما فيه مبدأ حركة مستديرة فهو لا يقبل الحركة المستقيمة أصلاً، وبالعكس، إلا بالقسر. فالفلكيات لا تنخرق ولا تلتئم، ولا تنمو ولا تذبل، ولا تتخلخل ولا تتكاثف، ولا تشتد في حركاتها ولا تضعف؛ ولا يكون لها رجوع ولا انعطاف ولا وقوف ولا خروج من حيز ولا اختلاف حال غير حركتها المستديرة المتشابهة في جميع الأوقات.

/1/ ...] M , β = ويصير] α .

BOOK II

The Configuration of the Celestial Bodies

Fourteen Chapters

CHAPTER ONE

On the Sphericity of the Sky and the Earth; On the Earth Being in Relation to the Sky As the Center of a Sphere to Its Circumference; and on [the Earth] Being Completely Stationary

[1] The sphericity of the sky is evidenced by: the movement of the fixed stars along parallel circles about a stationary point; the occurrence of those [stars] nearest this [point] upon [relatively] small circuits of permanent visibility, while those further away occur on [progressively] larger circuits until those stars are reached that just touch the horizon but do not disappear and then are followed by those that disappear fleetingly, keeping one and the same rising and setting place; the proportional increase in the periods of invisibility thereafter according to the increase in distance until those [stars] are reached whose periods of visibility and invisibility are equal and then are followed by those whose period of invisibility exceeds their period of visibility; the increase as well in the period of invisibility until those [stars] are reached that are visible briefly and then are followed by those that just touch the horizon once every revolution but do not rise; the converse equality of the periods of visibility and invisibility of those [stars] equidistant and on the two sides of that circuit upon which the periods of visibility and invisibility are equal; the ascent of whatever rises occurring little by little up to a certain maximum that is at the midpoint of the visible segment of its circuit and then its descent little by little until it disappears; the [successive] rising of one part of its body after another, as also occurs during setting; the observed equality of size of a [celestial body] at all distances along its revolution except at the horizon [which can be explained by] the accumulation of vapors rising from the Earth causing objects behind them to appear larger than they should just as we find to be the case for things that we observe at one time in air and in another in water. Therefore size increases

الباب الثاني
في هيئة الأجرام العلوية
أربعة عشر فصلاً

الفصل الأول
في استدارة السماء والأرض
وكون الأرض عند السماء كمركز الكرة عند محيطها
وكونها غير متحركة بالجملة

[١] تحركُ الثوابت على دوائر متوازية حول نقطة لا تتحرك ؛ وكون ما هو أقرب منها على مدار أصغر أبدي الظهور ، وما هو أبعد على مدار أكبر إلى أن ينتهي إلى ما يماس الأفق ولا يخفى ، ثم إلى ما يخفى زماناً يسيراً حافظاً لمطلع ومغيب بعينهما ؛ وتزايُدُ أزمنة الخفاء بعد ذلك بحسب تزايد البعد على نسبة إلى أن ينتهي إلى ما يتساوى زمانا ظهوره وخفائه ، ثم إلى ما يزيد زمان خفائه على زمان ظهوره ؛ وتزايُدُ أزمنة الخفاء أيضاً إلى أن ينتهي إلى ما يظهر قليلاً ، ثم إلى ما يماس الأفق في دورة مرة ولا يطلع ؛ وتساوي زماني الظهور والخفاء للمتساوية الأبعاد عن المدار الذي يتساوى زمانا ظهوره وخفائه عن الجنبتين على التبادل ؛ وارتفاعُ ما يطلع يسيراً يسيراً إلى غاية ما عند منتصف القطعة الظاهرة من مداره ، ثم انحطاطه يسيراً يسيراً إلى أن يخفى ؛ وطلوعُه شيئاً بعد شيء من جرمه وكذلك غروبه ؛ وتساوي مقداره في النظر في جميع أبعاده في دورته إلا عند الأفق ، فإنّ تراكُم الأبخرة المرتفعة من الأرض يُري ما وراءها من الأشخاص أكبر مما يجب أن يُرى كما نشاهد فيما نرى تارة في الهواء وتارة في الماء ولذلك يزداد

when the air becomes denser and the opposite holds; the permanent visibility of half the sky—or thereabouts—to everyone on Earth whatever his location; and, in addition to the above, other characteristics that are peculiar to sphericity.

[2] The general sphericity of the Earth is evidenced by: the earlier rising and setting of the stars for those persons in the east relative to their rising and setting for those in the west, and the increase and decrease of that [priority] according to farness and nearness; the increasing altitude of the pole and the northern stars and the decreasing altitude of the southern ones for those who advance northward and the opposite for those who advance southward, this being in both cases according to the extent of their advance; and the combining of these two variations for those traveling on a course between the two [cardinal] lines [*i.e.* east-west and north-south]. The [Earth's] undulations that occur due to mountains and depressions do not cause it to depart from being basically spherical since there is no appreciable ratio between them and the [Earth] taken as a whole. For a mountain that rises to a height of half a parasang is to the [Earth] as $\frac{1}{5}$ of $\frac{1}{7}$ the width of a barleycorn is to a sphere whose diameter is approximately one cubit. This will be elucidated when investigating the area of the Earth.

[3] The sphericity of the surface of the water standing on the face of the Earth is evidenced, in addition to what has been stated above regarding the Earth, by the fact that the curvature of the sea waters conceals the lower parts of mountains rising up from them but not their higher reaches and that these [lower parts] appear little by little to someone approaching them.

[4] That the Earth is in the middle of the Universe at the center is evidenced by: the equality of the periods of ascent and descent of the stars during the time of their visibility; the permanent visibility of half the celestial sphere; the coincidence on a single straight line of the sun's shadows at the time of its rising and setting when it is on the [day-] circle on which its periods of visibility and invisibility are equal or when it is at opposite parts of the circle that it describes with its proper motion; and the eclipse of the moon at its true oppositions to the sun.

[5] The permanent visibility of half the ecliptic orb as well as the orbs below it down to the orb of the sun is evidence that the Earth does not have a perceptible size with respect to the orb of Mars and all orbs beyond; indeed, it is as a point since there is no difference

الكبر إذا صار الهواء أغلظ وبالضد ؛ وظهورُ النصف أو قريب منه دائماً لكل من على الأرض في أي موضع يكون ؛ إلى غير ذلك من الأعراض الخاصة بالاستدارة : يدل على استدارة السماء .

[٢] وتقدمُ طلوع الكواكب وغروبها للمشرقيين على طلوعها وغروبها للمغربيين ، وزيادة ذلك ونقصانه بحسب بُعد المسافة وقربها ؛ وازديادُ ارتفاع القطب والكواكب الشمالية وانحطاط الجنوبية للواغلين في الشمال وبالعكس للواغلين في الجنوب بحسب وغولهما ؛ وتركُبُ الاختلافين للسائرين على سمت بين السمتين : يدل على استدارة الأرض جملة . وتضاريسها التي تلزمها من جهة الجبال والأغوار لا تُخرِجها عن أصل الاستدارة إذ لا نسبة محسوسة لها إلى جملتها ، فإنّ جبلاً يرتفع نصف فرسخ يكون عندها كخُمس سُبع عرض شعيرة عند كرة قطرها ذراع بالتقريب ــ يتبين ذلك عند الوقوف على مساحة الأرض .

[٣] وسترُ تقبيب مياه البحار أسافلَ الجبال الطالعة منها دون أعاليها المرتفعة وظهورها قليلاً قليلاً للمتقارب إليها ، مضافاً إلى ما مرّ في الأرض ، يدل على استدارة سطح الماء الواقف على وجه الأرض .

[٤] وتساوي زماني ارتفاع الكواكب وانحطاطها مدة ظهورها ؛ وظهورُ النصف من الفلك دائماً ؛ وتطابُقُ أظلال الشمس في وقتي طلوعها وغروبها عند كونها على المدار الذي يتساوى زمانا ظهوره وخفائه على خط واحد مستقيم أو عند كونها في جزأين متقابلين من الدائرة التي تقطعها بسيرها الخاص بها ؛ وانخسافُ القمر في مقاطراته الحقيقية للشمس : يدل على كون الأرض في وسط الكل عند المركز .

[٥] وظهور النصف من فلك البروج ومما تحته من الأفلاك إلى فلك الشمس دائماً يدل على أنّ الأرض ليست بذات قدر محسوس عند فلك المِرّيخ وما وراءه من الأفلاك ، بل هي كالنقطة إذ لا فرق

between the plane passing over the face of the Earth and separating the visible from the invisible part of these orbs, and the plane parallel to it passing through the center of the Universe. It does, however, have an appreciable size with respect to the moon's orb whose visible segment is therefore less than half; we shall make this clear in the appropriate place.

[6] The reliability of all the evidence we have adduced establishes that these bodies are according to the stated configuration. It is not possible to attribute the primary motion to the Earth. This is not, however, because of what has been maintained, namely that this would cause an object thrown up in the air not to fall to its original position but instead it would necessarily fall to the west of it, or that this would cause the motion of whatever leaves the [Earth], such as an arrow or a bird, in the direction of the [Earth's] motion to be slower, while in the direction opposite to it to be faster. For the part of the air adjacent to the [Earth] could conceivably conform to the Earth's motion along with whatever is joined to it, just as the aether conforms to the orb as evidenced by the comets, which move with its motion. Rather, it is on account of the [Earth] having a principle of rectilinear inclination that it is precluded from moving naturally with a circular motion.

[7] The sphericity of the Earth, / sky, / and water having been established, let it be known that the inclination of all heavy things is to the center of the Earth, which is the center of the Universe, and that the inclination of light things is to the circumference. Thus "up" from all sides of the Earth is what is toward the sky, while "down" is what is toward the Earth's center. Objects are situated on the Earth at the endpoints of its diameters; hence, the distance between their apices is greater than the distance between their bases. The amount of water a filled vessel contains when it is closer to the center, such as at the bottom of a well, is greater than what it contains when it is further away from it, such as at the top of a lighthouse. This is because in the former case the water is more greatly cupolaed than in the latter. This is one of the many things that those not acquainted with these matters find strange.

[8] The above proofs are *inniyya*, which convey existence; those which convey the necessity of that existence are *limmiyyāt* proofs and are given in the book *De caelo* of the Natural Philosophy.

/14/ sky] β = –α, –M.

بين السطح المارّ بوجه الأرض الفاصل بين الظاهر والخفي من تلك الأفلاك وبين السطح المارّ بمركز الكل الموازي لذلك السطح . وأما عند فلك القمر فلها قدر محسوس ، ولذلك تكون القطعة الظاهرة من فلكه أقل من النصف ، وسنبيّن ذلك في موضعه .

[٦] وثبات جميع ما ذكرنا من الدلائل يدل على ثبات تلك الأجرام على الهيئة المذكورة . ولا يمكن إسناد الحركة الأولى إلى الأرض ـ لا لِما قيل من أنّ ذلك يوجب أن لا يقع المرمي في الهواء على موضعه الأول بل يجب أن يقع في الجانب الغربي منه أو يوجب / أن الحركة / لِما انفصل منها كالسهم والطائر إلى جهة حركتها أبطأ وفي خلافها أسرع ، فإنّ المتصل بها من الهواء يمكن أن يشايعها بما يتصل بها[!] كما يشايع الأثير الفلك بدلالة حركات ذوات الأذناب بحركته ـ بل لكونها ذات مبدأ ميل مستقيم ، فيمتنع أن تتحرك على الاستدارة بالطبع .

[٧] وإذا ثبت استدارة الأرض / والسماء / والماء فليُعلم أنّ ميل الأثقال جميعاً إلى مركز الأرض الذي هو مركز الكل ، وميل ما هو خفيف إلى المحيط . فالفوق من جميع جوانب الأرض ما يلي السماء ، والتحت ما يلي مركز الأرض . والأشخاص تقوم على الأرض على أطراف أقطار لها ، فيكون البعد بين رؤوسها أكثر من البعد بين قواعدها . والإناء المملوء ماءً يحوي من الماء وهو أقرب إلى المركز كقعر بئر مثلاً أكثر مما يحويه وهو أبعد منه كرأس منارة مثلاً ، وذلك لكونه هناك أشد تقبيباً من هاهنا . وهذا من جملة ما يستغربه مَن لا يعرف هذه المسائل .

[٨] وهذه الأدلة إنّية تُفيد الوقوع ، والتي تُفيد وجوب الوقوع من اللِميات ما يذكر في كتاب السماء والعالم من العلم الطبيعي .

/9/ ...] أن تكون الحركة = β , M [α . /14/ ...] −M = β , −α .

CHAPTER TWO

On the Arrangement and Order of the Bodies

[1] An observer of the two luminaries and the stars finds that they all move with the daily motion: those that rise do so in the east, travel westward and become invisible there. After a period of invisibility, they return to the east once more, rising there as they did before and continuing thus in the same way always. Those stars that do not rise have a parallel motion.

[2] With a more precise observation than the first, he will then find that they [also] move with a slower motion and in a direction opposite to the first, seemingly from west to east. Actually this latter motion may be distinguished from the former because of a difference in their equators and poles, as will be discussed more fully below, since it would be impossible to perceive two distinct motions occurring on the same sphere and sharing the same equator and poles. In such a case one would only perceive them as a single motion, either as the sum of the two motions if they are in the same direction or as the excess of the faster over the slower if they are in [opposite] directions. The same rule applies as well if more [than two motions] are involved. Both of these [primary] motions are uniform in and of themselves, and both embrace the entirety of the stars and bodies that are perceived above.

[3] [The observer] will then find that the two luminaries and the five planets have irregular motions that are neither uniform in themselves nor in relation to one another. Therefore the practitioners of this science established nine orbs upon their initial consideration—two for the above-mentioned motions and seven for the seven planets. Since the rest of the stars did not have any motion other than the two primary ones, they were satisfied with using one of these two orbs as the place for them even though their being on manifold orbs was conceivable. In addition, the attribution of one of these two primary motions to the whole rather than to a specific orb would not be precluded / if it were not for the other motion. They did not do this, however, because of its existence. /

/22–23/ if it were not…existence] β = they did not do this, however] α, M.

الفصل الثاني
في ترتيب الأجرام ونضدها

[١] الناظر في النيّرين والكواكب يجدها بأسرها متحركة بالحركة اليومية ، يطلع ما يطلع منها من المشرق ويسير إلى المغرب ويخفى فيه ، وبَعد خفائه مدة يعود إلى المشرق ثانياً ويطلع كما طلع أولاً ، وهكذا دائماً . ويتحرك ما لا يطلع منها على موازاته .

[٢] ثم يجدها بنظر أدق من الأول متحركة حركة بطيئة مخالفة للأولى كأنها من المغرب إلى المشرق . وإنما امتازت هذه الحركة من الأولى باختلاف المنطقتين والأقطاب كما سيجيء شرحه ، وذلك لأنّ الإحساس بحركتين مختلفتين في كرة واحدة على منطقة وقطبين بأعيانها ممتنع ، بل إنما يُحسّ منهما بحركة واحدة هي مركبة من مجموعهما إن كانتا إلى جهة أو حاصلة من فضل أسرعهما على أبطئهما إن كانتا إلى جهتين ، وكذلك الحكم فيما زاد على ذلك . وهاتان الحركتان متشابهتان في أنفسهما شاملتان لجميع ما يحس به علواً من الكواكب والأجرام .

[٣] ثم إنه يجد النيّرين والخمسة من الكواكب ذوي حركات مختلفة غير متشابهة لا في أنفسها ولا بقياس بعضها إلى بعض . فلذلك أثبت أهل هذا العلم تسعة أفلاك في بادئ نظرهم _ اثنين منها للحركتين المذكورتين وسبعة للسيّارات السبعة . ولمّا لم تكن لباقي الكواكب حركة غير الأوليين اكتفوا بأحد فلكيهما مكاناً لها وإن كان كونها على أفلاك شتّى جائزاً . وأيضاً إسناد إحدى الأوليين إلى المجموع لا إلى فلك خاص به لم يكن ممتنعاً / لو لا الحركة الأخرى ، لكنهم لم يذهبوا إلى ذلك لوجودها / .

/22–23/ ...] = β [لكنهم لم يذهبوا إلى ذلك] α , M .

[4] They then made the highest orb, in view of its being without stars, for the most conspicuous motion and called it the **orb of orbs** and the ***aṭlas* orb**. The next was for the less conspicuous motion, and they made it the place for the rest of the stars and called it the **orb of the ecliptic** [*lit.*, of the signs] and the **orb of the fixed stars**. Its stars they called "fixed" either because of the minuteness of their second motion or else because their [relative] positions are permanently fixed. The seven remaining orbs were for the seven wandering planets, arranged according to the occultation of one by another. The most remote of these [orbs] was reserved for Saturn, followed by the one for Jupiter and then Mars. The lowest was for the moon, and above it were the [orbs] of Mercury and then Venus. They placed the sun in the medial orb between the former and the latter even though it was not occulted except by the moon, deeming this the most elegant arrangement and the most excellent structure inasmuch as the six were connected to it—the upper [planets] in a certain way, the lower in another and the moon in yet another. In addition, the known distance of the [sun] from the Earth corresponded with this positioning. There are also reports that Venus has been observed at its farthest and nearest distances occulting the [sun], appearing like a mole on its surface. Each of these seven orbs must be further divided into other orbs so that its planet's compound motion results from them, consistent with what is observed. A discussion of this will be forthcoming. These nine [orbs] then were the least number they could allow; as for a maximum, there was no upper limit.

[5] The orbs terminate with the orb of the moon. Below it are the elements and they are likewise divided into various levels: [a] a level of pure fire; [b] a level composed of fire and hot air in which the smoke rising from the lower region dissolves. Comets, meteors [*nayāzik*], and the like are formed in this [level], and they are sometimes found to move in conformity with the motion of the orb; [c] a level that is predominantly air in which shooting stars [*shuhub*] occur; [d] a level of intense cold [*zamharīr*] that is the place of origin of clouds, thunder, lightning, and lightning bolts; [e] a level of dense air that is contiguous to the earth and water; [f] a level of water, part of which has been drawn aside, uncovering

[٤] فجعلوا أعلى الأفلاك للحركة الأظهر على أنه غير مكوكب ، وسموه فلك الأفلاك والفلك الأطلس ؛ وتاليه للحركة الأخفى وجعلوه مكاناً لسائر الكواكب ، وسموه فلك البروج وفلك الثوابت ، وسموا كواكبه الثوابت إما لقلة حركتها الثانية أو لثبات أوضاعها أبداً ؛ والسبعة الباقية للسيّارة السبعة على ترتيب خسف بعضها بعضاً ، أقصاها لزُحَل وما يليه للمُشْتَري ثم للمِرّيخ والأدنى للقمر والذي فوقه لعُطارِد ثم للزُهَرة ، وجعلوا الشمس في الفلك الأوسط بين هذه وتلك ، وإن / لم / تنكسف إلا بالقمر ، استحساناً لما في ذلك من حسن الترتيب وجودة النظام إذ الستة مربوطة عليها ـ العلوية بوجه والسفليان بوجه آخر والقمر بوجه غيرهما . وكان أيضاً بُعدها المعلوم من الأرض مناسباً لهذا الوضع ، وقد قيل إنّ الزهرة رُئيت في بعدينا الأبعد والأقرب كاسفة إياها كخالة في صفحتها . ويجب أن ينقسم كل واحد من الأفلاك السبعة إلى أفلاك تتألف حركة كوكبه المركبة منها مطابقة لما يوجد ، وسيأتي ما قيل فيه . فهذه التسعة هي التي لم يجوّزوا أن يكون أقل منها ، وأما في جانب الكثرة فلا قطع .

[٥] وبفلك القمر تتناهى الفلكيات ويكون ما دونه العنصريات وهي أيضاً طبقات : طبقة للنار الصرف ؛ ثم طبقة لما يمتزج من النار والهواء الحار التي تتلاشى فيها الأدخنة المرتفعة من السفل ، تتكوّن فيها الكواكب ذوات الأذناب والنَيازِك وما يشبهها ، وربما توجد متحركة بحركة الفلك تشييعاً له ؛ ثم طبقة الهواء الغالب التي تحدث فيها الشُهُب ؛ ثم طبقة الزَمْهَرِير التي هي منشأ السحب والرعد والبرق والصواعق ؛ ثم طبقة الهواء الكثيف المجاور للأرض والماء ؛ ثم طبقة الماء وبعض هذه الطبقة منكشفة عن

/8/ ... [β = لم تكن] M , α .

the earth; [g] a level of earth intermixed with other [elements] in which is generated mountains, minerals, and much plant and animal life; and, finally, [h] a level of pure earth surrounding the center.

CHAPTER THREE

On the Well-Known Great Circles

[1] When mathematicians wish to assign values to circles and their diameters, it is customary for them to divide [the circle] into 360 **parts** and the diameter into 120 **parts**. These parts are then subdivided into **minutes, seconds**, and so forth. Thus one-fourth of a revolution will be 90. For every arc less than [90], its complement is what remains from one-fourth [of a revolution] after subtracting that [arc] from it.

[2] Let us now proceed with the main purpose [of this chapter]. We say: the most conspicuous of the great circles is the equator of the first motion, *i.e.* the daily universal motion. It is called the **equinoctial orb** and the **equinoctial circle**, their use of the name "orb" [*falak*] to designate its equator being permissible. It is named "equinoctial" because of the equality of night and day at all locations when the sun occurs on it. Its poles are called the **poles of the first motion**, one of them being northern and the other southern. This [circle's] parts are in units of time because time is initially measured by its motion. Every given point on the orb produces by virtue of its daily motion a circle that is parallel to the equinoctial, and all [the circles thus produced] are called the **day-circles**.

[3] The equator of the slower second motion is called the **equator of the ecliptic** and the **orb of the ecliptic**, its poles being those of the ecliptic. It intersects the equinoctial at non-right angles in all the orbs that have these two motions, and there occur between the two equators two facing intersection points called the two **equinox points**. The sun

الأرض ؛ ثم طبقة الأرض المخالطة / بغيرها التي فيها تتولد / الجبال والمعادن وكثير من النباتيات والحيوانات ؛ ثم طبقة الأرض الصرفة المحيطة بالمركز .

الفصل الثالث
في الدوائر العظمى المشهورة

[١] من عادة الحسّاب إذا أرادوا تقدير الدوائر وأقطارها تجزئتُها بثلاثمائة وستين جزءاً وتجزئة القطر بمائة وعشرين جزءاً ، ثم تجزئة الأجزاء إلى دقائقها وثوانيها وما يتلوها . فيكون الربع من الدور تسعين ، وكل قوس أقل منه فتمامها ما يبقى من الربع بعد نقصانها عنه .

[٢] ولنشرع في المقصود فنقول : أظهر الدوائر العظمى منطقة الحركة الأولى أعني حركة الكل اليومية وتسمى فلك مُعدِّل النهار ودائرة مُعدِّل النهار ، وقد يطلقون اسم الفلك على منطقته تجوّزاً . وسُميت معدل النهار لتعادل الليل والنهار في جميع البقاع عند كون الشمس عليها . ويسمى قطباها قطبي الحركة الأولى : أحدهما شمالي والآخر جنوبي ، وأجزاؤها أزماناً لأنّ الزمان يتقدر أولاً بحركتها . وكل نقطة تُفرض على الفلك فهي تفعل بحركتها اليومية دائرة موازية لمعدل النهار ويسمى جميعها المدارات اليومية .

[٣] ومنطقة الحركة الثانية البطيئة تسمى منطقة البروج وفلك البروج وقطباها قطبي البروج . وهي تُقاطِع معدّل النهار في جميع الأفلاك التي تتحرك بالحركتين على زوايا غير قائمة . ويحدث بين المنطقتين تقاطُعان متقابلان يسميان نقطتي الاعتدال . والشمس

/1/ ...] β = بغيره التي تتولد فيها] M = لغيره التي تتولد فيها] α .

adheres to this second equator, and the intersection point which is such that the [sun], when crossing it, comes to be north of the equinoctial is the **vernal** [equinox]; the other is **autumnal**. The maximum distance between the two equators is the distance between their two poles that are in the same direction and is called the **total obliquity** [*lit.*, the complete declination].

[4] Now let there be imagined a **great circle passing through the four poles**, and it will be called by this name.* It is perpendicular to each one of the two equators, and its two poles are the equinox points. It passes through two points of the ecliptic orb at which occurs the maximum inclination from the equinoctial and which mark quarter points of the [ecliptic] equator. They are called the **solstice points**: the northern being the summer one, the southern being the winter. The arc that occurs on the **solstitial colure** between the two equators or between their two poles is the total obliquity whose value is found by observation. Its complement is that [arc] occurring on this [circle] between the pole of one of them and the other equator.

[5] The ecliptic equator is divided into 12 equal parts, each of which is called a **zodiacal sign**. Their 12 names are well-known; they were derived from forms fancied from among those fixed stars that happened to be in alignment with the [zodiacal signs] at the time they were named. Since these [forms] are no longer aligned with their [respective signs], it is up to those who name these things to change their names. A [zodiacal sign's] units are called **degrees**, and each zodiacal sign consists of 30 degrees. Every point describes by virtue of its second motion a circle parallel to the ecliptic orb, which is its circuit; all [the circles thus produced] are called **parallels of latitude**.

[6] If a circle is imagined to pass through an arbitrary part of the ecliptic orb, or through a given star, and through the poles of the equinoctial, it is a **circle of declination**. The arc occurring on it between that part [of the ecliptic] and the equinoctial is the declination of that part; this is known as a **particular declination**. The [arc] occurring between the star and the equinoctial is that star's **distance from the equinoctial**. For each [arc], its complement is the distance of [the part or the star] from the pole. The plane of this circle intersects the plane of the equinoctial at right angles.

[7] If a circle is imagined to pass through an arbitrary part of the ecliptic orb, or through a given star, and through the poles of the ecliptic orb, it is a **circle of latitude**.

*Henceforth *al-dā'ira al-mārra bi-'l-aqṭāb al-arbaᶜa* (the circle passing through the four poles) will be translated "the solstitial colure."

تلازم هذه المنطقة فالتقاطع الذي إذا جازته صارت شمالية عن معدل النهار ربيعي والآخر خريفي . وغاية البعد بين المنطقتين هي البعد بين قطبيهما اللذين في جهة ويسمى الميل الكلي .

[٤] فتتوهم دائرة عظيمة تمرّ بالأقطاب الأربعة وتسمى بهذا الاسم . وهي تقوم على كل واحدة من المنطقتين على زوايا قائمة ، ويكون قطباها نقطتي الاعتدالين . وتمرّ بنقطتين من فلك البروج عندهما غاية الميل من معدل النهار ، تُربَّع المنطقة بهما وتسميان نقطتي الانقلابين ــ الشمالية صيفية والجنوبية شتوية . والقوس الواقعة من الدائرة المارّة بالأقطاب الأربعة بين المنطقتين أو بين القطبين هي الميل الكلي ومقدارها يُعرف بالرصد ، وتمامها ما يقع منها بين قطب إحداهما والمنطقة الأخرى .

[٥] وتقسم منطقة البروج باثني عشر قسماً متساوية يسمى كل قسم بُرْجاً . وأسماؤها الاثنا عشر مشهورة وهي مأخوذة من صور تُوهِّمت من كواكب وقعت وقت التسمية بحذائها من الثوابت . وإذا انتقلت عن محاذاتها فللمُسمِّين أن يسموها بغيرها . وأجزاؤها تسمى درجاً فكل برج ثلاثون درجة . وكل نقطة تفعل بحركتها الثانية دائرة موازية لفلك البروج هي مدارها ويسمى الجميع بالمدارات العرضية .

[٦] وإذا توهمت دائرة تمرّ بجزء من فلك البروج ــ أي جزء كان ــ أو بكوكب ما وبقطبي معدّل النهار فهي دائرة الميل . والقوس الواقعة منها بين ذلك الجزء وبين معدل النهار هي ميل ذلك الجزء وهي من الميول الجزئية ، والواقعة بين الكوكب وبين معدل النهار هي بعد ذلك الكوكب من معدل النهار . وتماماهما بعداهما من القطب . وسطح هذه الدائرة يقطع سطح معدل النهار على زوايا قائمة .

[٧] وإذا توهمت دائرة تمرّ بجزء من فلك البروج ــ أي جزء كان ــ أو بكوكب ما وبقطبي فلك البروج فهي دائرة العرض .

The arc occurring on it between that part [of the ecliptic] and the equinoctial is the **latitude** of that part. An [arc] on the circle of declination is sometimes called the **first declination**, while this latter is the **second declination**. At maximum declination, they are the same since the circles of declination and latitude [thereupon] coincide becoming none other than the solstitial colure. The arc occurring on the [latitude circle] between a star and the ecliptic orb is the **star's latitude**; its **colatitude** is the [arc] between it and the ecliptic pole.

[8] The **longitude** of a star is that arc along the ecliptic orb measured sequentially from the vernal equinox point to the star if it is on the ecliptic with no value in latitude or, if it has a value in latitude, to the point at which its latitude circle intersects the ecliptic orb. The longitude is sometimes called ***taqwīm***. Taking the vernal equinox point instead of some other starting point is by convention.

[9] If six circles of latitude pass through the beginning of each of the twelve zodiacal signs—one of these circles necessarily being the solstitial colure—they will divide the orb into twelve parts that are the zodiacal signs, each of which [extends] in latitude from pole to pole and in longitude 30 degrees. Everything falling in one of these divisions is considered to be in that zodiacal sign. The ecliptic equator passes through the midpoints of the zodiacal signs and hence is also called the **orb of the middle of the zodiacal signs**.

[10] These then are five circles that can be conceived without reference to the lower regions. Three of them—the equinoctial, the ecliptic orb, and the solstitial colure—are each individual circles. The other two—the circle of declination and the circle of latitude—are classes, each consisting of an unlimited number of individual members.

[11] By referring to the lower regions we have (among others):

[12] [a] the **horizon circle**. This is the great circle that divides the visible from the invisible on the celestial sphere. One of its poles is the **zenith** and the other is directly opposite it from below.

والقوس الواقعة منها بين ذلك الجزء وبين معدل النهار هي عرض ذلك الجزء . وقد تسمى التي تكون من دائرة الميل ميلاً أوّلَ وهذه ميلاً ثانياً . وعند غاية الميل يتحدان لأنّ دائرتي الميل والعرض تتحدان فتصيران المارّة بالأقطاب الأربعة بعينها . والقوس الواقعة منها بين الكوكب وبين فلك البروج عرض الكوكب ، والتي بينه وبين قطب البروج تمام عرضه .

[٨] وطول الكوكب هو قوس من فلك البروج على التوالي تقع بين نقطة الاعتدال الربيعية وبين الكوكب إن كان على فلك البروج عديم العرض ، أو بين النقطة التي تقطع دائرةُ عرضه فلكَ البروج عليها إن كان ذا عرض . وقد يسمى الطول تقويماً . وإنما اعتبر نقطة الاعتدال الربيعية دون غيرها لأنّها جعلت مبدأً اصطلاحاً .

[٩] وإذا مرّت ستّ من دوائر العروض بأوائل البروج الاثني عشر – وتكون إحداها لا محالة المارة بالأقطاب الأربعة – قسمت الفلك باثني عشر قسماً هي البروج ، كل قسم في العرض من القطب إلى القطب وفي الطول ثلاثون درجة . وكل ما يقع في كل قسم منها يكون في ذلك البرج . ومنطقة البروج تمرّ بأوساط البروج ولذلك تسمى أيضاً فلك أوساط البروج .

[١٠] فهذه خمس دوائر تتوهم من غير ملاحظة السفليات : ثلاث منها أشخاص بأعيانها وهي معدل الهار وفلك البروج والمارّة بالأقطاب / الأربعة / ، واثنتان نوعان لهما أشخاص بلا نهاية وهما دائرة الميل ودائرة العرض .

[١١] وأما التي تكون بملاحظة السفليات فمنها :

[١٢] دائرة الأفق وهي العظيمة الفاصلة بين الظاهر والخفي من الفلك . وأحد قطبيها سمت الرأس والآخر ما يحاذيه من تحت .

/20/ …] M , β = α– .

The circles parallel to it above the Earth are called **almucantars of altitude**, while those below are **almucantars of depression**;

[13] [b] the **meridian circle**. This divides the eastern from the western half of the celestial sphere as well as the ascending from the descending with respect to the primary motion. It passes through the poles of the horizon and those of the equinoctial and is perpendicular to both the horizon and the equinoctial. It bisects the visible and the invisible segments of the day-circles as well as those day-circles that are visible or invisible in their entirety. Since it passes through the poles of the equinoctial and of the horizon, these two will pass through the [meridian's] poles, which are the points of their intersection. These poles are the places of rising and setting of the equinoxes and are called the **east and west points**. The arc occurring on the [meridian] between the pole of the equinoctial and the horizon circle (or [alternatively] between the pole of the horizon and the equinoctial circle) is called the **local latitude**; its complement is the [arc] between the two poles or between the two equators;

[14] [c] the **east-west circle**. It passes through the poles of the horizon and through the poles of the meridian. Its poles are the points of intersection of the horizon and the meridian, which are called the **north and south points**. This circle is also called the **circle of the initial azimuth** [prime vertical]; the meaning of "azimuth" will be forthcoming.

[15] These three circles divide the celestial sphere into eight equal parts—triangles whose sides are quarter revolutions. Four of them are visible, while four are invisible;

[16] [d] the **ecliptic meridian circle** [*lit.*, midheaven circle of appearances]. It passes through the poles of the ecliptic orb and the poles of the horizon, bisecting the visible and invisible halves of the ecliptic orb. It is also called the **circle of the local ecliptic latitude** [*lit.*, the clime latitude circle of appearances]. The arc occurring on it between the pole of the ecliptic and the horizon circle (or [alternatively] between the pole of the horizon and the ecliptic equator) is the **local ecliptic latitude**;

وتسمى الدوائر الموازية لها فوق الأرض مُقَنْطَرات الارتفاع والتي تحتها مقنطرات الانحطاط ؛

[١٣] ودائرة نصف النهار وهي الفاصلة بين النصف الشرقي والنصف الغربي من الفلك بل الصاعد والهابط بقياس الحركة الأولى . وهي المارّة بقطبي الأفق وقطبي معدل النهار ، وتقوم على الأفق وعلى معدل النهار على زوايا قائمة ، وتنصّف القِطَع الظاهرة الخفية من المدارات اليومية والمدارات الظاهرة والخفية / بأسرها أيضاً / . ولكونها مارّة بأقطاب معدل النهار والأفق فهما تمرّان بقطبيها فيكون قطباها نقطتي تقاطعيهما وهما مطلع الاعتدالين ومغيبهما وتسميان نقطتي المشرق والمغرب . والقوس الواقعة منها بين قطب معدل النهار ودائرة الأفق أو بين قطب الأفق ودائرة معدل النهار تسمى عرض البلد ، والتي بين القطبين أو المنطقتين تمامه ؛

[١٤] ودائرة المشرق والمغرب وهي المارة بقطبي الأفق وبقطبي نصف النهار . ويكون قطباها نقطتي تقاطُع الأفق ونصف النهار وتسميان نقطتي الشمال والجنوب . وتسمى هذه الدائرة أيضاً دائرة أول السموت وسيجيء معنى السمت .

[١٥] وهذه الدوائر الثلاث تقسم الفلك بثمانية أقسام متساوية مثلّثات أضلاعها أرباع الدور ـ أربعة ظاهرة وأربعة خفية ؛

[١٦] ودائرة وسط سماء الرؤية وهي المارّة بقطبي فلك البروج وبقطبي الأفق . وهي تنصّف النصفين الظاهر والخفي من فلك البروج . وتسمى دائرة عرض إقليم الرؤية . والقوس الواقعة منها بين قطب فلك البروج ودائرة الأفق أو بين قطب الأفق ومنطقة البروج هي عرض إقليم الرؤية ؛

/7–8/ [...] β = أيضاً بأسرها] M , α .

[17] [e] the **altitude circle**. This passes through any given point on the celestial sphere and through the poles of the horizon circle. If the point is above the Earth, then the [arc] between it and the horizon is its **altitude**; if below, the [arc] is the point's **depression**. The [arc] between this circle and the prime vertical along the horizon circle is the point's **azimuth**. An azimuth may be northeast, southeast, northwest, or southwest. The altitude circle coincides with the meridian circle if the star is at the middle of the period of its visibility or of its invisibility, with the prime vertical if the star has zero azimuth, and with the ecliptic meridian circle if the star is at quadrature to the ascendent.

[18] Each of these five circles is a class with numerous individual members.

[19] These then are the best-known great circles.

CHAPTER FOUR

On the Circumstances Occurring Due to the Two Primary Motions, and the Situation of the Fixed Stars

[1] The total obliquity as found by ancient and modern observations is not the same; rather, what the ancients found is greater than that found by the moderns. It has been supposed that the value found by a more recent observer will be less than that of one more ancient, even though the maximum that was found did not come to 24°, while the minimum was not less than $(23 + \frac{1}{2} + (\frac{1}{2} \text{ of } \frac{1}{10}))°$. Most have accepted the value of $(23 + \frac{1}{3} + \frac{1}{4})°$.

[2] Because of this difference, some have maintained that the ecliptic equator moves in latitude and approaches the equinoctial. If this were true, then another orb would need to be established / whereby the ecliptic orb would move with that motion. / Now the [ecliptic] equator, if

/23/ whereby...that motion] β = to move the ecliptic orb with that motion] α, M.

[١٧] ودائرة الارتفاع وهي التي تمرّ بأي نقطة تُفرض على الفلك وبقطبي دائرة الأفق. فإن كانت النقطة فوق الأرض فما بينها وبين الأفق ارتفاعها وإن كانت تحتها فهو انحطاطها . وما بين هذه الدائرة ودائرة أول السموت من دائرة الأفق سمتها ، فمن السمت شرقي شمالي ومنه شرقي جنوبي وكذلك غربي شمالي وغربي جنوبي . وهي تتّحد بدائرة نصف النهار إذا كان الكوكب في منتصف زمان ظهوره أو خفائه ، وبدائرة أول السموت إذا كان عديم السمت ، وبدائرة وسط سماء الرؤية إذا كان على تربيع الطالع .

[١٨] وهذه الدوائر الخمس وحدتها نوعية وتتكثر بالأشخاص.

[١٩] فهذه هي المشهورة من العظام .

الفصل الرابع
في الأوضاع التي تحدث بسبب الحركتين الأوليين وأحوال الكواكب الثابتة

[١] الميل الكلي الموجود بالأرصاد القديمة والحديثة ليس شيئاً واحداً بل كان ما وجده القدماء أكثر مما وجده المحدثون . وقد يظن أنّ ما وجده من هو أحدث زماناً كان أقل مما وجده من هو أقدم مع أنّ أكثر ما وجدوه لم يبلغ أربعة وعشرين جزءاً ، وأقله لم ينقص من ثلاثة وعشرين جزءاً ونصف جزء ونصف عُشر جزء . والجمهور على أنه ثلاثة وعشرون جزءاً وثُلث ورُبع جزء .

[٢] فلهذا الاختلاف زعم بعضهم أنّ منطقة البروج تتحرك في العرض فتقرب من معدل النهار . وإن كان ذلك حقاً فيجب أن يُثبت فلكٌ آخر / يتحرّك فلكُ البروج بتلك الحركة /. ثم المنطقة إن

/23/ ...] = β [يُحرِّك فلكَ البروج تلك الحركة] α , M .

it moves, may complete a revolution or it may not complete it but instead move to a certain limit then return. / This limit may be after it has coincided twice with the equinoctial equator, or at the second coincidence, or between the two coincidences. If between, then the limit may be after half a revolution, or exactly at the halfway mark, or before it. If the equator does not reach the region between the two coincidences, then it will either return upon arriving at the first coincidence or else return before this [limit is reached].

[3] These then are eight possibilities. On the first five assumptions, the northern and southern halves of the surface of the ecliptic orb, along with whatever conditions pertain to them, would interchange. In the first three cases only, each of the two halves of the equator of the ecliptic orb would coincide with each of the two halves of the equinoctial equator. For the remaining three assumptions (those after the first five), only part of the surfaces would interchange. On the basis of the first seven assumptions, one half of the ecliptic equator would coincide with that half of the equinoctial equator adjacent to it; during each coincidence, day and night would be of equal length in all localities and the seasons of the year would cease to occur. On the eighth assumption, this would not be the case; however, altitudes [of stars] and the extent of days and nights would increase and decrease in a given locality. /

[4] Furthermore there was a divergence regarding the rate of the second motion. The ancients found it to describe one degree every 100 years, while the moderns observed it to be every 66 years. One reliable group among the latter found it to be

---

/2–16/ This limit...a given locality] β = This limit may be after it has coincided with the equinoctial and departed from it; it may be at its coincidence; or, it may be before its coincidence. On the first assumption, the halves of the ecliptic orb, *i.e.* the northern and the southern, could completely interchange. On the second assumption, this could partially occur. On the third assumption, this could not occur; however, day and night would become equal during the coincidence under all circumstances and the seasons of the year would cease to occur. On the fourth assumption, this would not be the case; however, altitudes [of stars] and the extent of days and nights would increase and decrease in a given locality] α, M.

تحركت فيمكن أن تتمّم الدورة ويمكن أن لا تتمّم بل تتحرك إلى غاية ما ثم تعود . / وتلك الغاية يمكن أن تكون بعد انطباقها على منطقة معدل النهار مرّتين أو حال انطباقها الثاني أو فيما بين الانطباقين ، وذلك إما بعد قطع نصف دورتها أو حال قطع النصف سواءً أو قبله ، وإن لم تصل إلى ما بين الانطباقين فإما أن تعود حال انطباقها الأول أو قبل ذلك .

[٣] وهذه ثمانية احتمالات . وعلى التقديرات الخمسة الأُوَل يتبادل نصفا سطح فلك البروج الشمالي والجنوبي مع ما يتبعهما من الأحكام . وفي الثلاثة الأُوَل منها فقط ينطبق كل واحد من نصفي منطقة فلك البروج على كل واحد من نصفي منطقة معدل النهار . وعلى التقديرات الثلاثة الباقية بعد الخمسة الأُوَل لا يتبادل غير البعض من السطح . وعلى التقديرات السبعة الأُوَل ينطبق النصف من منطقة البروج على النصف المجاور إياه من منطقة معدل النهار وعند كل انطباق يتساوى النهار والليل في جميع البقاع وتبطل فصول السنة . وعلى التقدير الثامن لا يكون ذلك إلا أنّ الارتفاعات ومقادير الأيام والليالي تزيد وتنقص في بقعة بعينها . /

[٤] وأيضاً وقع الاختلاف في مقدار الحركة الثانية . وذلك أنّ القدماء وجدوها تقطع جزءاً واحداً في كل مائة سنة والمحدثون وجدوها في كل ست وستين سنة ، وقوم من محققيهم وجدوها في

/2–16/ ...] β = وتلك الغاية يمكن أن تكون بعد انطباقها على معدل النهار ومفارقتها إياه ويمكن أن تكون حال انطباقها ويمكن أن تكون قبل انطباقها وعلى التقدير الأول يمكن تبادل نصفي فلك البروج أعني الشمالي والجنوبي بالتمام وعلى التقدير الثاني يمكن ذلك في البعض وعلى التقدير الثالث لا يمكن ذلك إلا أن النهار والليل يصيران متساويين عند الانطباق في جميع الأحوال وتبطل فصول السنة وعلى التقدير الرابع لا يمكن ذلك إلا أن الارتفاعات ومقادير الأيام والليالي تزيد وتنقص في بقعة بعينها] M , α .

every 70 years. Some of the practitioners of [the art of] talismans have claimed that the orb undergoes accession and recession, the maximum extent of each being 8 degrees, which is completed in / every / 640 years. Some of the practitioners of this discipline have entertained this idea and held that the [second] motion becomes slower due to the recession during which the vernal equinox, which is the starting point, shifts counter-sequentially from its position, and that the motion becomes faster due to the accession during which the vernal equinox shifts sequentially from its position. This would, if they are correct, also require that another mover be established other than those already mentioned.

[5] One [some?] of them came to be satisfied with one mover for both divergences. This mover would cause the ecliptic orb to move in such a way that every point on it moves about a small circle. Accession would then result from the motion in one of its halves, while recession would occur in the other half. [In addition], there would occur a decrease in the obliquity during the motion from the midpoint of one of these halves to the midpoint of the other half, while there would be an increase during the motion in the other half. This then is what has been said concerning this matter. But to establish conclusively the authenticity of a mover and its configurations depends upon ascertaining the true state of affairs; let us then put this matter aside.

[6] It should be noted that when one orb moves another, the moved retains its position with respect to the mover; its relation to it is as a part to the whole. Thus it along with its poles and the rest of its parts moves due to the motion of [the mover] just as an occupant on a ship moves by virtue of the ship's motion. At the same time, the [moved orb] has its own proper motion just as is the case with the occupant of a ship who moves back and forth on the ship—sometimes in the direction of the ship's motion, at other times opposite that motion.

[7] This having been determined, let there be imagined a similar situation with respect to the eighth orb, which is moved by the motion of the ninth orb. And it should be noted that the fixed stars—as indeed all given points on the eighth orb—do not depart from their latitude circles in any way nor do their positions relative to one another or relative to the ecliptic equator or its poles change. However, their positions relative to the equinoctial do change. Every star on the ecliptic equator crosses the equinoctial twice in

/2/ every] $\beta = -\alpha$, –M.

كل سبعين سنة . وقد زعم بعض أهل الطِّلسمات أنّ للفلك إقبالاً وإدباراً غاية كل واحد منهما ثمانية أجزاء يتم في / كل / ستمائة وأربعين سنة . فسمع ذلك بعض أهل هذا العلم فظن أنّ تلك الحركة تبطؤ بسبب الإدبار وانتقال النقطة الربيعية التي هي المبدأ من موضعها إلى خلاف التوالي ، وتسرع بسبب الإقبال وانتقالها من موضعها إلى التوالي . وذلك أيضاً إن كان كما ظنوا محوج إلى إثبات محرك آخر غير ما مرّ .

[٥] وذهب بعضهم إلى الاكتفاء بمحرّك واحد للاختلافين يحرّك فلكَ البروج فتتحرك كل نقطة منه حول دائرة صغيرة . فيكون من الحركة في أحد نصفيه[!] الإقبال ومن الحركة في النصف الآخر الإدبار . ومن الحركة من منتصف أحد النصفين إلى منتصف النصف الآخر انتقاص الميل ومن الحركة في النصف الآخر ازدياده . فهذا ما قيل فيه والقطع بإثبات محرّك وهيئاته موقوف على تحقّق الحال ، فلنُعرض عنه .

[٦] وآعلم أنّ تحريك فلك فلكاً يكون بملازمة المتحرك لمكانه من المحرك ، وكونه منه كالجزء من الكل . فيتحرك مع قطبيه وسائر أجزائه بحركته مثل حركة ساكن السفينة بحركة السفينة . ثم إنه مع ذلك يتحرك بنفسه حركته الخاصة به كساكن السفينة إذا تردّد في السفينة تارة إلى جهة حركتها وتارة إلى خلاف تلك الجهة .

[٧] وإذا تقرر ذلك فليتصوّر في الفلك الثامن المتحرك بحركة الفلك التاسع مثل ذلك . وليعلم أنّ الثوابت بل جميع النقط المفروضة على الفلك الثامن لا تفارق مداراتها العرضية البتة ولا تختلف أوضاعها بقياس بعضها إلى بعض ولا بقياسها إلى منطقة البروج وقطبيها ، لكنّ أوضاعها بالقياس إلى معدل النهار تختلف . فكل كوكب يكون على منطقة البروج فهو يقطع معدل النهار في

/2/ …] β = M− , α− .

a revolution of the second motion; on one half of its circuit it is north of it and on the other half south. Every star having a latitude less than the total obliquity will also cross the celestial equator twice; however, the north and south segments of its circuit are unequal, the larger being in the direction of the star's latitude. Every star whose latitude is equal to the total obliquity will not cross the equinoctial; rather, it will touch it once every revolution at the solstice point that is in the direction of its latitude. Every star whose latitude exceeds the total obliquity neither crosses the equinoctial nor touches it [at a point]; instead, it will approach and recede from it. If the star's latitude is equal to the complement of the total obliquity, once every revolution it will reach the equinoctial pole that is in its direction. On account of this change [relative to the equinoctial], the day-circles for each star will vary. No star remains on the same circuit but will instead shift to a larger circuit if it approaches the equinoctial or shift to a smaller one if the opposite is the case.

[8] The positions of the stars also vary with respect to the inhabitants of the climes. Those stars having a greater altitude come to have a lesser one and conversely. Some of them will transit the zenith after not having done so previously. This is so when their distance and direction from the equinoctial becomes the same as the local latitude. Some stars will become permanently visible or permanently invisible after not being so. This will occur when the complement of their distance from the equinoctial becomes equal to the local latitude in the direction of the visible pole or the invisible one after having been greater than this amount. Some stars will come to rise and set after having been permanently visible or permanently invisible. This will take place when the complement of their distance from the equinoctial becomes greater than the local latitude after having been less than or equal to it. Polaris is one of those stars that will eventually reach the northern pole. Rigil Centaurus and Canopus are two of the stars that will become permanently invisible in the Fourth Clime.

دورة من الحركة الثانية مرّتين ، ويكون في أحد نصفي مداره شمالياً عنه وفي النصف الآخر جنوبياً . وكل كوكب ذي عرض يكون عرضه أقل من الميل الكلي فهو يقطع / أيضاً معدل النهار / مرّتين ، لكن تختلف قطعتا مداره الشمالية والجنوبية ويكون أعظمهما ذات جهة العرض . وكل كوكب يساوي عرضه الميل الكلي فهو لا يقطع معدل النهار ولكن يماسه على نقطة الانقلاب التي في جهة عرضه في دورة مرة واحدة . وكل كوكب يفضل عرضه الميل الكلي فهو لا يقطع معدل النهار ولا يماسه بل يقرب منه ويبعد عنه . فإن كان عرضه مساوياً لتمام الميل الكلي فهو ينتهي في دورة إلى قطب معدل النهار الذي في جهته مرة واحدة . وبحسب هذا الاختلاف تختلف المدارات اليومية لكل كوكب ولا يبقى كوكب على مدار واحد بل ينتقل إلى مدار أكبر إن كان يقرب من معدل النهار أو إلى مدار أصغر إن كان بالضد .

[٨] وتختلف أيضاً أوضاع الكواكب بالقياس إلى سكّان الأقاليم ، فيصير ما هو أكثر ارتفاعاً أقل وبالعكس . ويحدث لبعضها مرور بسمت الرأس بعد ما لم يكن ، وذلك عند صيرورة بعده عن معدل النهار بقدر عرض البلد وفي جهته . ويصير بعضها أبدي الظهور أو أبدي الخفاء بعد أن لم يكن ، وذلك عند صيرورة تمام بعده عن معدل النهار مساوياً لعرض البلد في جهة القطب الظاهر أو الخفي بعد أن كان أكثر من ذلك . ويحدث لبعضها طلوع وغروب بعد أن كان أبدي الظهور أو الخفاء ، وذلك عند ازدياد تمام بعده عن معدل النهار على عرض البلد بعد أن كان أقل منه أو مساوياً له . والجُدَيّ مما سينتهي إلى القطب الشمالي ، ورِجْل قَنْطُورِس وسُهَيل مما يصيران أبدي الخفاء في الإقليم الرابع .

/3/ ...] M, β = معدل النهار أيضاً] α .

[9] The **fixed stars** cannot possibly be enumerated, but 1022 of them have been observed and their positions in longitude and latitude are known. They arranged their magnitudes into six grades, the first being the greatest. In order to identify them, they imagined certain constellations [*lit.*, forms] that the stars were either on or near; thus one could say "that star on the head of such and such a constellation" or "that one near the foot of such and such a constellation." The constellations are 48 in number, of which 21 are in the north, namely: Ursa Minor, Ursa Major, Draco, Cepheus, Boötes, Corona Borealis, Hercules, Lyra, Cygnus, Cassiopeia, Perseus, Auriga, Aquila, Delphinus, Sagitta, Ophiuchus, Serpens, Equuleus, Pegasus, Andromeda, and Triangulum. Twelve are on the [ecliptic] equator; they are the zodiacal signs and their names are well-known. There are 15 southern [constellations] and they are: Cetus, Orion, Eridanus, Lepus, Canis Major, Canis Minor, Argo Navis, Hydra, Crater, Corvus, Centaurus, Lupus, Ara, Corona Australis, and Piscis Austrinus. Three hundred sixty stars were observed in the northern constellations, 346 in the ecliptic constellations, and 316 in the southern constellations.

[10] The **Milky Way**, *i.e.* the galaxy, is made up of a very large number of small, tightly-clustered stars, which, on account of their concentration and smallness, seem to be cloudy patches. Because of this, it was likened to milk in color.

[11] As for the **mansions of the moon**, they are made up of the stars that are near the ecliptic equator. The Arabs used them as signs for the 28 divisions into which the [ecliptic] equator was divided so as to coincide with the number of days of the moon's revolution. Thus every night it was seen "residing" near one of these mansions whose names are well-known.

[12] Knowledge of the fixed stars and that which concerns them being a separate discipline, it is best that we confine ourselves to just what has been presented.

[٩] والكواكب الثابتة لا يمكن أن تُحصى كثرة . وقد رُصد منها ألف واثنان وعشرون كوكباً فعرف مواضعها في الطول والعرض . ورتّبوا أقدارها في ست مراتب أُولاها أعظمها ، وتوهموا لتعريفها صوراً تكون هي عليها أو بقربها فيقولون : الذي على رأس الصورة الفلانية أو بقرب رجل الصورة الفلانية . وكانت الصور ثمانية[!] وأربعين منها إحدى وعشرون في الشمال وهي : الدبُ الأصغر ، والدبّ الأكبر ، والتِنّين ، وقَيْقاوُس ، والعَوّاء ، والفَكّة ، والجاثي على رُكْبَتَيه ، وشلْياق ، والدَجاجة ، وذات الكُرْسي ، وحامِل رأس الغُول ، ومُمسِك العِنان ، والعُقاب ، والدُلْفِين ، والسَهْم ، والحوّاء ، والحيّة ، وقِطْعة الفَرَس ، والفَرَس الأعظم ، والمرأة المُسَلْسَلة ، والمُثَلَّث ؛ واثنتا عشرة على المنطقة وهي البروج وأسماؤها مشهورة ؛ وخمس عشرة في الجنوب وهي : قِيْطُس ، والجَبّار ، والنَهْر ، والأرْنَب ، والكَلْب الأكبر ، والكَلْب الأصغر ، والسَفينة ، والشُجاع ، والباطِيَة ، والغُراب ، وقَنْطُورِس ، والسَبُع ، والمِجْمَرَة ، والإكْليل الجنوبي ، والحُوت الجنوبي . وكان من المرصودة ثلاثمائة وستون على الصور الشمالية ، وثلاثمائة وستة وأربعون على صور المنطقة ، وثلاثمائة وستة عشر على الصور الجنوبية .

[١٠] والدائرة اللَبَنِيّة أعني المَجَرّة مؤلفة من كواكب صغار متقاربة متشابكة كثيرة جداً صارت من تكاثفها وصغرها كأنها لطخات سحابية ، ولذلك شُبّهت باللبن لوناً .

[١١] وأما منازل القمر فهي من الكواكب القريبة من منطقة البروج جعلته[!] العرب علامات الأقسام الثمانية والعشرين التي قسمت المنطقة بها لتكون مطابقة لعدد أيام دور القمر ، فيرى كل ليلة نازلاً بقرب أحدها . وأسماؤها مشهورة .

[١٢] ومعرفة الثوابت وأحوالها فنّ مفرد ، فالأَولى أن نقتصر هاهنا على هذا القدر .

CHAPTER FIVE

On Basing Some of the Apparently Irregular Motions Upon Models That Bring About Their Uniformity

[1] If a celestial motion is irregular from our perspective, we must require that it have a model according to which that motion is uniform; this model should also bring about its irregularity with respect to us. For irregular [motion] does not arise from the celestial bodies.

[2] One of these models is such that the motion is uniform about a point different from the center of the World, which is in our proximity. The circumference upon which the moved thing—a planet, for example—moves about that point must either enclose the center of the World or else not enclose it. In the first case it is called an **eccentric**, while in the second it is called an **epicycle**.

[3] If an eccentric is taken by itself and a planet is assumed to move upon it about its center with a simple, uniform motion, it will cause that motion, with respect to the center of the World (as well as to points other than the [eccentric] center) to be irregular. Thus in the segment that is farther away, the motion will be slower, while in the nearer segment it will be faster. This is so because for equal arcs whose distances vary, those that are farther away appear smaller than those that are nearer. If a line is produced that passes through the [eccentric] center and through the center of the World (or through some given point other than these two), it will pass through the **farthest distance**, which is the midpoint of the far segment, and through the **nearest distance**, which is the midpoint of the near segment. Then if a line is produced perpendicular to it passing through the center of the World (or through that [other given] point) and reaching the circumference on both sides, it will pass through the two **mean distances**, which are the common part between the two segments. At these two points the motion will be at its mean between the faster and slower motions.

[4] As for the epicycle, if it is taken by itself and the planet moves on its circumference, equal arcs will likewise vary with respect to the center of the World. The line joining the two centers passes through the **farthest and nearest distances** of the epicycle.

الفصل الخامس
في إسناد بعض الحركات المختلفة في الرؤية إلى أصول تقتضي تشابهها

[١] إذا اختلفت حركة فلكية عندنا وجب أن نطلب لها أصلاً تتشابه تلك الحركة بحسبه ويقتضي ذلك الأصل أيضاً اختلافها بالقياس إلينا ، فإنّ المختلفة لا تصدر عن الفلكيات .

[٢] فمن الأصول كون الحركة متشابهة حول نقطة خارجة عن مركز العالم الذي نحن بقربه . ولا يخلو من أن يكون المحيط الذي يتحرك عليه ذلك المتحرك ـ وليكن كوكباً مثلاً ـ حول تلك النقطة إما محيطاً بمركز العالم وإما غير محيط به ، والأول يسمى الخارج المركز والثاني يسمى التدوير .

[٣] والخارج المركز إذا فُرض وحده وفُرض الكوكب متحركاً عليه حول مركزه حركة بسيطة متشابهة صيّر الحركةَ بالقياس إلى مركز العالم وغيره من النقط التي هي غير ذلك المركز مختلفة . فتكون في القطعة التي هي أبعد منه بطيئة وفي القطعة التي هي أقرب سريعة . وذلك لأن القسي المتساوية المختلفة بالبعد والقرب يُرى البعيد منها أصغر من القريب . وإذا أُخرج خط يمرّ بمركزه وبمركز العالم أو بالنقطة المفروضة التي هي غيرهما مرّ بالبعد الأبعد وهو منتصف القطعة البعيدة وبالبعد الأقرب وهو منتصف القطعة القريبة . ثم إذا قام عليه عمود يمرّ بمركز العالم أو بتلك النقطة ووصل إلى المحيط في الجانبين مرّ بالبعدين الأوسطين ، وهما الفصل المشترك بين القطعتين وعندهما تكون الحركة متوسطة بين السرعة والبطء .

[٤] وأما التدوير فإذا فُرض وحده وتحرك الكوكب على محيطه كانت القسي المتساوية أيضاً مختلفة بالقياس إلى مركز العالم ، وكان الخط الواصل بين المركزين ماراً بالبعدين الأبعد والأقرب منه ،

The two lines produced from the center of the World tangent to the epicycle on both sides separate the far and near segments. [As distinct from the eccentric], however, the planet is observed to retrograde in one segment from the direction in which it had been proceeding in the other until it reaches the starting point of its motion. [Also] it does not traverse by that motion all parts of the orb that surrounds the center of the World. This is their illustration:

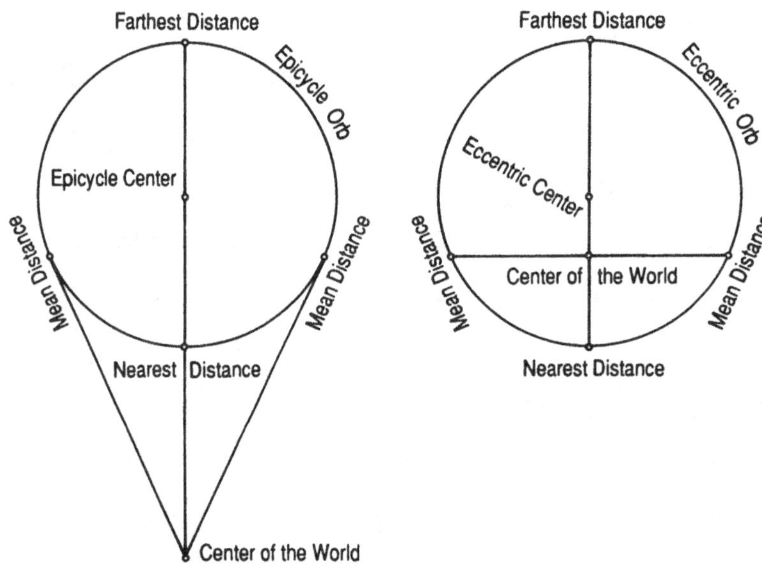

[Fig. T1]

[5] Now if an epicycle is taken to be on a concentric orb that is its deferent such that the ratio of the deferent's radius to the radius of the epicycle is the same as the ratio of the eccentric radius to the eccentricity, and if the motion of the deferent is made the same as the motion of the eccentric and in the same direction so that they complete their revolutions simultaneously (with the epicycle center then also moving with this same motion), and if the epicycle is also made to move with a motion the same as the previous two in such a way that it is in the direction opposite the motion of the deferent in the far segment and in the same direction in the near segment, then the motion of the planet in the far segment is observed to be in the amount of the excess of the motion of the deferent over

والخطّان الخارجان من مركز العالم الماسّان للتدوير من جانبيه يفصلان بين القطعتين البعيدة والقريبة ؛ إلا أنّ الكوكب يُرى في إحدى القطعتين راجعاً عن السمت الذي يقصده في القطعة الأخرى إلى أن يصل إلى المبدأ الذي تحرك منه . ولا يقطع أجزاء الفلك المحيط بمركز العالم جميعاً بتلك الحركة . وهذه صورتهما :

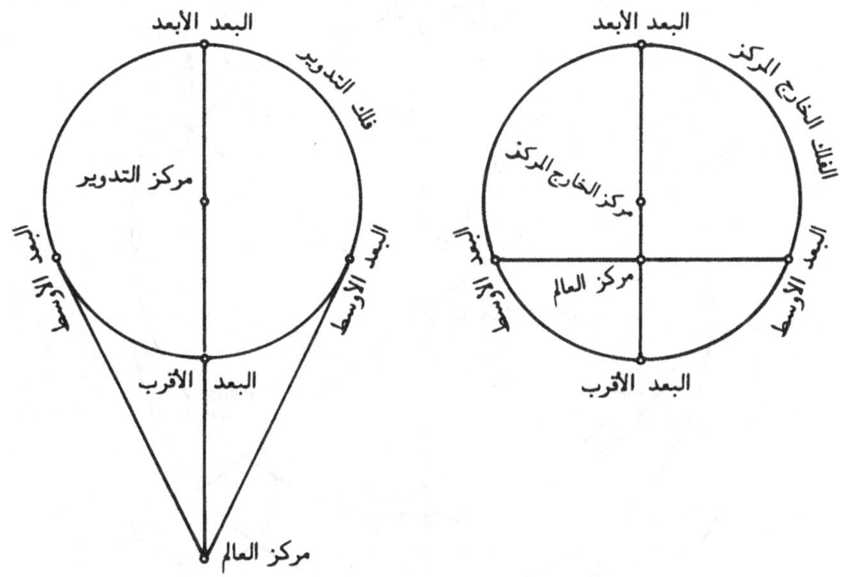

[شكل ١]

[٥] أما إن فُرض التدوير على فلك آخر حامل له موافق المركز على أنّ نسبة نصف قطر الحامل إلى نصف قطر التدوير كنسبة نصف قطر الخارج المركز إلى ما بين المركزين ، وجُعلت حركة الحامل شبيهة بحركة الخارج المركز وفي جهته بحيث يتمّان الدورتين معاً فيتحرك مركز التدوير بتلك الحركة ، وجُعل التدوير متحركاً أيضاً بحركة شبيهة بهما على وجه تكون في القطعة البعيدة إلى خلاف جهة حركة الحامل وفي القطعة القريبة إلى جهتها ، رُئيت حركة الكوكب في القطعة البعيدة بقدر فضل حركة الحامل على

that of the epicycle and in the near segment to be their sum. Thus the apparent motion is exactly the same as that observed in the previously mentioned eccentric model, there being no difference whatsoever. The planet will describe by this compound motion an eccentric circuit that is the same as with the eccentric orb. This is its illustration:

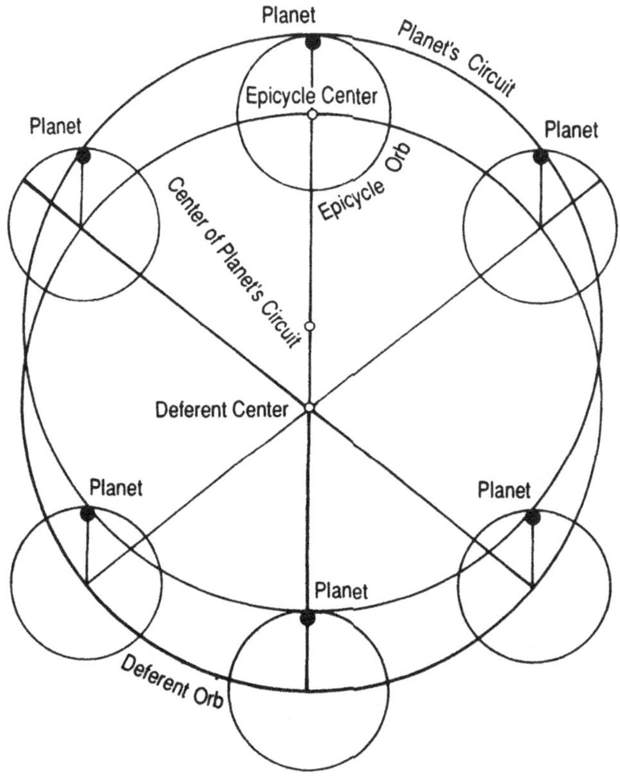

[Fig. T2]

[6] The difference between the two models in this case is twofold: first, the eccentric model is realized by means of a single motion, whereas the epicyclic model is realized with two motions; second, the epicycle entails a circuit that is eccentric, while the eccentric does not entail an epicycle. Therefore Ptolemy considered the eccentric in this case to be simpler than the epicycle.

[7] If the epicycle is assumed to move in such a way that in the far segment it is in the direction of the motion of the deferent, the faster motion will be attained in that segment and the slower in the near segment, which is the opposite of what occurred in the first case. However, the period

حركة التدوير وفي القطعة القريبة بقدر مجموعهما ، فصارت الحركة المرئية مثل ما يُرى في أصل الخارج المركز المذكور بعينه من غير تفاوت أصلاً . ويفعل الكوكب بحركته المركبة مداراً خارجَ المركز شبيهاً بالفلك الخارج المركز . وهذه صورته :

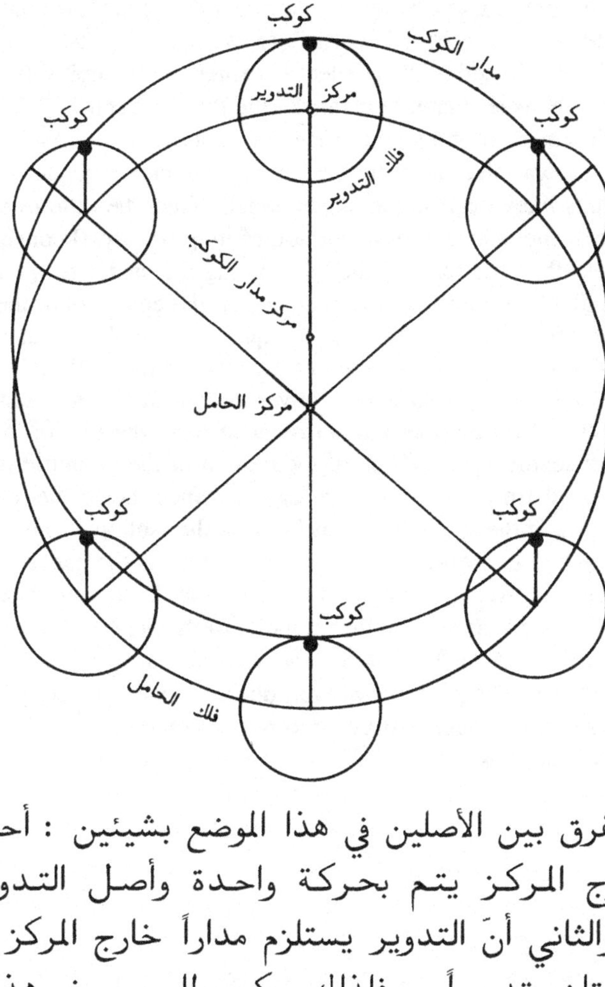

[شكل ٢]

[٦] والفرق بين الأصلين في هذا الموضع بشيئين : أحدهما أنَ أصل الخارج المركز يتم بحركة واحدة وأصل التدوير يتم بحركتين، والثاني أنّ التدوير يستلزم مداراً خارج المركز والخارج المركز لا يستلزم تدويراً . فلذلك حكم بطلميوس في هذا الموضع بأنّ الخارج المركز أبسط من التدوير .

[٧] وإن فُرض التدوير متحركاً على وجه يكون في القطعة البعيدة إلى جهة حركة الحامل حصلت السرعة في تلك القطعة والبطء في القطعة القريبة بخلاف ما كان في الأول ؛ إلا أنّ زمان

of the faster motion will be longer than the period of the slower motion in this figure, whereas previously it was shorter. This is because the far segment is larger than the near one since the dividing [line] between them cannot pass through the center; hence, it will not bisect the epicycle but rather divide it into two unequal segments, the smaller of which is the one nearest the center of the deferent.

[8] Related to this discussion is the following: let us assume the eccentric also to have a concentric mover; / let us make the ratio of the line connecting the concentric center and the eccentric perigee to the eccentric radius the same as the ratio of the line connecting the concentric center and the epicyclic perigee to the radius of the epicycle; / and, let us make the two concentrics move, say, sequentially with equal motion, the eccentric counter-sequentially, and the epicycle in such a way that at its farthest distance it moves sequentially (these latter two motions likewise being assumed equal). Then the ratio of the motion of the eccentric or the epicycle to the motion of its [respective] concentric must either be: [a] less than the ratio of the line joining the center of the concentric and the nearest distance on either [the eccentric or the epicycle] to the radius of the eccentric or epicycle, respectively; [b] equal to it; or [c] greater than that [ratio]. If it is less, the planet on account of the two motions will only undergo a faster motion in the far segment and a slower motion in the near segment. For the eccentric this is because the amount reduced from the motion of the concentric in the far segment on account of the motion of the eccentric is less than the amount reduced from it in the near segment since the former arcs appear smaller. In the case of the epicycle, this is because the motion in the far segment is the sum of the two motions, while in the near one it is the excess of the motion of the concentric over the motion of the epicycle. If the ratio is equal to it, the planet will become stationary at the midpoint of the period of the slower motion, this occurring when it is at the nearest distance, which is located on the above-mentioned line. The planet, however, does not undergo retrogradation. If the ratio is greater, the planet will undergo retrograde motion in the near segment between two stations.

/6–8/ β = let us make the ratio of the eccentric radius to the [distance] between the two centers the same as the ratio of the [epicycle's] deferent radius to the radius of the epicycle] α, M.

السرعة يكون في هذه الصورة أطول من زمان البطء وهناك كان أقصر . وذلك لأنّ القطعة البعيدة تكون أكبر من القريبة ، فإنّ الفاصل بينهما لا يمكن أن يمرّ بالمركز فهو لا ينصّف التدوير بل يقطعه بمختلفين [!] أصغرهما التي تلي مركز الحامل .

[٨] ومما يتّصل بهذا البحث أنّا إذا فرضنا لخارج المركز أيضاً محركاً موافق المركز ، / وجعلنا نسبة الخط الواصل بين مركز الموافق وحضيض الخارج إلى نصف قطر الخارج كنسبة الخط الواصل بين مركز الموافق وحضيض التدوير إلى نصف قطر التدوير / ، وجعلنا الموافقي المركز متحركين إلى التوالي مثلاً حركتين متشابهتين والخارجَ المركز إلى خلافه والتدويرَ على وجه يكون في بعده الأبعد إلى التوالي وحركتاهما أيضاً متشابهتان ، فنسبة حركة الخارج المركز أو التدوير إلى حركة موافقيهما لا يخلو من أن تكون إما أصغر من نسبة الخط الواصل بين مركز الموافق وبين البعد الأقرب من كل واحد منهما إلى نصف قطر الخارج المركز أو التدوير كل إلى صاحبه ، وإما مساوية لها ، وإما أكبر منها . فإن كانت أصغر فلا يحدث للكوكب بسبب الحركتين إلا السرعة في القطعة البعيدة والبطء في القطعة القريبة : أما في الخارج المركز فلأنّ ما ينقص في القطعة البعيدة بسبب حركة الخارج المركز من حركة الموافق المركز يكون أقل مما ينقص في القطعة القريبة لكون تلك القسي أصغر في الرؤية ، وأما في التدوير فلأنّ الحركة في القطعة البعيدة مجموع الحركتين وفي القريبة فضل حركة الموافق على حركة التدوير . وإن كانت مساوية حدث للكوكب في منتصف زمان البطء وقوف وهو عند كونه في البعد الأقرب على الخط المذكور ولا يكون له رجوع . وإن كانت أكبر حدث للكوكب رجوع في القطعة القريبة بين وقوفين .

/6–8/ ...] β = وجعلنا نسبة نصف قطر الخارج المركز إلى ما بين المركزين كنسبة نصف قطر الحامل إلى نصف قطر التدوير] α , M .

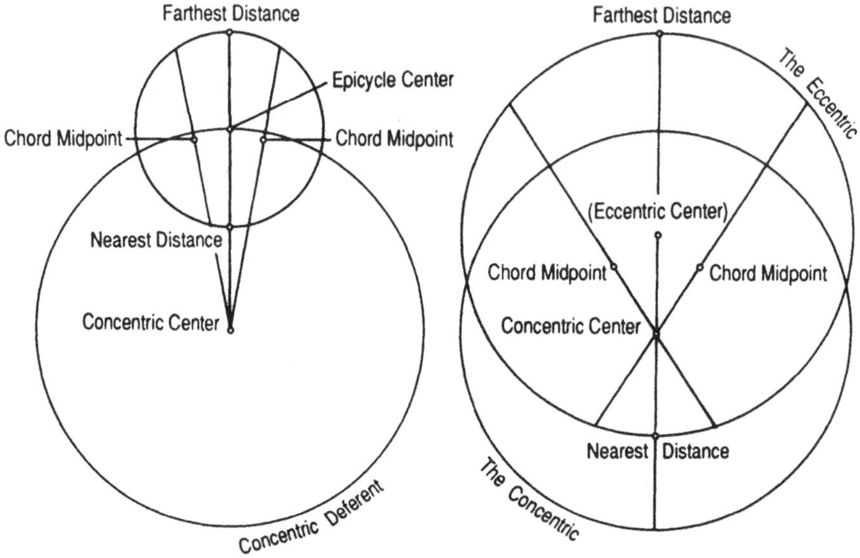

[Fig. T3]

[9] Let two lines be produced from the two concentric centers on both sides of the above-mentioned line, *i.e.* that joining the concentric center and the nearest distance for each of the two orbs, to the circumferences of the eccentric and the epicycle on both sides in such a way that the ratio of the motion of the eccentric or the epicycle to the motion of its respective concentric is equal to the ratio of that part of each of those two lines lying between the concentric center and the circumference of the eccentric or epicycle that is on the nearer side to half the chord, likewise from that line, that divides each of the orbs, respectively, into two segments. This is possible for this type of eccentric or epicycle [in which the ratio is greater] but not for the first two cases. The planet will thereby, after a gradual slowing down, become stationary when it reaches the first of the two lines in the near segment. From there until it reaches the second line, it will retrograde, gradually going from a slower to a faster speed whose maximum occurs at the nearest distance and then slowing until it reaches the second line whereupon it once more becomes stationary. After that, its motion becomes

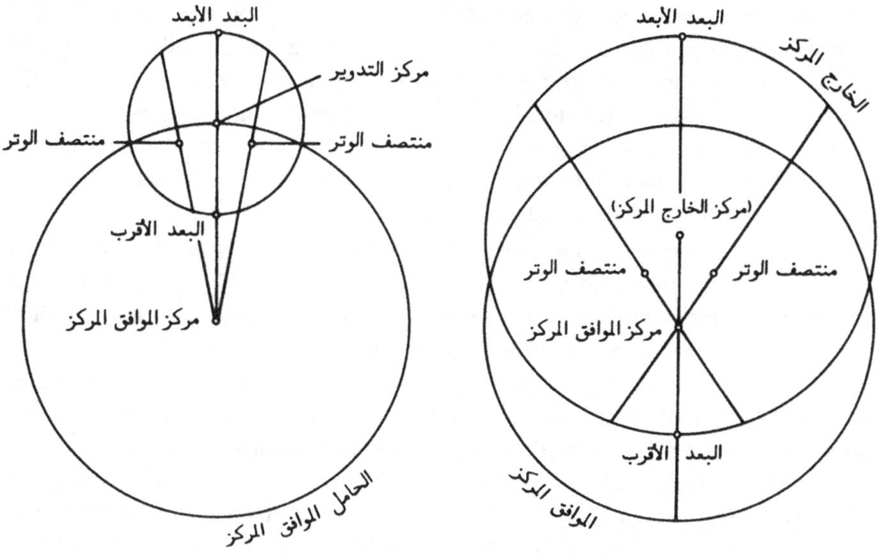

[**شكل ٢**]

[٩] وليخرج خطّان عن مركزي الموافق عن جنبتي الخط المذكور – أعني الواصل بين مركز الموافق وبين البعد الأقرب في كل واحد من الفلكين – إلى محيطي الخارج المركز والتدوير في الجانبين بحيث تكون نسبة حركة الخارج المركز أو التدوير إلى حركة الموافقين ، كل إلى صاحبه ، مساوية لنسبة ما وقع من كل واحد من ذينك الخطّين بين مركز الموافق ومحيط الخارج المركز أو التدوير من الجانب الأقرب إلى نصف الوتر الفاصل لكل واحد من الفلكين إلى قطعتين أيضاً من ذلك الخط ، كل إلى صاحبه . وذلك يكون في مثل هذا الخارج المركز والتدوير ممكن[!] دون الأولين . فيكون الكوكب عند وصوله إلى أول الخطّين في القطعة القريبة واقفاً بعد بطء متدرج إلى الوقوف ، ومنه إلى وصوله إلى الخط الثاني راجعاً رجوعاً متدرجاً من بطء إلى سرعة غايته في البعد الأقرب ، ثم منها إلى بطء ينتهي عند الخط الثاني ، وعند وصوله إلى الخط الثاني واقفاً وقوفاً ثانياً ، وبعد ذلك يستقيم

direct as it gradually goes from being stationary to its fastest speed. The two mean speeds between the slower and faster speeds occur at the two mean distances. This speed / in the case of the epicycle / is that of the motion of the concentric alone. If the concentric and eccentric motions are set opposite the directions we have assumed previously, while the deferent's motion is as it was before and the motion of the epicycle is such that at the farthest distance it is counter-sequential, and all the rest of the conditions remain unchanged, then the situations of the near and far segments become reversed.

[10] These then are models and rules that should be known. We have only stated them here; their geometric proofs are given in the *Almagest*. Restricting oneself to circles is sufficient in the entirety of this science for whoever studies the proofs. However, one who attempts to understand the principles of the motions must know the configuration of the bodies, which move with these motions in such a way that these motions manifest themselves on their equators. He should conceive of the **concentric** and the **deferent** each as an orb bounded by two parallel surfaces with the same center and the eccentric as an orb within the thickness of the concentric and bounded by two parallel surfaces with the same center, which is eccentric to the concentric center by an amount dictated by the anomaly. The convex of the two surfaces is tangent to the convex surface of the concentric at a single point that is the farthest point on it from the center of the concentric. Its concave surface is tangent to the concave surface of the concentric at a single point that is opposite the first and is the nearest point on it to the [concentric center]. Its thickness is such that it is large enough to include in it an epicycle or planet so that the latter's convex surface is tangent to the [eccentric's] two surfaces at two points. Its **[inner] equator** is the circuit of the epicycle's center or the planet's center. The **[inner] equator of the concentric** is a circle whose center is the concentric center and is equal to the [inner] equator of the eccentric, intersecting it in two points. One group makes it a circle that is tangent to the eccentric's [inner] equator at a point facing the farthest distance. The epicycle [should be conceived of] as a sphere in the thickness of its deferent with its convex surface tangent to the [deferent's] two surfaces at two points that are the farthest and nearest points on it

/3/ in the case of the epicycle] $\beta = -\alpha, -M$.

متدرجاً من وقوف إلى سرعة سير . ويكون السيران المتوسطان بين البطء والسرعة عند البعدين الأوسطين ، وذلك السير هو حركة الموافق وحدها / في التدوير / . وإن جُعل حركتا الموافق المركز والخارج المركز مخالفتين في الجهة لما فرضنا ، وحركة الحامل كما / كان / ، لكنّ حركة التدوير على وجه تكون في البعد الأبعد إلى خلاف التوالي ، وسائر الشروط بحالها ، تبادلت حالتا القطعتين القريبتين والبعيدتين .

[١٠] فهذه أصول وقوانين لا بدّ من معرفتها أوردناها ههنا على سبيل الحكاية وبراهينها مذكورة بالخطوط في المجسطي . والاقتصار على الدوائر كافٍ للناظر في البراهين في جميع هذا العلم . أما لمن يحاول تصور مبادئ الحركات فلا بدّ من معرفة هيئة الأجسام المتحركة بتلك الحركات على وجه تظهر تلك الحركات في مناطقها . وعليه أن يتصور كلاً من الموافق المركز والحامل فلكاً يحيط به سطحان متوازيان مركزاهما واحد ؛ والخارجَ المركز فلكاً في ثخن الموافق المركز يحيط به سطحان متوازيان مركزاهما واحد خارج عن مركز الموافق بقدر ما يوجبه الاختلاف ، والمحدّب من سطحيه مماسّ لمحدّب الموافق على نقطة واحدة هي أبعد نقطة عليه من مركز الموافق ومقعّره مماسّ لمقعّر الموافق على نقطة واحدة مقابلة للأولى هي أقرب نقطة عليه منه ، وثخنه بحيث يسع ما يجب أن يكون فيه من تدوير أو كوكب بحيث يماسّ محدّبه سطحيه على نقطتين ، ومنطقته مدار مركز التدوير أو مركز الكوكب ، ومنطقة الموافق دائرة مركزها مركز الموافق مساوية لمنطقة الخارج مقاطعة إياها في نقطتين ، وقوم يجعلونها دائرة تماسّ منطقة الخارج على نقطة محاذية للبعد الأبعد ؛ وفلكَ التدوير كرة في ثخن حامله محدّبها مماسّ لسطحيه على نقطتين هما أبعد نقطة عليه وأقربها

/3/ ...] β = M– , α– . /5/ ...] β = كانت] M , α .

from the center of its deferent. The planet is embedded in it in such a way that its outer surface is tangent to the convex surface of the epicycle at a point. For neither [planet nor epicycle] is their concave surface taken into account. The **[inner] equator of [the epicycle]** is a circle that is the circuit of the planet's center. The **[inner] equator of the deferent** is a circle that is the circuit of the epicycle center. After removing the eccentric from the concentric, there are left two solid, curved bodies that are thick in the middle, becoming more narrow until the thickness disappears altogether at a point opposite the maximum thickness. The two bodies bound the eccentric in accordance with the alternating position of their two thicknesses. They are called the **complementary bodies**.

[11] The farthest distance for the eccentric is called the **apogee**; for the epicycle, it is called the **apex**. For both, the nearest [distance] is called the **perigee**. The eccentric is sometimes referred to as the **orb of the apogee**. A moving object on either orb is [called] **descending** from the farthest distance to the nearest and **ascending** from the [nearest] to the farthest. This is their illustration:

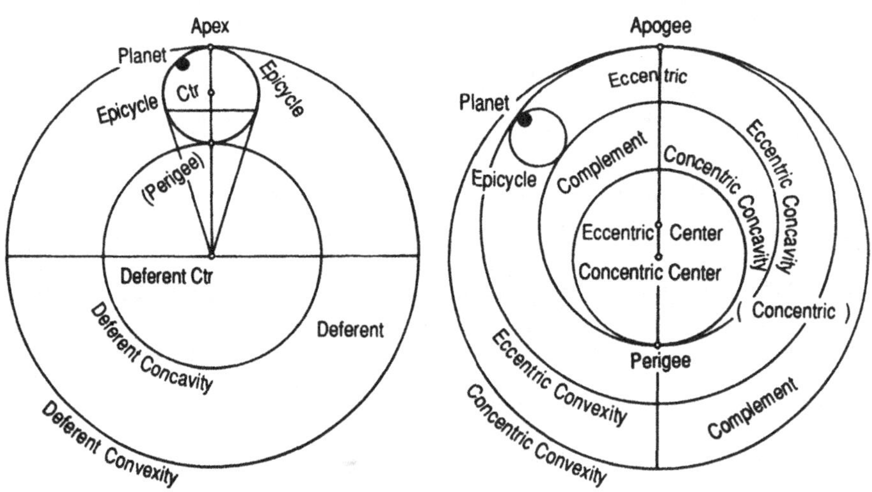

[Fig. T4]

من مركز حامله ، والكوكب مركوز فيه بحيث يماس سطحه الخارج محدّب التدوير على نقطة ولا يعتبر مقعرهما ، ومنطقته دائرة هي مدار مركز الكوكب ، ومنطقة الحامل دائرة هي مدار مركز التدوير . ويفضل من الموافق المركز بعد انفصال الخارج المركز منه جسمان مستديران ثخينان غليظا الوسط يستدق ذلك الغلظ إلى أن ينعدم عند نقطة مقابلة لغاية الغلظ يحيطان بالخارج المركز على تبادل وضع غلظيهما ، ويسميان المُتمِّمين .

[١١] والبعد الأبعد في الخارج المركز يسمى الأوج وفي التدوير يسمى الذروة ، والأقرب فيهما يسمى الحضيض . وقد يسمى الخارج المركز بفلك الأوج . والمتحرك في الفلكين من البعد الأبعد إلى الأقرب هابط ، ومنه إلى الأبعد صاعد . وهذه صورتهما :

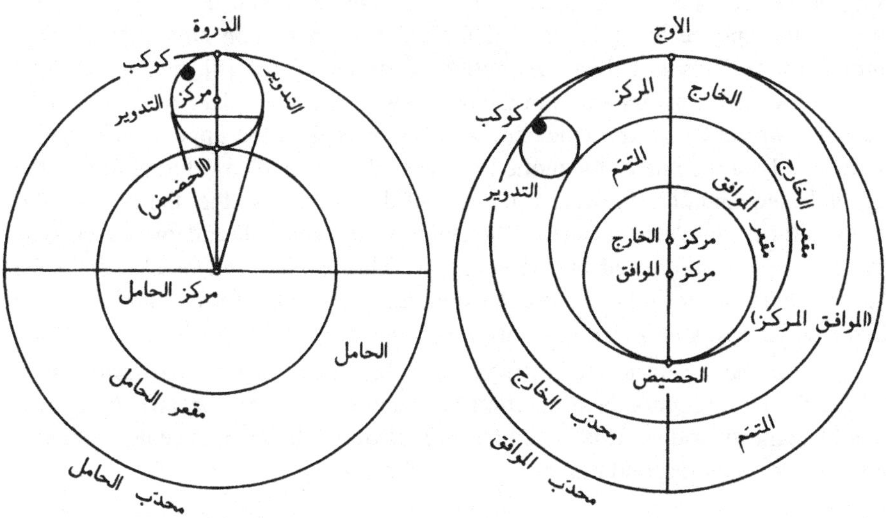

[شكل ٤]

CHAPTER SIX

On the Orbs and Motions of the Sun

[1] When the situation of the sun was considered, its motion was found to vary in different parts of the ecliptic equator: it was slower in a certain half, faster in the other. The center of its body was found to adhere always to the ecliptic equator, deviating neither north nor south of it; for this reason, the ecliptic is sometimes known as the solar circuit. By careful examination of solar eclipses, the solar body was found during the middle of the period of slower motion to be somewhat smaller than during the middle of the period of faster motion. It was inferred from this that during its slower motion the sun was farther away from the center of the World, while during the faster motion it was nearer. The moderns have found the midpoints of its slower and its faster motions—and, indeed, every position whatever the circumstance—to have a movement through the parts of the ecliptic equator in the sequence [of the signs] that is approximately the movement of the fixed stars due to the second motion. This was something Ptolemy did not find.

[2] The above required that there be established for the sun either: [a] an eccentric whose equator would be in the plane of the ecliptic equator. The sun would be within its thickness, and the [eccentric] would move, thus moving the sun, in the sequence of the signs at the rate of the sun's mean motion (the motion of the apogee being subtracted from the mean motion among those who propound it). This is called the motion of its center; or [b] an epicycle and deferent whose equators are as in the previous case [in the plane of the ecliptic]. The sun would be on the epicycle, which moves it in the upper half countersequentially at the rate of the motion of the sun's center. The deferent moves the epicycle sequentially, likewise at the rate of that motion so that the two revolutions will be completed together. The center of the sun will undergo exactly the same motion as produced by the eccentric. This motion will be slower in the apogeal half, faster in the perigeal half. Ptolemy chose the former model—there being no necessity to do so—because it is simpler.

[3] For the eccentric model, there must be established a concentric orb in whose thickness occurs the eccentric and which exceeds the eccentric by its two complementary bodies. It is called the **pareccliptic orb** since its center, equator, and two poles correspond to those of the [ecliptic orb].

الفصل السادس
في أفلاك الشمس وحركاتها

[١] لمّا تؤمّل في أحوال الشمس وُجدت حركتها مختلفة في أجزاء منطقة البروج بأن كانت بطيئة في نصف بعينه ، سريعة في النصف الآخر . ووُجد مركز جرمها دائماً ملازماً لمنطقة البروج غير مائل عنها إلى الشمال ولا إلى الجنوب ، ولذلك ربّما تعرّف بمدار الشمس . ووُجد بالنظر الدقيق في الكسوفات جرمها في أواسط زمان البطء أصغر قليلاً منه في أواسط زمان السرعة . فاستدلوا من ذلك على كونها في البطء أبعد من مركز العالم وفي السرعة أقرب . والمتأخرون وجدوا لمنتصفي بطئها وسرعتها ــ بل لكل موضع حال من أحوالها ــ انتقالاً في أجزاء منطقة البروج على التوالي قريباً من انتقالات الثوابت بالحركة الثانية . وبطلميوس لم يجد ذلك .

[٢] فاقتضى ذلك أن يثبت لها إما خارج مركز ، منطقته في سطح منطقة البروج ، تكون الشمس في ثخنه وهي[!] تتحرك وتحرك الشمس على توالي البروج بقدر حركة وسط الشمس إذا نقص منها حركة أوجها عند من يقول بها ، وتسمى حركة مركزها ؛ وإما تدوير وحامل ، منطقتاهما كذلك ، تكون الشمس على التدوير وهو يحرّكها في النصف الأعلى إلى خلاف التوالي بقدر حركة مركز الشمس ، والحامل يحرك التدوير إلى التوالي أيضاً بقدر تلك الحركة لتتم الدورتان معاً . وتحدث لمركز الشمس حركة كما أحدثها الخارج المركز بعينها ، وتكون تلك الحركة في النصف الأوجي بطيئة وفي النصف الحضيضي سريعة . وبطلميوس اختار الأول من غير ضرورة لكونه أبسط .

[٣] ويلزم على أصل الخارج المركز إثبات فلك موافق المركز يكون الخارج المركز في ثخنه ويفضل عليه بمتمِّميه . ويسمى الفلك المُمَثَّل بفلك البروج لكونه بالمركز والمنطقة والقطبين موافقاً له ، وهو

According to the moderns, it has the motion of the fixed stars and moves the apogee and perigee. For the epicyclic model, the eighth orb suffices for the moving of the apogee and the perigee since it moves everything below it; thus the deferent is the parecliptic. Since the sun is always in the plane of the eccentric or epicyclic equator, which are themselves in the plane of the parecliptic, [the sun] does not have any latitude. We have set forth the illustration of the sun's two orbs according to the eccentric model as was Ptolemy's preference.

[4] The sun has a single **anomaly** that is in the amount by which the sun's observed motion differs from its mean motion, it being an angle that is formed at the center of the sun from two lines produced from the centers of its two orbs to it. This angle attains its greatest possible value at the two mean distances and disappears at the two other distances; it depends upon the amount of the eccentricity. According to Ptolemy, it is 2;30; according to more recent observational [astronomers], it is about 2;05 (the radius of the eccentric being 60). The position of the apogee, according to Ptolemy, is 24½° ahead of the summer solstice point. According to the moderns, the value varies, as they have stated in their *zījes*, depending on the date. This is the illustration of the orbs of the sun:

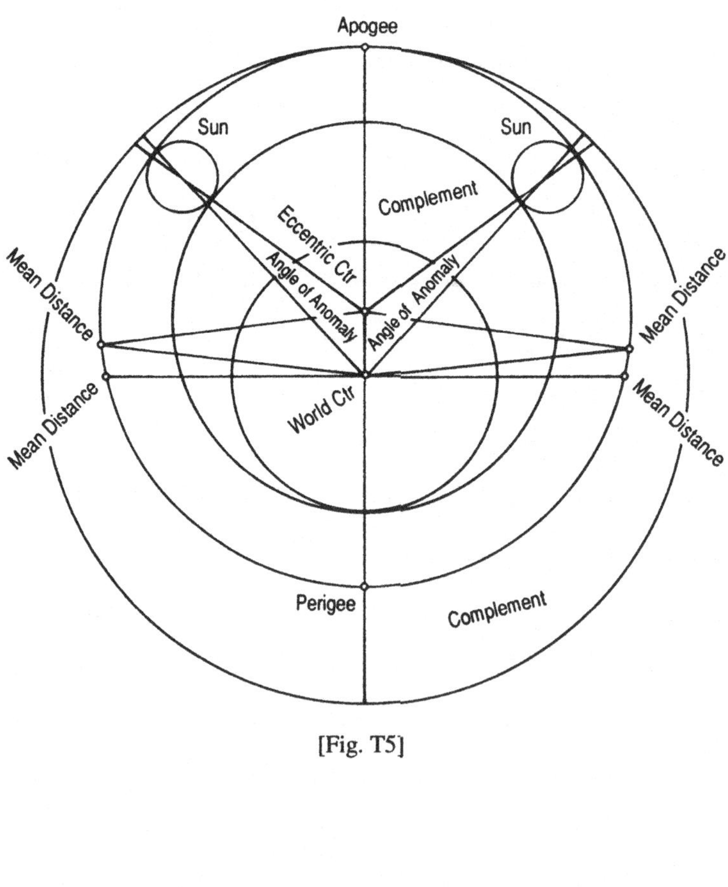

[Fig. T5]

يتحرك حركة الثوابت فيحرك الأوج والحضيض وذلك عند المتأخرين . وأما على أصل التدوير فالفلك الثامن كافٍ في تحريك الأوج والحضيض إذ هو محرك لجميع ما دونه فالحامل هو الممثل . ولكون الشمس دائماً في سطح منطقة الخارج أو التدوير ـ وهما في سطح الممثل ـ لا يكون لها عرض . ونحن أوردنا صورة / فلكيها / على أصل الخارج كما مال إليه بطلميوس .

[٤] ويلزم للشمس اختلاف واحد / بقدره / تخالف حركتها المرئية حركتها الوسطى ، وهو زاوية تحدث عند مركز الشمس من خطّين يخرجان من مركزي / فلكيها / إليه . ويصير أعظم ما يمكن في البعدين الأوسطين وينعدم عند البعدين الآخرين ، ويكون بقدر ما بين المركزين وهو عند بطلميوس بَلَ وعند أصحاب الأرصاد من المتأخرين قريب من بَهَ على أن يكون نصف قطر الخارج المركز ستين . وموضع الأوج عند بطلميوس متقدم على نقطة الانقلاب الصيفية بأربعة وعشرين جزءاً ونصف ، وعند المتأخرين مختلف فيه كما ذكروه في زيجاتها بقيد التأريخ . وهذه صورة أفلاك الشمس :

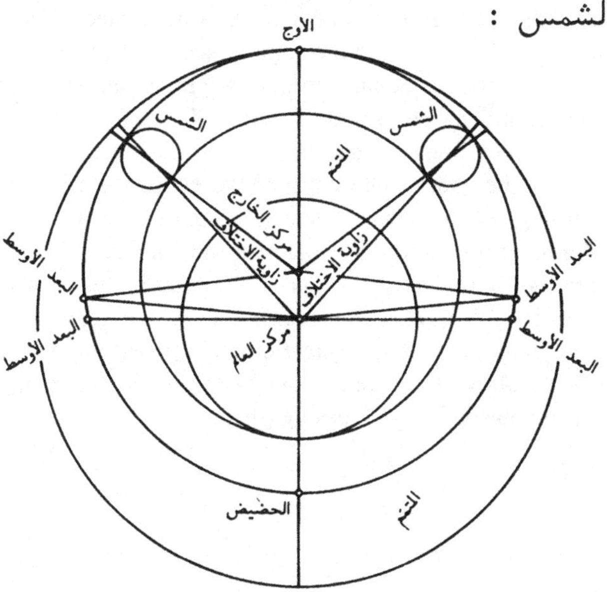

[شكل ٥]

/5/ ...] β = فلكيه] M , α . /7/ ...] M , β = بقدر ما] α . /9/ ...] β = فلكيه] M , α .

[5] One group has made the **mean distance** so that the two produced lines from the two centers to the sun are equal. This [mean] distance is based on distance, whereas what we have previously stated is based on motion. The above having been determined, it should be known that:

[a] the **solar apogee** designates the [arc] measured sequentially occurring on the parecliptic between the first of Aries and the apogee point;

[b] the **solar center** designates the [arc] measured sequentially occurring on the eccentric between the apogee and the center of the sun;

[c] the **mean** designates the sum of [a] and [b];

[d] the **true position** [*taqwīm*] designates the [arc] occurring on the parecliptic between the first of Aries and the endpoint of the line produced from the center of the World to the sun's body. This will be less than the mean by the amount of the anomaly as long as the sun is descending, greater than the mean when it is ascending.

[6] Thus the situation of the sun has been ordered by means of two orbs and two motions, and this is what we intended.

CHAPTER SEVEN

On the Orbs and Motions of the Moon

[1] The moon was found to move along a circuit different from the solar circuit that intersected the latter in two places facing one another. These [points] were not fixed, however, but shifted in the counter-sequence [of the signs]. Thus the moon is north of the ecliptic equator in half its circuit and south of it in the other half, the maximum distance in both directions being the same amount. Its motion on that circuit was found to be nonuniform, having anomalous speeds [that did not recur] in the exact same parts of the ecliptic but instead shifted, each anomaly returning not to its exact same value but rather, a short time after the completion of a lunar revolution, to what was comparable to it. The distance of the [moon] from the Earth was also found to vary; during the slower motion, it was sometimes nearer and at other times farther away, and likewise during the faster motion. It was found at mean conjunction and opposition to the sun, [this being] at a farther distance, to increase and decrease [in distance], it being slower with increasing, faster with decreasing [distance];

[٥] وقوم يجعلون البعد الأوسط حيث يتساوى الخطّان الخارجان من المركزين إليه ، وهذا بُعد بحسب المسافة وما ذكرناه أولاً هو بحسب الحركة . وإذا تقرر هذا فآعلم أنَ أوج الشمس يقال لِما يقع من الممثل بين أول الحمل ونقطة الأوج على التوالي ؛ ومركز الشمس لِما يقع من الخارج المركز بين الأوج ومركز الشمس على التوالي ؛ والوسط لمجموعهما ؛ والتقويم لِما يقع من الممثل بين أول الحمل وطرف الخط الخارج من مركز العالم إلى جرم الشمس ، وهو ناقص من الوسط بقدر الاختلاف ما دامت الشمس هابطة ، زائد عليه ما دامت صاعدة .

[٦] فإذن انتظم أمر الشمس بفلكين وحركتين ، وذلك ما أردناه .

الفصل السابع
في أفلاك القمر وحركاته

[١] وُجد القمرُ متحركاً على مدار غير مدار الشمس مقاطع إياه في موضعين متقابلين غير ثابتين بل منتقلين إلى خلاف التوالي ، فيكون القمر في نصف مداره شمالياً عن منطقة البروج وفي النصف الآخر جنوبياً عنها ، وغاية البعد في الجهتين بمقدار واحد ؛ وحركتُه على ذلك المدار غير متشابهة بل مختلفة بالبطء والسرعة في أجزاء لا بأعيانها من فلك البروج بل منتقلة ، عائدٌ كلُ اختلاف لا إلى مثله بعينه بل إلى ما يشبهه بعد تمام دور القمر بزمان قليل ؛ وبعدُه عن الأرض أيضاً مختلفاً اختلافاً يكون في البطء تارة قريباً وتارة بعيداً وكذلك في السرعة . ووُجد في مقارنة الشمس ومقابلتها الوُسطيين في بعد أبعد يزيد وينقص ، فيكون أبطأ كلما زاد وأسرع كلما نقص ، وتختلف مقادير جرمه في

/ similarly, / its body size varies during lunar and solar eclipses. At its quadrature to the sun, [this being] at a nearer distance, it increases and decreases as well. The lunar body was found to have different illuminated shapes according to its position from the sun. Its markings were found to be fixed.

[2] They therefore established four orbs and four simple motions for the [moon].

[3] The first orb is the parecliptic whose convex surface is contiguous with the concave surface of Mercury's parecliptic.

[4] The concave surface of the [moon's parecliptic] is contiguous with the convex surface of the second of its orbs, which is called **the inclined orb**. The concave surface of the inclined orb is contiguous with the sphere of fire from among the four elements. It is called inclined because its equator has a fixed inclination to the parecliptic equator whose maximum value as found by observation is five degrees. Its center is the center of the World.

[5] The third orb is an eccentric in the thickness of the inclined orb. Its equator is in the plane of the inclined equator.

[6] The fourth orb is an epicycle in the thickness of the eccentric, which is its deferent. The moon is embedded in the epicycle and adheres always to its equator that occurs in the plane of the eccentric equator.

[7] The equators of the parecliptic and inclined orbs intersect at two points facing one another called the **nodes** and the *jawzahr*. One of them, the northern crossing point or the **head**, is such that the moon when crossing it comes to be north. The other node is the southern crossing point or the **tail**.

[8] Turning to the motions, the first is the motion of the parecliptic with the motion of the nodes; it is three minutes plus a fraction daily in the countersequence [of the signs] about the center of the World. All the orbs of the moon move with this motion, and hence the head and tail shift as well; the motion, for this reason, is attributed to them. The motion of the fixed stars is indistinguishable from the other lunar motions but not because of what is [sometimes] said, namely that it is imperceptible since its ratio to these very fast motions is so small. For the small over a long period multiplies,

---

/1/ similarly] β = because of that] α, M.

الخسوفات والكسوفات / كذلك / ، وفي تربيعه للشمس في بعد أقرب يزيد وينقص أيضاً ؛ وجرمه مختلفَ الأشكال في النور بحسب أوضاعه من الشمس ؛ ومَحْوه ثابتاً .

[٢] فأثبتوا له أربعة أفلاك وأربع حركات بسيطة .

[٣] الفلك الأول هو المثل بفلك البروج محدّبه يماسّ مقعّر المثل لعطارد .

[٤] ومقعّره يماسّ محدّب الفلك الثاني من أفلاكه وهو المسمى بالفلك المائل ، ومقعّر المائل يماسّ كرة النار من العناصر الأربعة . وإنما سُمي مائلاً لكون منطقته مائلة عن منطقة المثل ميلاً ثابتاً ، غايته على ما وُجد بالرصد خمسة أجزاء . ومركزه مركز العالم .

[٥] والفلك الثالث فلك خارج المركز في ثخن المائل ، ومنطقته في سطح منطقة المائل .

[٦] والفلك الرابع فلك تدوير في ثخن الخارج المركز وهو حامله . والقمر مركوز في التدوير ملازم أبداً لمنطقته الكائنة في سطح منطقة الخارج المركز .

[٧] ومنطقتا المثل والمائل تتقاطعان على نقطتين متقابلتين تسميان العقدتين والجَوْزَهْر ، إحداهما التي إذا جاوزها القمر أخذ في الشمال هي المجاز الشمالي والرأس ، والأخرى هي المجاز الجنوبي والذنب .

[٨] وأما الحركات فالأولى حركة المثل بحركة الجوزهر ، وهي كل يوم ثلاث دقائق وكسر إلى خلاف التوالي حول مركز العالم . وبها تتحرك جميع أفلاك القمر ، فينتقل الرأس والذنب ولذلك تُنسَب إليهما . وأما حركة الثوابت فغير متميّزة عن غيرها في القمر ـ لا لِما قيل من أنها غير محسوسة لقلة نسبتها إلى هذه الحركات السريعة جداً ، فإنّ القليل في المدد الطويلة يتكثر

/1/ ...] β = لذلك] M , α .

and the models of the moon cannot tolerate much variation since the matter of the solar and lunar eclipses would thereby become disturbed. Rather, the motion [of the fixed stars] is indistinguishable from the motion of the nodes since their circumstances are the same in all respects. Hence the perceptible motion of the nodes is, in actuality, composed, being the excess of the motion of the nodes over the simple motion [of the fixed stars].

[9] The second motion is that of the inclined [orb] in the counter-sequence [of the signs], also about the center of the World, and it is 11°9′ daily. The eccentric is moved by this / motion, / which is called the **motion of the apogee** because it manifests itself there.

[10] The third motion is that of the eccentric in the sequence [of the signs], also about the center of the World, and it is 24°23′ daily. It is called the **motion of the center** because the epicycle center is moved by it this amount.

[11] Because the center of the epicycle is moved by the motions of the parecliptic and inclined [orbs] counter-sequentially, these motions being 11°12′, and sequentially by the above amount [*i.e.* 24°23′], its daily elongation from the apogee is this amount [24°23′] and from a fixed point on the ecliptic orb in the amount of the excess of the motion of the center over the sum of the first two motions, this being 13°11′. This is called the **mean motion of the moon**. The mean sun is always aligned with the center of the epicycle when the latter is at the apogee. Since [the mean sun] moves sequentially 0;59° daily, its elongation from the lunar apogee will be 12°11′, and its elongation from the epicycle center will by subtraction be the same. The sun then, after the epicycle center has departed from the apogee, will always be midway between the apogee and the center until the apogee is directly opposite the center at the [sun's] quadrature. The [center] will once more meet the [apogee] at the [sun's] opposition, will be opposite it at the other quadrature, and will then return to conjunction with the apogee. The motion of the center is therefore called the **double elongation**, *i.e.* the elongation of the epicycle center from the sun is doubled.

/8/ motion] $\beta = -\alpha$, –M.

وأصول القمر لا تحتمل كثير تفاوت لأنّ أمور الكسوفات والخسوفات تختل بذلك ـ بل لأنّ تلك الحركة لا تتميز عن حركة الجوزهر لاتّحاد موضوعيهما من جميع الوجوه . فإذن الحركة المحسوسة من الجوزهر مركبة في الحقيقة ، أعني أنها فضل حركة الجوزهر على تلك الحركة البسيطة .

[٩] والحركة الثانية حركة المائل إلى خلاف التوالي حول مركز العالم أيضاً ، كل يوم إحدى عشرة درجة وتسع دقائق . ويتحرك الخارج المركز بتلك / الحركة / وتسمى حركة الأوج لظهورها فيه .

[١٠] والثالثة حركة الخارج المركز إلى التوالي حول مركز العالم أيضاً ، كل يوم أربعاً وعشرين درجة وثلاثاً وعشرين دقيقة . وتسمى حركة المركز لانتقال مركز التدوير به[!] ذلك القدر .

[١١] ولكون مركز التدوير متحركاً بحركتي الممثل والمائل إلى خلاف التوالي ، وهما إحدى عشرة درجة واثنتا عشرة دقيقة ، وإلى التوالي هذا القدر ، يكون بعده عن الأوج كل يوم هذا القدر وعن النقطة الثابتة من فلك البروج بقدر فضل حركة المركز على مجموع الأوليين ، وهو ثلاث عشرة درجة وإحدى عشرة دقيقة . وتسمى هذه حركة وسط القمر . والشمس بوسطها تكون أبداً مع مركز التدوير عند كونه في الأوج ـ وهي تتحرك كل يوم تسعاً وخمسين دقيقة إلى التوالي ـ فيصير بعدها عن أوج القمر اثنتي عشرة درجة وإحدى عشرة دقيقة ، ويبقى بعدها عن مركز التدوير مثله . فتكون الشمس ، بعد مفارقة مركز التدوير الأوجَ ، متوسطة دائماً بين الأوج والمركز إلى أن يقابل الأوجُ المركزَ عند تربيعها ، ويلاقيه مرة أخرى عند استقبالها ، ويقابله في التربيع الآخر ، ويعود إلى الاجتماع مع الأوج . ولذلك تسمى حركةُ المركز البعدَ المضعَّف ، يعني بُعد مركز التدوير من الشمس مضعَّفاً .

/8/ ...] β = M– , α– .

In this way, the center will be at the eccentric apogee at mean conjunction and opposition, and at its perigee at the quadratures.

[12] Because all the above motions are about the center of the World, they are all uniform with respect to us.

[13] The fourth motion is that of the epicyclic orb. The moon, due to its motion, moves 13°4′ per day such that it is counter-sequential in the upper half [of the epicycle]. This is called the [moon's] **proper motion**.

[14] Because the ratio of this last motion to the mean motion is less than the ratio of the line joining the center of the World and the epicyclic perigee to the [epicyclic] radius, the moon does not have a station or retrogradation; rather, / its motion / will be slower in the apex half [of the epicycle], faster in the perigeal half. At conjunction, opposition and the two quadratures, the moon will have a slower speed with increasing distance and a faster one with decreasing distance. And because the motion of the epicycle is less than the mean motion, the slower and faster motions will not [recur] in the exact same parts of the ecliptic orb but instead their positions will shift; the return to the same anomaly will occur after the return to the same part of the ecliptic orb. For this reason as well, an eccentric alone cannot substitute for this epicycle. Because the radius of the epicycle varies in size with respect to the center of the World due to the varying distance of [the epicycle] from it in the two orbs, the rates of the slowest and fastest speeds are not always the same but will also vary; thus the slowest speed will sometimes return to a less slow, sometimes to a more slow speed, and the same holds for the fastest speed as well as the other anomalous speeds.

[15] These then are the motions of the moon.

/10/ its motion] β, M = –α.

فعلى هذا الوجه يكون المركز في الاجتماع والاستقبال الوَسطيّين في الأوج من خارج المركز ، وفي التربيعين في الحضيض منه .

[١٢] ولكون جميع هذه الحركات حول مركز العالم تكون الجميع عندنا متشابهة .

[١٣] والحركة الرابعة حركة فلك التدوير . ويتحرك القمر بحركته إلى غير التوالي في النصف الأعلى كل يوم ثلاث عشرة درجة وأربع دقائق . وتسمى حركته الخاصة .

[١٤] فلكون نسبة هذه الحركة إلى حركة الوسط أصغر من نسبة الخط الواصل بين مركز العالم وحضيض التدوير إلى نصف قطره لا يكون للقمر وقوف ولا رجوع ، بل تصير / حركته / بطيئة في نصف الذروة ، سريعة في نصف الحضيض . ويكون للقمر في الاجتماع والاستقبال والتربيعين بطءٌ مع زيادة بعد وسرعةٌ مع نقصانه . ولكون حركة التدوير أقل من حركة الوسط لا يكون البطء والسرعة في أجزاء بأعيانها من فلك البروج ، بل تنتقل مواضعها ، ويكون العود إلى اختلاف بعينه بعد العود إلى جزء بعينه من فلك البروج . ولا يقوم خارج مركز وحده بدل هذا التدوير لهذا السبب أيضاً . ولكون نصف قطر التدوير مختلف المقادير بالقياس إلى مركز العالم لاختلاف أبعاده منه في الفلكين تكون أقدار البطء والسرعة غير متشابهة بل مختلفة ، فيعود البطء تارة إلى بطء أقل وتارة إلى بطء أكثر ، وكذلك السرعة وغيرهما من الاختلافات .

[١٥] فهذه حركات القمر .

. −α = β , M [… /10/

[16] Turning now to the anomalies that inhere in the [moon] as a result of these motions, / the first is that anomaly effected by the radius of the epicycle at conjunctions and oppositions. It is an angle occurring at the center of the World from the production of two lines from it, one to the epicycle center and the other to the lunar body. / Its maximum is based upon the radius of the epicycle being at [the epicycle's] two mean distances whose size has been found by observation to be $5\frac{1}{4}$ parts, the radius of the inclined orb being 60 parts. The [anomaly] disappears at the apparent apex and perigee; it is subtractive from the mean as long as the moon is descending on the epicycle, additive when it is ascending. This is called the **independent equation**.

[17] The second anomaly is that resulting from an increase in the previous one when the epicycle is at a distance other than the farthest distance. Its maximum occurs when the epicycle is at the two quadratures, *i.e.* at the perigee, and it is $2\frac{2}{3}^\circ$ for this [epicycle] radius if the first anomaly is at its maximum. When the [first anomaly] decreases, the [second anomaly] will decrease accordingly. It is additive when the first anomaly is additive, subtractive when it is subtractive. It is called the **anomaly of the nearest distance**.

[18] The moon has another anomaly whose maximum occurs when the center of the epicycle is at the sextiles or trines with respect to the sun. It is caused by the apex of the epicycle, which is the starting point of its proper motion, and its perigee, which is opposite the apex, not being aligned with either the eccentric center or the center of the World except when the epicycle center is at the apogee or perigee. At that time they will be aligned with them since the diameter passing through [the apex and epicyclic perigee] will coincide with the diameter passing through the apogee, the [eccentric] perigee, and the centers. At other times, they will always be aligned with a point that is in the direction of the [eccentric] perigee whose distance from the center of the World

/2–4/ …] β, M = the first is that anomaly effected by the radius of the epicycle. It is an angle occurring at the center of the World from the production of two lines from it, one to the epicycle center and the other to the lunar body at conjunctions and oppositions] α.

[١٦] وأما الاختلافات التي تلزمه بسبب هذه الحركات / فالاختلاف الأول الذي بسبب نصف قطر التدوير في الاجتماعات والاستقبالات ، وهو زاوية تحدث على مركز العالم من خروج خطّين منه _ أحدهما إلى مركز التدوير والآخر إلى جرم القمر / . وتكون غايته بحسب نصف قطر التدوير في البعدين الأوسطين منه ، وقد وُجد بالرصد مقداره خمسة أجزاء / وربعاً / على أنّ نصف قطر المائل ستون جزءاً . وينعدم في الذروة والحضيض المرئيين ، وهو ناقص من الوسط ما دام القمر هابطاً في التدوير ، زائد ما دام صاعداً . ويسمى التعديل المفرد .

[١٧] والاختلاف الثاني هو الذي يكون بسبب زيادة الاختلاف المذكور عند كون التدوير في بعد غير الأبعد . وتكون غايته عند كون التدوير في التربيعين _ أعني في الحضيض _ وهي لنصف القطر جزءان وثُلثا جزء ، وذلك إذا كان الاختلاف الأول في الغاية . ولما نقص منه يكون بحسب نقصانه ، ويكون زائداً مع زيادة الاختلاف الأول ، ناقصاً مع نقصانه . ويسمى اختلاف البعد الأقرب .

[١٨] وللقمر اختلاف آخر تكون غايته عند كون مركز التدوير على تسديس الشمس أو / تثليثها / . وسببه أنّ ذروة التدوير التي هي مبدأ حركته الخاصة وحضيضه المقابل لها لا يحاذيان مركز الخارج ولا مركز العالم إلا عند كون مركز التدوير في الأوج أو الحضيض ، فإنهما حينئذ يحاذيانهما لانطباق القطر المارّ بهما على القطر المارّ بالأوج والحضيض والمراكز . أما في غير ذلك الوقت فيحاذيان أبداً نقطة مما يلي الحضيض ، بُعدها عن مركز العالم

/2–4/ [... β , M = فالاختلاف الأول الذي بسبب نصف قطر التدوير وهو زاوية تحدث على مركز العالم من خروج خطّين منه أحدهما إلى مركز التدوير والآخر إلى جرم القمر يكون في الاجتماعات والاستقبالات] α. /6/ [... β = وربع] α , M. /17/ [... β = تثليثه] α , M.

is the same as the distance of the eccentric center, which is in the direction of the apogee, from the [center of the World]. This point is called the **point of alignment**. Each of these distances, according to what has been found by the observational [astronomers], amounts to 10 parts and 19 minutes, the radius of the inclined orb being 60. Because of this alignment, the **mean apex**, from which is the beginning of the proper motion, will always differ from the **apparent apex** at which the first two anomalies disappear. The situation is the same for the two [epicyclic] perigees. Thus the moon will have an anomaly when it may be thought to have none, and it will not have one when it is thought to occur. The maximum of this anomaly is based upon the above-mentioned distance. It disappears when the center is at the [eccentric] apogee or perigee. It is additive when the center is descending and subtractive when it is ascending. This is called the **equation of the proper** [motion].

[19] In addition, the moon has another anomaly that is the difference between the distance from the nodes of its two positions on the parecliptic and inclined equators. This anomaly is taken into account if one wishes to convert from one position to the other.

[20] All the above are concerned with longitude.

[21] As for latitude, it should be clear from what has already been stated. The return of the moon in both directions is always to the maximum [latitude]. The moon is north from the head to the tail and south from the tail to the head. It is ascending from its maximum latitude in the south to its maximum in the north and descending in the other half.

[22] As regards the changes in the illuminated shapes of the lunar body according to its position from the sun, this will be forthcoming in a separate chapter.

[23] The variation in the parts of the lunar surface in receiving light, which is called the **lunar markings**, is due to an anomaly whose true character has yet to be ascertained. Most likely, there exist various bodies occurring with the moon in its epicycle that do not accept illumination equally, either because of a difference in type or because of a difference in position.

[24] This then is the situation of the moon.

[25] There is a difficulty that arises regarding the motion of the epicycle center on the circumference of the eccentric about the center of the World and also regarding the alignment of its diameter with a point other than the center of the deferent. To see this, [we note] that when the deferent moves the epicycle with a simple, uniform motion:

كبعد مركز الخارج مما يلي الأوج عنه ، تسمى تلك النقطة نقطة المحاذاة . ومقدار كل واحد من البعدين عشرة أجزاء وتسع عشرة دقيقة على أنّ نصف قطر المائل ستون بحسب ما وجده أهل الرصد. وبسبب هذه المحاذاة تخالف الذروة الوسطى التي منها مبدأ الحركة الخاصة أبداً الذروة المرئية التي عندها ينعدم الاختلافان الأوّلان وكذلك الحضيضان ، فيوجد للقمر اختلاف عندما يظن عدمه ويعدم اختلافه عندما يظن وجوده . وغاية هذا الاختلاف بحسب البعد المذكور ، وينعدم عند كون المركز في الأوج أو الحضيض .ويكون زائداً ما دام المركز هابطاً ، وناقصاً ما دام صاعداً . ويسمى تعديل الخاصة .

[١٩] وأيضاً له اختلاف آخر وهو التفاوت بين بعد موضعيه في منطقتي الممثل والمائل عن العقدتين . ويعتبر ذلك إذا أريد تحويل أحدهما إلى الآخر .

[٢٠] وهذه الأمور كلها تتعلق بالطول .

[٢١] وأما العرض فقد تبيّن مما مرّ . ويكون عود القمر في الجهتين دائماً إلى غايته . فيكون شمالياً من الرأس إلى الذنب وجنوبياً من الذنب إلى الرأس ، وصاعداً من غاية عرضه في الجنوب إلى غايته في الشمال وهابطاً في النصف الآخر .

[٢٢] وأما اختلاف التشكّلات النورية في جرمه بحسب وضعه من الشمس فسيجيء في باب مفرد .

[٢٣] وأما اختلاف أجزاء سطحه في قبول النور المسمى بالمَحْو فلاختلاف فيه لم يوقف على حقيقته . والأشبه وجود أجرام مختلفة معه في تدويره غير قابلة للإنارة بالتساوي إما لاختلاف نوعي أو لاختلاف وضعي .

[٢٤] فهذه أحوال القمر .

[٢٥] وورد على حركة مركز التدوير في محيط الخارج المركز حول مركز العالم ومحاذاة قطره نقطة غير مركز الحامل إشكالٌ . وبيان ذلك أنّ الحامل إذا حرك التدوير حركةً بسيطة متشابهة

[a] the distances of the epicycle center from the [deferent] center must be equal under all circumstances; [b] the angles [formed] at the [deferent center] must be equal in equal times; and [c] the diameter passing through the apex and the [epicyclic] perigee must be aligned with the [deferent center] under all circumstances. If any one of these three were not to hold, this would be due to the motion being composed. In fact, we do find these things violated in the case of the moon. For while the distances of its epicycle center *are* equal with respect to the center of the eccentric, the equality of angles occurs at the center of the World and the alignment of the diameter is to the point of alignment. The practitioners of the profession have not yet explained in what way this motion is composed; in fact, they have not ventured any explanation of this at all. I shall present below what I have regarding this matter, God willing.

[26] The moon has another anomaly called parallax whose description will be forthcoming.

[27] This is the illustration of the orbs of the moon:

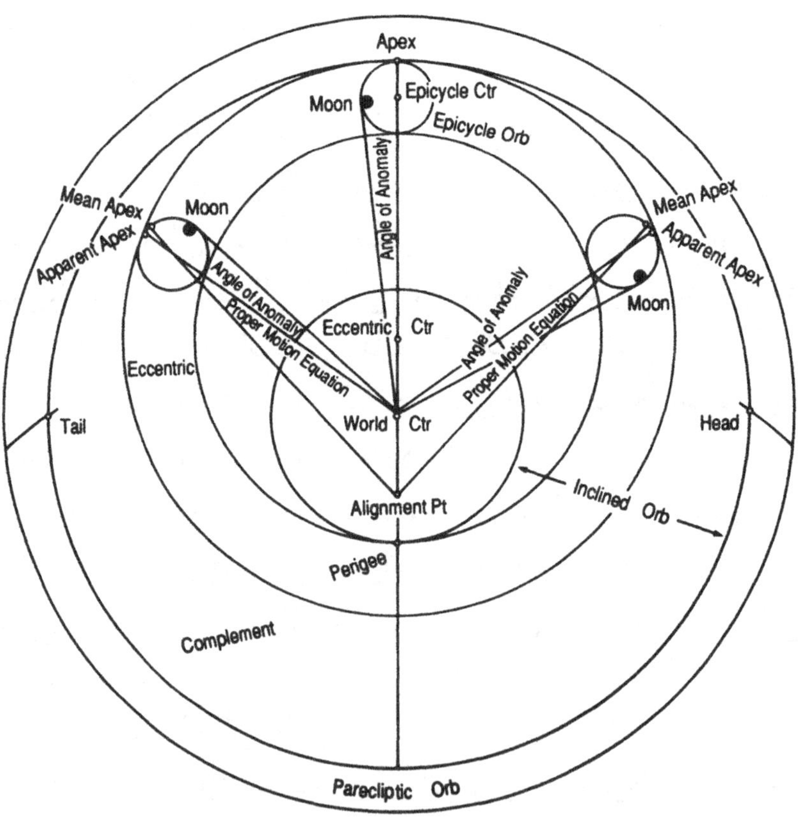

[Fig. T6]

وجب تساوي أبعاد مركز التدوير عن مركزه في جميع الأحوال ، وتساوي الزوايا حوله في الأزمنة المتساوية ، وكون القطر المارّ بالذروة والحضيض محاذياً له في جميع الأحوال . فإن اختلف بعض هذه الأمور الثلاثة فذلك يكون لتركُب في الحركة . ثم إنّا نجد هذه الأمور مختلفة في القمر ، فإنّ تساوي أبعاد مركز تدويره إنما يكون عند مركز الخارج المركز ، وتساوي الزوايا عند مركز العالم ، ومحاذاة القطر لنقطة المحاذاة . وأهل الصناعة لم يبيّنوا الوجه في كيفية هذا التركيب ، بل لم يتعرضوا لبيان شيء من ذلك . وسأورد ما عندي فيه إن شاء الله تعالى .

[٢٦] وللقمر اختلاف آخر يسمى اختلاف المنظر سيجيء وصفه .

[٢٧] وهذه صورة أفلاك القمر :

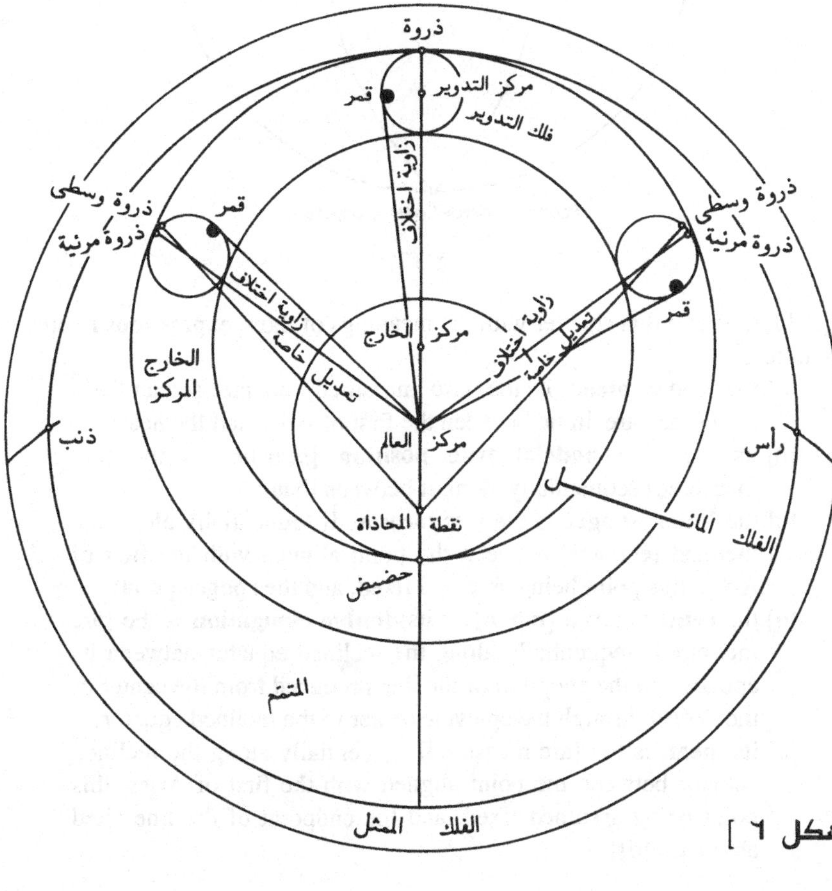

[شكل ٦]

[28] Those who limit themselves to circles make the parecliptic and the inclined equators intersect one another, the eccentric equator tangent to the inclined [equator] at the apogee point, and the equator of the epicycle such that its center is on the eccentric equator. If the motion of the sun is disregarded, the circuit of the center of the epicycle, which reaches each of the apogee and the perigee twice during its revolution, would be thus:

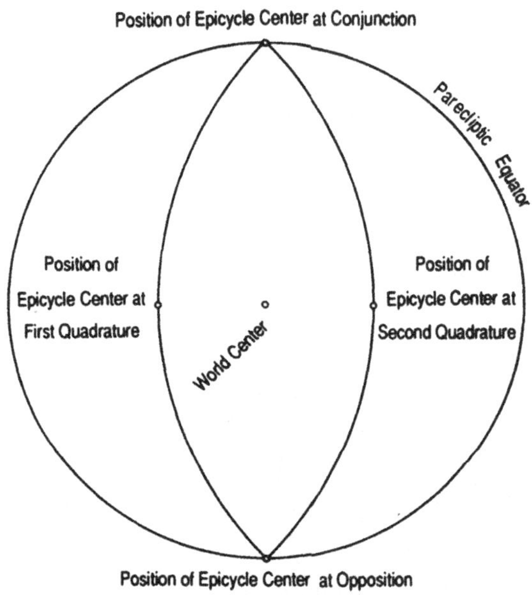

[Fig. T7]

[29] Let us end this chapter with the meanings of those expressions related to the moon:

[a] the **nodal mean** is the [arc measured] counter-sequentially along the parecliptic between the first of Aries and the head;

[b] its [*i.e.* the **node's**] **true position** [*taqwīm*] is the [arc measured] sequentially along it between them;

[c] the **lunar apogee** is the [arc measured] sequentially along the inclined [equator] between the point aligned with the first of Aries, this point being assumed fixed, and the apogee point;

[d] the **center of the [moon]** or its **double elongation** is the [arc measured] sequentially along the inclined equator between its apogee and the endpoint of the line produced from the center of the World through the epicycle center to the inclined equator;

[e] its **mean** is the [arc measured] sequentially along the inclined equator between the point aligned with the first of Aries, this point being assumed fixed, and the endpoint of the line cited above [in [d]];

[٢٨] ومَن اقتصر على الدوائر أورد منطقتي المثل والمائل متقاطعتين ، ومنطقةَ الخارج المركز مماسّةً للمائل على نقطة الأوج ، ومنطقةَ التدوير على أنّ مركزها على منطقة الخارج المركز . ومدار مركز التدوير – لو لا حركة الشمس – في وصوله إلى الأوج في الدورة مرتين وإلى الحضيض مرتين يكون هكذا :

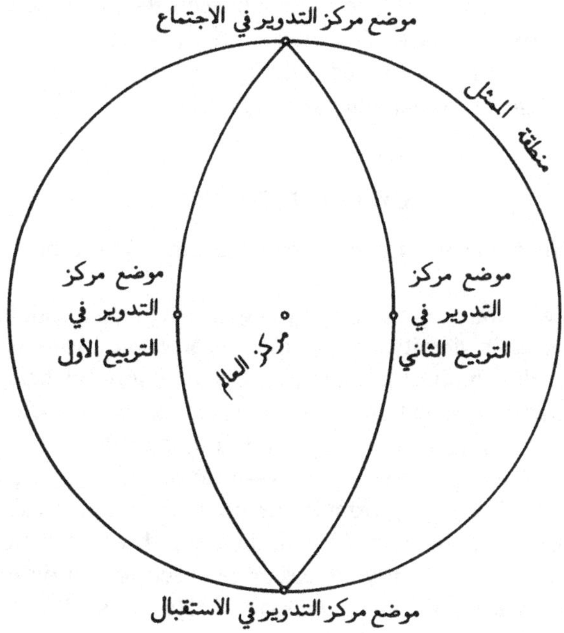

[شكل ٧]

[٢٩] ولنختم الفصل بمعاني ألفاظ تتعلق بالقمر فنقول : وسط الجوزهر ما بين أول الحمل ونقطة الرأس من الممثل على خلاف التوالي ؛ وتقويمه هو ما بينهما منه على التوالي ؛ وأوج القمر هو ما بين النقطة المحاذية لأول الحمل ، على أنها لا تتغير ، ونقطة الأوج من المائل على التوالي ؛ ومركزه أو بعده المضعَف هو ما بين أوجه وطرف الخط الخارج من مركز العالم إلى مركز التدوير ومنه إلى منطقة المائل من منطقة المائل على التوالي ؛ ووسطه ما بين النقطة المحاذية لأول الحمل ، على أنها لا تتغير ، وطرف الخط المذكور

[f] its **mean proper anomaly** [*khāṣṣa*] is the [arc measured] sequentially (as defined for the [epicycle]) along the equator of its epicycle between the mean apex and the center of the [lunar] body.

Among those whose motion is irregular:

[g] the [moon's] **apparent proper anomaly** is the [arc] along the equator of its epicycle between its apparent apex and the center of its body;

[h] its **true position** [*taqwīm*] is the [arc measured] sequentially along the parecliptic equator between the first of Aries and the point at which its latitude circle intersects the parecliptic;

[i] its **argument of latitude** [*ḥiṣṣat* *ᶜarḍihi*] is the [arc measured] sequentially along [the parecliptic equator] between the head and the intersection point cited above [in [h]].

CHAPTER EIGHT

The Orbs and Longitudinal Motions of Mercury

[1] Mercury was found not to move in longitude precisely along the ecliptic equator but rather in its vicinity, approaching it sometimes from the north and at other times from the south, the limits [north and south] not being exactly the same. It will increase in speed, thus moving ahead of the sun after having been in conjunction with it, and appear in the west. It then gradually slows until it becomes stationary. Thereupon it retrogrades—disappearing, coming into conjunction with the sun, and departing from it; the sun will then move ahead of it, [the planet now] appearing in the east. / It will thereafter become stationary and undergo a direct motion, / gradually increasing in speed until it disappears. It will then reach the sun and be in conjunction with it. Thus it is with [the sun] during the middle of its periods of direct and of retrograde motion. Its elongation ahead of or behind [the sun] does not exceed 27°. If one compares one retrogradation to another retrogradation, a direct motion to another direct motion, a slower motion to another slower motion, or a faster motion to another faster motion, one will not find them to be identical; instead, in some parts of the ecliptic they will be less in terms of both / extent / and time, while in other parts they will be greater. That part in which the slower motion is at its slowest and the time periods are at their shortest is not fixed but shifts with the movement of the fixed stars. The opposite of these circumstances does not occur

/15/ It will thereafter become stationary and undergo a direct motion] β, M = It will become stationary and thereafter undergo a direct motion] α. /20/ extent] β, M = revolution] α.

من منطقة المائل على التوالي ؛ وخاصته الوسطى ما بين ذروته الوسطى ومركز جرمه من منطقة تدويره على التوالي المفروض فيه . ومما تختلف حركته : خاصته المرئية وهي ما بين ذروته المرئية ومركز جرمه من منطقة تدويره ؛ وتقويمه وهو ما بين أول الحمل والنقطة التي تقاطع دائرة عرضه المثل من منطقة المثل على التوالي ؛ وحِصة عرضه / وهي / ما بين نقطة الرأس ونقطة التقاطع المذكورة منه على التوالي .

الفصل الثامن
في أفلاك عطارد وحركاته الطولية

[١] وُجد عطارد متحركاً في الطول لا على نفس منطقة البروج بل / حواليها ، يقرب منها تارة في شمالها وتارة في جنوبها / لا إلى حدّين بعينهما . وهو يسرع في سيره فيسبق الشمسَ بعد مقارنتها ويظهر مُغرّباً ، ثم يأخذ في البطء متدرجاً إلى أن يقف ، ثم يرجع ويختفي ويقارن الشمسَ ويفارقها فتسبقه الشمسُ ويظهر مشرّقاً ، / ثم يقف ويستقيم / ويتدرج إلى السرعة إلى أن يختفي ، ثم يُدرِك الشمسَ ويقارنها . فيكون معها في منتصف زماني استقامته ورجوعه ، ولا يبعد عنها من قدامها وخلفها أكثر من سبعة وعشرين جزءاً . وإذا قيس رجوعٌ إلى رجوع أو استقامة إلى استقامة أو بطء إلى بطء أو سرعة إلى سرعة لم توجد متشابهة بل كانت في بعض أجزاء البروج أقل / قدراً / وزماناً وفي بعضها أكثر . والجزء الذي يوجد البطء فيه أشد والزمان أقل لا يكون ثابتاً بل منتقلاً انتقال الثوابت . وأضداد تلك الأحوال ليست في

/6/ ...] M , β = وهو] α . /11/ ...] β = حواليه يقرب منه تارة في شماله وتارة في جنوبه] M , α . /15/ ...] M , β = ويقف ثم يستقيم] α . /20/ ...] M, β = دوراً] α .

directly facing that part [of the ecliptic] but at the two trines. Directly facing that, one finds a situation similar to what occurs at this part but not to that maximum extent.

[2] They therefore established for [Mercury] four orbs and four motions.

[3] The first orb is the pareccliptic whose convex surface is contiguous with the concave surface of Venus's orb. Its concave surface is contiguous with the convex surface of the moon's pareccliptic.

[4] The second orb is an eccentric and is called the **dirigent**. It is in the thickness of the pareccliptic just as we have described an eccentric being in the thickness of a concentric. Its equator is not in the plane of the pareccliptic equator but has an inclination to it that is not fixed; its description will be forthcoming. The apogee of the [dirigent] occurs at the place of the maximum inclination. The plane of its equator intersects the plane of the pareccliptic equator at acute and obtuse angles. There thus occurs a great circle in the pareccliptic orb whose center is the center of the World and which intersects the pareccliptic in two places that are called the nodes—the "head" and the "tail"—of this planet. This great circle is called [Mercury's] **inclined orb**.

[5] The third orb is another eccentric called the deferent for the epicycle. It is in the thickness of the dirigent just as the dirigent is in the thickness of the pareccliptic. The [deferent's] equator is in the plane of the [dirigent's] equator.

[6] On account of its two eccentric orbs, this planet has four complementary bodies: two for the dirigent from the pareccliptic and two for the deferent from the dirigent.

[7] The fourth orb is the epicycle, and it is in the thickness of the deferent. Its equator is not fixed in the [deferent's] equator as will be explained below. Mercury is on the epicycle, being embedded in it, and moves along its equator.

[8] Turning now to the motions, the first is the motion of the pareccliptic, [which moves] with the motion of the fixed stars sequentially about the center of the World. It manifests itself in the apogee and perigee of the dirigent and in the head and tail.

[9] The second is the motion of the dirigent and is the same as the motion of center of the mean sun, *i.e.* the excess of its mean motion over the motion of its apogee; it is counter-sequential about its own center. This motion manifests itself in the apogee and perigee of the deferent.

مقابلة ذلك الجزء بل في تثليثيه . وفي مقابلة ذلك يوجد مثل ما يوجد في ذلك الجزء ولكن لا في تلك الغاية .

[٢] فأثبتوا له أربعة أفلاك وأربع حركات .

[٣] الفلك الأول المثل بفلك البروج محدّبه مماسّ لمقعّر فلك الزهرة ومقعّره مماسّ لمحدّب ممثل القمر .

[٤] والفلك الثاني خارجُ مركزٍ يسمى بالمُدير . ويكون في ثخن المثل كما وصفنا في كون الخارج المركز في ثخن الموافق المركز . ومنطقته ليست في سطح منطقة المثل بل مائلة عنها غير ثابتة الميل ، وستجيء صفتها ، وأوجُه عند موضع غاية الميل . وسطح منطقته يقاطع سطح منطقة المثل على زوايا حادّة ومنفرجة ، فتَحدث في الفلك المثل دائرةٌ عظيمة مركزها مركز العالم مقاطعة للممثل في موضعين يسميان عقدتي الرأس والذنب لهذا الكوكب . وتسمى تلك العظيمة فلكه المائل .

[٥] والفلك الثالث خارج مركز آخر يسمى الحامل للتدوير . ويكون في ثخن المدير مثل كون المدير في ثخن المثل ، ومنطقته في سطح منطقته .

[٦] وتكون لهذا الكوكب بحسب فلكيه الخارجي المركز أربعةُ متمّمات : اثنان للمدير من المثل ، واثنان للحامل من المدير .

[٧] والفلك الرابع فلك التدوير وهو في ثخن الحامل . ومنطقته ليست بثابتة في منطقته على ما سيجيء بيانه . وعطارد على التدوير مركوز فيه يتحرك على منطقته .

[٨] وأما الحركات فالأولى حركة المثل بحركة الثوابت حول مركز العالم على التوالي . وتظهر في أوج المدير وحضيضه وفي الرأس والذنب .

[٩] والثانية حركة المدير وهي مثل حركة مركز الشمس الوسطى ، أعني فضل حركة وسطها على حركة أوجها إلى خلاف التوالي حول مركزه . وتظهر هذه الحركة في أوج الحامل وحضيضه.

On account of it, the deferent center comes to have a circuit about the dirigent center called the **deferent orb for the deferent orb's center**.

[10] The third is the motion of the deferent; it is equal to twice the sun's motion of center and is sequential. [It moves] not about its own center nor about the center of the World nor about the dirigent center; it is instead about a point we will discuss below. This motion manifests itself in the epicycle center.

[11] The center of the epicycle is always in conjunction with the position of the mean sun. When it is at the dirigent apogee, it will be at the deferent apogee as well. The [center] and the [deferent apogee] then separate from the [dirigent apogee]: the deferent apogee will move counter-sequentially away from the dirigent apogee at the rate of the sun's motion of center, and the epicycle center will move sequentially away from the dirigent apogee at the rate of the excess of its motion over the motion of the deferent apogee, which is likewise equal to the sun's motion of center. Therefore the dirigent apogee is always at the midpoint between the deferent apogee and the epicycle center as was the case for the moon with respect to the center of the sun being midway between the apogee and the epicycle center. When each [of the deferent apogee and the epicycle center] describes a quarter [revolution], the center will reach the perigee of the deferent, and they will each be at quadrature with respect to the dirigent apogee. After describing another quarter, they will meet opposite the dirigent apogee and thus the center will be at the dirigent perigee and at the deferent apogee. Then [the center and deferent apogee] will separate, and they will be in opposition at the quadratures. They will once again meet at the dirigent apogee. Thus the epicycle center's farthest distance is when it is at both apogees simultaneously. Its nearest distance, however, is not directly opposite this position on account of it thereupon being at the deferent apogee and the dirigent perigee, nor is [this distance] at the quadratures since the two opposite distances [from the center of the World] occurring at the apogee [of the dirigent] and at the point directly opposite it are not equal. Instead, it occurs at two positions whose distance from the dirigent apogee is greater than from the point directly opposite it. These two positions are at the trines with respect to the apogee, a result dictated by the composition of the two perigees.

[12] This latter motion and the motion of the apogee together yield Mercury's **mean motion**.

ويظهر بسببها لمركز الحامل مدارٌ حول مركز المدير يسمى الفلكَ الحامل لمركز الفلك الحامل.

[١٠] والثالثة حركة الحامل وهي مثل ضعف حركة مركز الشمس إلى التوالي _ لا حول مركزه ولا حول مركز العالم ولا حول مركز المدير بل حول نقطة سنذكرها . وتظهر في مركز التدوير .

[١١] ومركز التدوير يقارن موضع الشمس الوسطى دائماً . وإذا كان في أوج المدير كان في أوج الحامل أيضاً . ثم يفارقانه فيتحرك أوج الحامل إلى خلاف التوالي ويبعد عن أوج المدير بقدر حركة مركز الشمس ، ويتحرك مركز التدوير إلى التوالي ويبعد عن أوج المدير بقدر فضل حركته على حركة أوج الحامل وهو أيضاً مثل حركة مركز الشمس . فيكون أوج المدير دائماً في المنتصف بين أوج الحامل ومركز التدوير كما مرّ في القمر من توسّط مركز الشمس بين الأوج ومركز التدوير . وإذا قطع كل واحد منهما الربع انتهى المركز إلى حضيض الحامل وهما في تربيعي أوج المدير ، وبعد قطع ربع آخر يتلاقيان في مقابلة أوج المدير فيكون المركز في حضيض المدير وأوج الحامل ، ثم يتفارقان ويتقابلان في التربيعين ويعودان إلى الملاقاة عند أوج المدير . فالبعد الأبعد لمركز التدوير يكون عند كونه في أوجيه معاً . ولا يكون بعده الأقرب في مقابلة ذلك الموضع لكونه في أوج الحامل وحضيض المدير هناك ، ولا في التربيعين لأنّ البعدين المتقابلين اللذين في الأوج ومقابله ليسا / بمتساويين / ؛ بل يكون في موضعين بعدهما من أوج المدير أكثر من مقابلته وهما تثليثا الأوج بحسب ما يقتضيه تركّب الحضيضين .

[١٢] وتجتمع من هذه الحركة وحركة الأوج حركةُ وسط عطارد .

/21/ ...] M , β = متساويين] α .

[13] The fourth motion is that of the epicycle orb, which is 3°6′ per day. The planet moves with this [motion] in such a way that in the far segment [of the epicycle] it is sequential. The planet will undergo retrogradation on this epicycle in the near segment since the ratio of the two motions is such so as to bring about retrogradation. The planet will not move away from the sun—either ahead or behind—except by the amount brought about by its epicycle radius. At the apex and perigee [of the epicycle], it will be in conjunction with the [sun] since its [epicycle] center is permanently in conjunction with it. The radius of the epicycle is according to observation 22½ parts, the deferent radius being 60 parts.

[14] The dirigent center is 6 parts removed from the center of the World, this also according to the above units. The point about which / the motion of the epicycle center and the motion of the deferent / is always uniform is at the midpoint of this distance on the diameter that passes through these two [centers]; this point is called the **equant center**. About it one may imagine a circle, called the **equant orb**, whose size is that of the deferent equator and which is in the same plane. Thus the center of the epicycle will cut off from its circumference equal arcs in equal times. It is as if there were a line extending from the center of the equant to the center of the epicycle so as to turn it with uniform motion. The mean apex and perigee of the epicycle are also aligned with this point. The distance of the deferent center from the dirigent center is likewise equal to the distance of the equant center from the latter. It therefore follows that the deferent center will encounter the equant center once every revolution, and this will occur when the center of the epicycle is in opposition to the dirigent apogee. At that time the deferent equator will coincide with the equant orb; afterwards, they will become separate. When the epicycle center is at the two apogees, the four centers will be equally spaced on the diameter passing through the centers.

[15] Turning now to Mercury's anomalies that follow from its motions, the first is its anomaly resulting from the diameter of its epicyclic orb when it is at the mean distance. It is an angle

---

/10–11/ the motion of the epicycle center and the motion of the deferent] β = the motion of the deferent] α, M.

[١٣] والحركة الرابعة حركة فلك التدوير ، كل يوم ثلاثة أجزاء وست دقائق . فيتحرك بها الكوكب على وجه يكون في القطعة البعيدة منه على التوالي . ويقع للكوكب في هذا التدوير رجوعٌ في القطعة القريبة لكون نسبة الحركتين على ما يقتضي الرجوع . ولا يبعد الكوكب من الشمس قدامها وخلفها إلا بقدر ما يقتضيه نصف قطر تدويره ، ويقارنها في الذروة والحضيض لكون / مركزه / مقارناً لها دائماً . ونصف قطر التدوير اثنان وعشرون جزءاً ونصف بالرصد على أنّ نصف قطر الحامل ستون جزءاً .

[١٤] ومقدار خروج مركز المدير عن مركز العالم ستة أجزاء بهذه الأجزاء أيضاً . وتكون النقطة التي تتشابه / حركة مركز التدوير وحركة الحامل / حولها أبداً عند منتصف هذا البعد على القطر المارّ بهما ، وتسمى مركز مُعدِّل المَسير . وتتوهم حولها دائرةٌ بقدر منطقة الحامل وفي سطحها تسمى فلك معدل المسير ؛ فإنّ مركز التدوير يقطع من محيطه في أزمنة متساوية قسياً متساوية كأنّ خطاً خرج من مركز معدل المسير إلى مركز التدوير ليُديره حركةً متشابهة . والذروة والحضيض الوسطيان من التدوير يحاذيان أيضاً هذه النقطة . ومقدار خروج مركز الحامل عن مركز المدير أيضاً بقدر بعد مركز معدل المسير عنه ؛ فلذلك يلزم أن يلاقي مركزُ الحامل في كل دورة مرةً مركزَ معدل المسير ، وذلك عند كون مركز التدوير في مقابلة أوج المدير ، وحينئذ تنطبق منطقة الحامل على فلك معدل المسير ثم يتفارقان . وعند كون مركز التدوير في الأوجين تكون المراكز الأربعة على القطر المار بالمراكز على أبعاد متساوية .

[١٥] وأما اختلافات عطارد اللازمة لحركاته فالأول اختلافه اللازم من جهة قطر فلك تدويره عند كونه في البعد الأوسط . وهو زاوية

/7/ ...] مركزها = β [M , α . /10–11/ ...] حركة الحامل = β [M , α .

that occurs at the center of the World from the production of two lines from it, one to the center of the epicycle and the other to the center of the planetary body. The maximum of this anomaly will be according to the size of the epicycle radius. It will be additive with respect to the position of the epicycle center in the descending half, subtractive in the ascending half. This anomaly is called the / **second** / **equation**.

[16] The second is the apparent increase [in size] of the epicycle radius when it is at a nearer distance over what is observed at the mean distance, and its decrease when it is at a farther distance. This anomaly will combine with the first anomaly according to the amount of the latter as determined by the radius, thus either decreasing or increasing the [first anomaly]. The resulting [quantity] will then follow [the above rule for the first anomaly] in being either additive or subtractive with respect to the center. This anomaly is called the **anomaly of the farthest and nearest distances**.

[17] The third is the anomaly that results from the motion of the epicycle center being uniform about a point other than the center of the World and from the difference between the apparent and the mean [epicyclic] apices. These two anomalies are one and the same since the diameter of the epicycle passing through the mean apex and perigee is aligned with this very same point. This [anomaly] is an angle occurring at the center of the epicycle from the production of two lines from it, one to the center of the World and the second to the equant center. The anomaly is subtractive from the center and additive to the proper motion as long as the epicycle center is descending on the dirigent; the opposite will hold when it is ascending. This anomaly is called the **equation of the center and of the proper** [motion].

[18] These then are [Mercury's] anomalies.

[19] The same difficulty that was mentioned in the chapter on the moon due to the uniformity of the motion of the epicycle center being about a point different from the center of its deferent is present here as well. However, the [difficulty] that was mentioned as arising on account of the anomaly in alignment is not present because the alignment is toward the point with respect to which the uniformity of motion occurs. There [also] results from the motions of the dirigent

/5/ second] $\beta = -\alpha$, –M.

على مركز العالم تَحدث من خروج خطّين عنه : أحدهما إلى مركز التدوير ، والآخر إلى مركز جرم الكوكب . وغاية هذا الاختلاف بقدر نصف قطر التدوير . ويكون زائداً على موضع مركز التدوير في النصف الهابط ، ناقصاً في النصف الصاعد . ويسمى هذا الاختلاف بالتعديل / الثاني / .

[١٦] والثاني زيادة نصف قطر التدوير في الرؤية على ما يُرى في البعد الأوسط إذا صار في بعد أقرب منه ، ونقصانه من ذلك إذا صار في بعد أبعد / منه / . وهذا الاختلاف يلحق الاختلاف الأول بقدر ذلك الاختلاف من نصف القطر فينقص منه أو يزيد عليه ، ويكون بعد ذلك في الزيادة على المركز أو النقصان منه تابعاً له . ويسمى هذا الاختلاف اختلاف البعد الأبعد والأقرب .

[١٧] والثالث الاختلاف اللازم بحسب تشابه حركة مركز التدوير حول نقطة غير مركز العالم وبحسب اختلاف الذروتين المرئية والوسطى . وهذان الاختلافان شيء واحد لكون قطر التدوير المارّ بالذروة والحضيض الوسطيين محاذياً لتلك النقطة بعينها . وهو زاوية تَحدث على مركز التدوير من خطّين يخرجان منه : أحدهما إلى مركز العالم ، والثاني إلى مركز معدل المسير . ويكون الاختلاف ناقصاً من المركز ، زائداً على الخاصة ، ما دام مركز التدوير هابطاً في المدير ، وبالعكس ما دام صاعداً . ويسمى هذا الاختلاف تعديل المركز والخاصة .

[١٨] فهذه اختلافاته .

[١٩] والإشكال المذكور في باب القمر بسبب تشابه حركة مركز التدوير حول نقطة خارجة عن مركز حامله واردٌ بعينه ههنا . وأما الذي ذُكر بحسب اختلاف المحاذاة فغير وارد لكون المحاذاة نحو النقطة التي بحسبها تتشابه الحركة . ويلزم من كون حركتي المدير

/5/ ...] β = M– , α– . /8/ ...] β = M– , α– .

and the deferent being about two different points an anomaly that was not mentioned with regard to the motion of the epicycle center, which is composed of these two.

[20] This is the illustration of the orbs of Mercury:

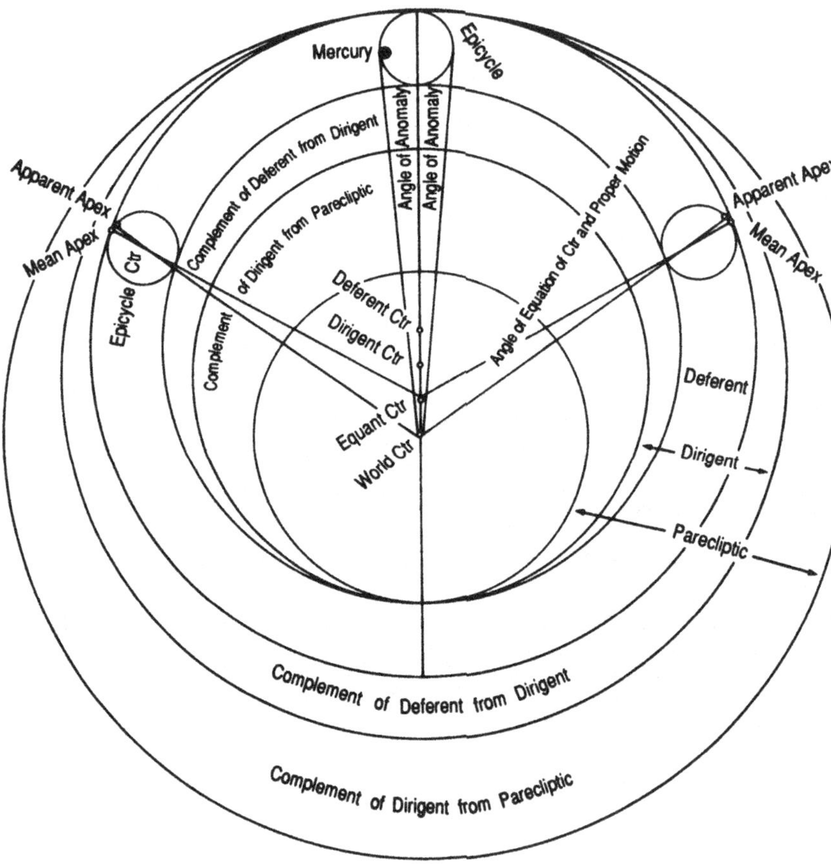

[Fig. T8]

والحامل حول نقطتين مختلفتين اختلافٌ لم يُذكر في حركة مركز التدوير المركبة عنهما .

[٢٠] وهذه صورة أفلاك عطارد .

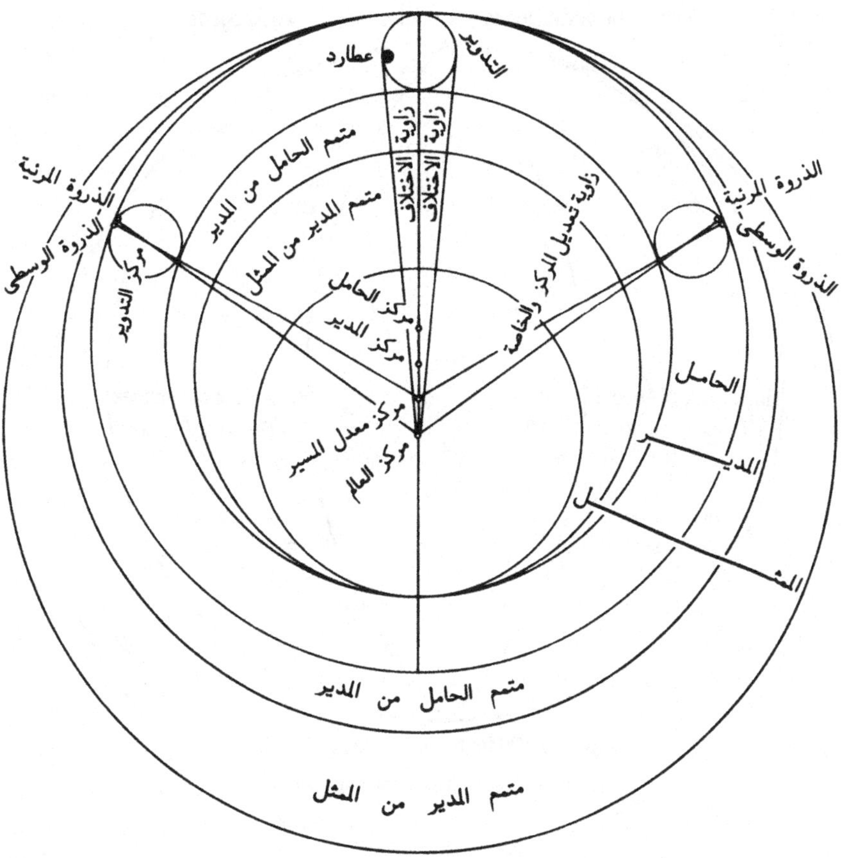

[شكل ٨]

[21] Those who limit themselves to circles set forth six orbs: the parecliptic, the inclined, the deferent of the epicycle, the equant, the deferent of the deferent's center, and the epicycle. The shape of the circuit of the epicycle center in relation to the inclined orb and to the center of the World is thus:

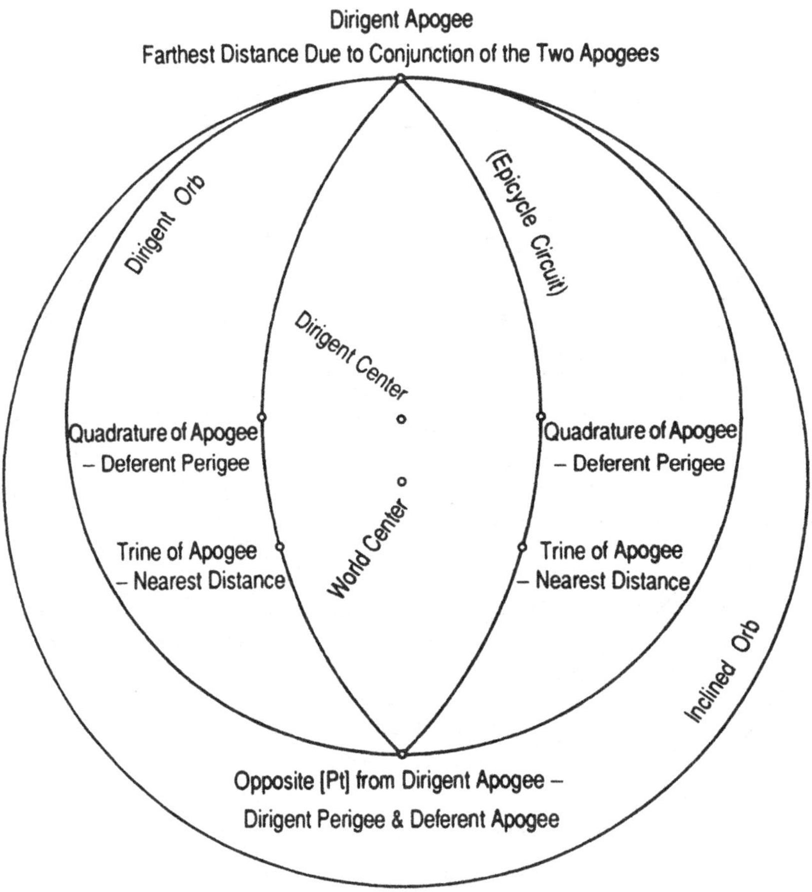

[Fig. T9]

[22] The explication of terms is analogous to what was given for the moon. A discussion of the latitudes will be forthcoming in a separate chapter.

[٢١] والمقتصر على الدوائر يُورد ستة أفلاك : المثل ، والمائل ، والحامل للتدوير ، ومعدل المسير ، وحامل مركز الحامل ، والتدوير . وشكل مدار مركز التدوير بالقياس إلى المائل وإلى مركز العالم يكون هكذا :

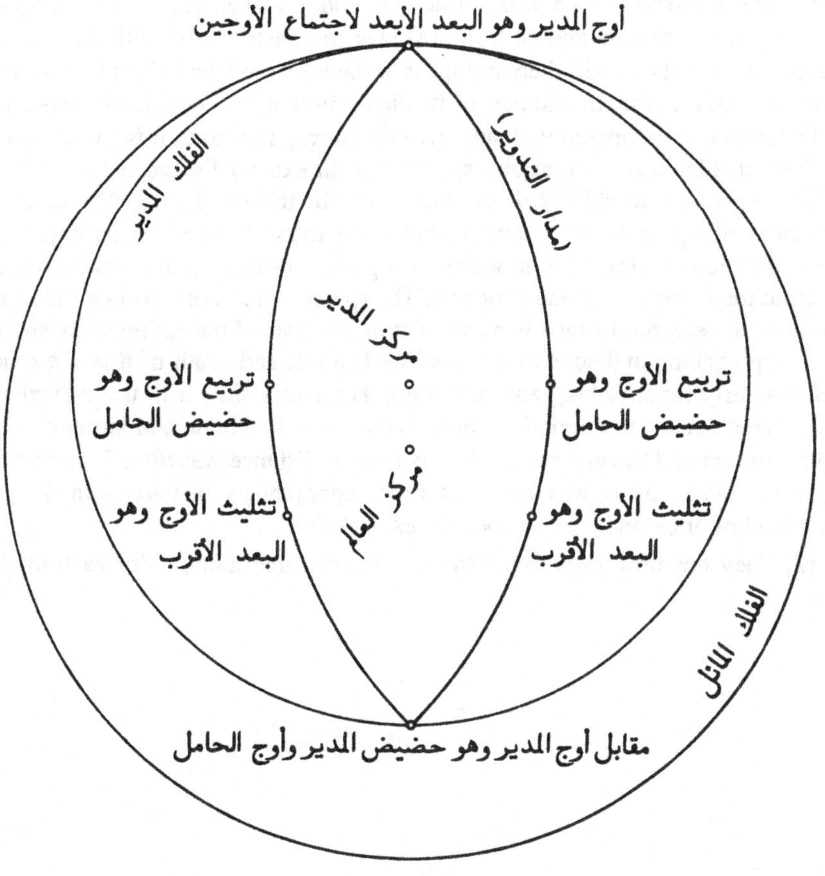

[شكل ٩]

[٢٢] وتفسير الألقاب يكون على قياس ما مرّ في القمر . والكلام في العروض يجيء في باب مفرد .

CHAPTER NINE

On the Orbs and Longitudinal Motions of the Remaining Planets

[1] They found that the three upper planets were slower in speed than the sun. So when the sun is in conjunction with one of them, it will move ahead of it; the [planet] will then appear in the east and be at its fastest speed. It then begins to slow until it stops when the sun is near its first trine or a little beyond. Thereupon the planet will retrograde and the sun will be in opposition to it at the middle part of its retrogradation. It will then stop a second time when the sun comes near the second trine or is a little / beyond it. / The [planet] will then proceed with direct motion and go from a slower to a faster speed until the sun approaches it; the planet will then disappear in the west, and the sun will be in conjunction with it at the middle part of its direct motion. If a certain circumstance of the [planet] is compared with another that corresponds to it, it is found to differ from it. Identical circumstances occurring in exactly the same parts of the ecliptic are found to shift with the movement of the fixed stars. The circumstances resulting from the nearest distance are found to occur in parts [of the ecliptic] directly facing those in which the opposite [set of circumstances] resulting from the farthest / distance / occurs. The [planet] does not travel on the same circuit as the sun but instead is north of it in one half of the ecliptic orb, sometimes approaching and sometimes receding from it, and south of it in the other half, similarly [approaching and receding]. The nodes shift with the movement of the fixed stars. They found Venus's situation in longitude and latitude to be similar to that of Mercury except that its nearest distance was directly facing its farthest distance, just as was the case for the upper planets. Its maximum elongation ahead of or behind the sun does not exceed 47°.

[2] They therefore established three orbs and three motions for each of the four.

/9/ beyond it] β, M = before it] α. /14/ distance] β, M = –α.

الفصل التاسع
في أفلاك الكواكب الباقية
وحركاتها الطولية

[١] وجدوا الكواكب الثلاثة العلوية أبطأ سيراً من الشمس . فإذا قارنتها الشمس سبقتها فظهرت مشرّقة وتكون في أسرع سيرها ، ثم تأخذ في البطء حتى إذا صارت الشمس إلى قريب من تثليثها الأول أو بَعده بقليل وقفت ، ثم رجعت وتقابلها الشمس في أواسط رجوعاتها ، ثم تقف ثانياً بقرب وصول الشمس إلى تثليثها الثاني أو / بعده / بقليل ، ثم تستقيم وتأخذ من البطء إلى السرعة إلى أن تقرب الشمس منها فتختفي مغرّبة وتقارنها الشمسُ في أواسط استقاماتها . وإذا قيست حال من أحوالها إلى نظير تلك الحال وُجدت مخالفة لها . والأحوال المتشابهة في أجزاء بأعيانها من فلك البروج تنتقل بانتقال الثوابت . ووُجدت الأحوال التي يقتضيها البعدُ الأقرب في أجزاء مقابلة للتي يقتضي فيها / البعدُ / الأبعد أضدادها . وهي لا تسير على مدار الشمس بعينه بل تكون شمالية عنه في نصف فلك البروج متقاربة إليه تارة ومتباعدة عنه أخرى ، وجنوبية عنه في النصف الآخر كذلك . والمَجازان ينتقلان انتقال الثوابت . ووجدوا الزُهَرَة شبيهة الأحوال بعطارد طولاً وعرضاً إلا أنّ أقرب أبعادها مقابل لأبعدها كما في العلوية . وغاية بعدها في الطول عن الشمس قدّاماً وخلفاً لا تتجاوز سبعاً وأربعين درجة .

[٢] فأثبتوا لكل من الأربعة ثلاثة أفلاك وثلاث حركات .

/9/ ...] M , β = قبله] α . /14/ ...] M , β = –α .

[3] The first orb is the parecliptic. For Saturn, its convex surface is contiguous with the concave surface of the eighth orb, and its concave surface is contiguous with the convex surface of Jupiter's parecliptic. The concave surface of Jupiter's parecliptic is contiguous with the convex surface of Mars's parecliptic. The concave surface of Mars's parecliptic is contiguous with the convex surface of the sun's parecliptic. The convex surface of Venus's parecliptic is contiguous with the concave surface of the sun's parecliptic, while its concave surface is contiguous with the convex surface of Mercury's parecliptic.

[4] The second is the eccentric deferent for the epicycle. It is located in the thickness of the parecliptic.

[5] The third is the epicycle, which is in the thickness of the deferent. The planet is embedded in the epicycle.

[6] The equator of the epicycle is not fixed in the plane of the deferent equator; rather, only its center is fixed therein. The deferent equator is inclined to the parecliptic equator, this inclination being fixed for the upper planets but not for Venus; its plane intersects the plane of the parecliptic equator, and there occurs on the parecliptic / a great circle whose center is the center of the World / called the **inclined orb** for the planet. It intersects the parecliptic equator in two places, which are the **head** and **tail** of that planet. The amounts of the inclinations are as we shall state in the chapter on latitudes.

[7] As for the motions, the first is the motion of the parecliptic [which moves] with the motion of the fixed stars. It manifests itself in the two [extreme] distances and / in the two nodes. /

[8] The second is the motion of the eccentric and is per day: 2 minutes for Saturn, 5 minutes for Jupiter, 31 minutes for Mars, and for Venus the same as the sun's mean motion of center. It manifests itself in the epicycle center and is therefore ascribed to it, thus being called the / epicycle's / **motion of center**.

[9] This motion is not uniform about the center of the World nor about the center of the eccentric but instead is uniform about a point removed from the eccentric center whose location is on the diameter passing through the two centers on the part that is toward the apogee and away from the eccentric center

/12/ a great circle whose center is the center of the World] β = a great circle] M = a great one] α. /17/ in the two nodes] β, M = the two nodes] α. /21/ epicycle's] β = planet's] α, M.

[٣] الفلك الأول المثل : محدبه لزُحَل يماس مقعرَ الفلك الثامن ومقعره لمحدب ممثل المُشْتَري ، ومقعر ممثل المشتري لمحدب ممثل المِرِيخ ، ومقعر ممثل المريخ لمحدب ممثل الشمس ، ومحدب ممثل الزهرة لمقعر ممثل الشمس ومقعره لمحدب ممثل عطارد .

[٤] والثاني خارج المركز الحامل للتدوير وهو في ثخن الممثل .

[٥] والثالث التدوير وهو في ثخن الحامل . والكوكب مركوز في التدوير .

[٦] ومنطقة التدوير لا تثبت في سطح منطقة الحامل بل يثبت مركزُه فيه فقط . ومنطقة الحامل مائلة عن منطقة الممثل ـ ثابتة الميل في العلوية ، غير ثابتة في الزهرة ؛ وسطحها يقاطع سطح منطقة الممثل وتَحدث في الممثل / دائرةٌ عظيمة مركزها مركز العالم / تسمى الفلك المائل لذلك الكوكب ، وتقاطع منطقةَ الممثل في موضعين هما الرأس والذنب لذلك الكوكب . ومقادير الميول على ما نوردها في باب العروض .

[٧] وأما الحركات فالأولى حركة الممثل بحركة الثوابت ، وتظهر في البعدين / وفي العقدتين / .

[٨] والثانية حركة الخارج المركز ، وهي كل يوم : لزحل دقيقتان ؛ وللمشتري خمس دقائق ؛ وللمرّيخ إحدى وثلاثون دقيقة ؛ وللزهرة مثل حركة مركز الشمس الوسطى . وهي تظهر في مركز التدوير ولذلك تُنسب إليه فتسمى حركةَ مركز / التدوير / .

[٩] وهذه الحركة لا تتشابه حول مركز العالم ولا حول مركز الخارج المركز ، بل تتشابه حول نقطة خارجة عن مركز الخارج المركز ، موضعها على القطر المارّ بالمركزين عمّا يلي الأوج من مركز

/12/ ...] دائرة عظيمة = β [عظيمة = M] α . /17/ ...] والعقدتين = β , M] α .

/21/ ...] الكوكب = β [α , M .

at a distance equal to the eccentricity. This [eccentricity] is: $(3 + \frac{1}{4} + \frac{1}{6})$ parts for Saturn, $2\frac{3}{4}$ parts for Jupiter, 6 parts for Mars, and for Venus approximately one-half the sun's eccentricity. All of these [values], which are based on the deferent radius for each planet being 60 parts, are known by observation. Twice this amount is the distance of that point from the center of the World. This point is called the **equant center**, and one may imagine a circle the size of the deferent equator with this point as center and it is called the **equant orb**.

[10] When the motion of the apogee is added to this [second] motion, there results the **mean motion** of the planet.

[11] The third motion is that of the epicycle orb. For the upper planets, it is equal to the excess of the mean motion of the sun over the mean motion of each one of the [planets]; for Venus it is 37 minutes per day. In the upper part of the epicycles, it is sequential. The starting point [for this motion] is the **mean apex**, which is aligned with the equant center as was the case for Mercury.

[12] Because the ratio of the two motions is a ratio that causes retrogradation in the epicycles, these planets will become retrograde in the segment nearest the Earth. The upper planets at the mean apices of their epicycles are always [aligned] with the mean sun. And on account of their motions on the epicycles being in the amount of the excess [of the motion] of the mean sun over their mean [motions], their distances along the epicycles from the apices are in the amount of the mean sun's distances along their orbs surrounding the Earth from the centers of their epicycles. Therefore the mean sun will be in opposition to the [planets] when they are at their mean [epicyclic] perigees during the middle of their periods of retrogradation; they will return to conjunction with [the mean sun] at the apices. As for Venus, the center of its epicycle is permanently in conjunction with the sun's center. For that reason, it will combust at the apex of its epicycle half-way through its period of direct motion and at its perigee half-way through its period of retrogradation. It will not move away from the [sun] beyond what is brought about by its epicycle radius.

الخارج على بعد مساوٍ لِما بين المركزين . وذلك لزحل ثلاثة أجزاء وربع وسدس جزء ؛ وللمشتري جزءان وثلاثة أرباع جزء ؛ وللمرّيخ ستة أجزاء ؛ وللزهرة قريب من نصف ما بين مركزي الشمس . جميع ذلك ، بحسب ما يكون نصف قطر حامل ذلك الكوكب ستين جزءاً ، عُرف بالرصد . وضعف هذا المقدار هو بعد تلك النقطة عن مركز العالم . وتسمى تلك النقطة مركز معدل المسير ، وتتوهم دائرةٌ بقدر منطقة الحامل مركزها هذه النقطة وتسمى فلك معدل المسير .

[١٠] وإذا أُضيفت حركةُ الأوج إلى هذه الحركة حصلت حركةُ وسط الكوكب .

[١١] والثالثة حركة فلك التدوير ، وهي للعلوية بقدر فضل حركة وسط الشمس على وسط كل واحد منها ، وللزهرة كل يوم سبع وثلاثون دقيقة . وهي تكون في أعالي التداوير إلى التوالي . ومبادئها الذروة الوسطى وهي محاذية لمركز معدل المسير كما في عطارد .

[١٢] ولكون نسبة الحركتين نسبة توجب الرجوع في التداوير تصير هذه الكواكب راجعة في القطعة القريبة من الأرض . والكواكب العلوية تكون في ذُرى تداويرها الوسطى مع وسط الشمس أبداً . ولكون حركاتها في التداوير بقدر فضل وسط الشمس على أوساطها تكون أبعادها في التداوير عن الذُرى بقدر أبعاد وسط الشمس عن مراكز تداويرها في أفلاكها المحيطة بالأرض . فإذن يقابلها وسطُ الشمس وهي في حضيضاتها الوسطى في أواسط أيام رجوعاتها ويعود إلى مقارنتها في الذرى . وأما الزهرة فمركز تدويرها مقارن لمركز الشمس أبداً ، ولذلك تحترق في ذروة تدويرها عند انتصاف مدة استقامتها وفي حضيضه عند انتصاف مدة رجوعها . ولا تبعد عنها فوق ما يقتضيه نصف قطر تدويرها .

[13] The size of the epicycle radius, by observation, is: 6½ parts for Saturn, 11½ parts for Jupiter, 39½ parts for Mars, and 43⅙ parts for Venus, which are based on the deferent radius being 60.

[14] It should be noted that the epicycles for Mars and Venus are very much larger than the remaining epicycles, and for this reason the difference between the sizes of their bodies at apex and perigee is greater than is the case for the remaining planets. It will be shown that the sphere of the epicycle of Mars is much larger than the parecliptic sphere of the sun including what is inside it. For this reason, they sometimes ask: how is it that Mars when it is at opposition to the sun and six zodiacal signs away from it is nearer to the [sun] than when it is at combust and in conjunction with it at the same minute [of arc]? This is so / because Mars / at combust is at the apex of its epicycle; thus the distance between them will be the diameter of its epicycle plus whatever occurs by way of the complementary bodies of their orbs. At opposition, Mars is at the perigee of its epicycle; thus the distance between them will be the diameter of the sun's parecliptic plus whatever occurs by way of the complementary bodies. This is another instance of something that is found to be strange in this science.

[15] As for the anomalies that result from these motions, they are three in number and are exactly the same as those for Mercury. The above-mentioned difficulty, which is due to the motion being uniform about a point other than the center of its equator but without there being the [difficulty] due to the alignment [of the epicycle diameter], occurs here just as it did for [Mercury].

/11/ because Mars] β = because it] α, M.

[١٣] ومقدار نصف قطر التدوير بالرصد : لزحل ستة أجزاء ونصف ؛ وللمشتري أحد عشر جزءاً ونصف ؛ وللمرّيخ تسعة وثلاثون جزءاً ونصف ؛ وللزهرة ثلاثة وأربعون جزءاً وسدس ، بحسب ما يكون نصف قطر الحامل ستين .

[١٤] وآعلم أنّ تدويري المرّيخ والزهرة أعظم جداً من سائر التدويرات ، ولذلك يكون الاختلاف بين جرميهما بالصغر والكبر في الذروة والحضيض أكثر مما يكون في سائر الكواكب . وسيتّضح أنّ كرة تدوير المرّيخ أعظم كثيراً من كرة ممثل الشمس بما فيها ، فلذلك ربما سألوا فقالوا : ما بال المرّيخ كان في مقابلة الشمس على بعد ستة بروج منها أقرب إليها منه في الاحتراق مجتمعاً معها في دقيقة واحدة ؟ وإنما يكون ذلك / لكون المريخ / في الاحتراق في ذروة تدويره فيكون البعد بينهما قطر تدويره مع ما يتّفق من متمّمات فلكيهما ، وفي المقابلة في حضيض تدويره فيكون البعد بينهما قطر ممثل الشمس مع ما يتّفق من المتمّمات . وهذا أيضاً مما يُستغرب في هذا العلم .

[١٥] وأما الاختلافات اللازمة لهذه الحركات فثلاثة وهي كما مرّ في عطارد بعينه . والإشكال المذكور بسبب كون الحركة متشابهة حول نقطة غير مركز منطقتها دون الذي بسبب المحاذاة واردٌ كما مرّ فيه .

/11/ ...] β = لكونه] M , α .

[16] This is the illustration of the orbs for each of the four planets:

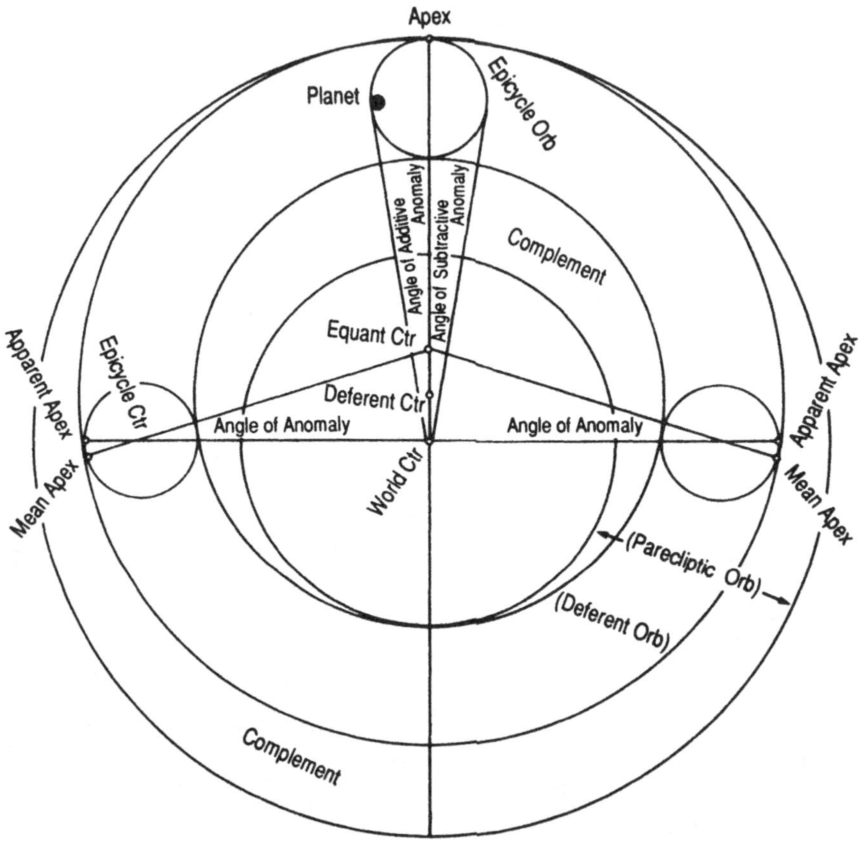

[Fig. T10]

[17] Those who limit themselves to circles set forth five orbs: the pareclīptic, the inclined, the deferent, the equant, and the epicycle. The explication of terms is analogous to what was given previously.

[18] This then is what the practitioners of this science have said concerning the orbs of the planets. The total number of solid orbs that they have established for the seven planets is 22; for those who limit themselves to circles, it is 32.

[١٦] وهذه صورة أفلاك كل كوكب من الأربعة :

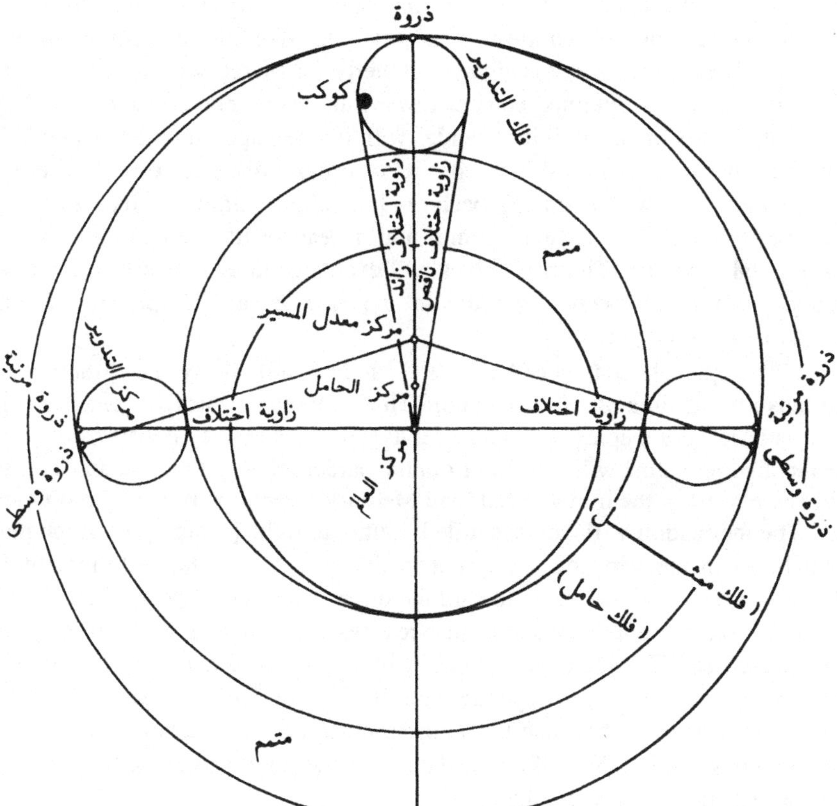

[شكل ١٠]

[١٧] والمقتصرون على الدوائر يوردون خمسة من الأفلاك : الممثل ، والمائل ، والحامل ، ومعدل المسير ، والتدوير . وتفسير الألقاب يكون على قياس ما مرّ .

[١٨] فهذا ما ذكر أهل هذا العلم في أفلاك الكواكب . وجميع الأفلاك المجسّمة التي أثبتوها للكواكب السبعة اثنان وعشرون ، وعند المقتصرين على الدوائر اثنان وثلاثون .

CHAPTER TEN

On the Latitudes of the Five Planets

[1] The maximum inclination of the inclined [orb] from the parecliptic is: 2½° for Saturn, 1½° for Jupiter, 1° for Mars, ⅙° for Venus, and (½ + ¼)° for Mercury. For the upper planets, this [inclination] is fixed in both directions, while in the case of the two lower planets it is not fixed. Rather, for Venus it is always northerly, whereas it is always southerly for Mercury. This is so because of the motion of the inclined equator toward the parecliptic equator; it will approach the [parecliptic] until it coincides with it whereupon it will separate from it in the other direction until it reaches its maximum distance from it. Then the [inclined equator] will again approach the [parecliptic] until it coincides with it once more. It will then separate from it until it reaches its maximum distance in the original direction. The two halves will exchange directions after each coinciding, the northern becoming southern and vice versa. This process is completed in each solar year.

[2] The epicycle centers of Venus and Mercury will always be at their heads / or tails / at the time of this coinciding. Thus if Venus's epicycle center is at its head and Mercury's epicycle center is at its tail and then they depart from them, the inclined [equator] will separate from the parecliptic / so that / Venus's center will come to be in the northern half and Mercury's center will be in the southern half. The inclination will increase little by little until the [centers] reach the midpoint between the two nodes whereupon the inclination attains its maximum. The two centers will then head toward the other node as the two inclinations begin to decrease until Venus's center reaches the tail and Mercury's center reaches the head. The inclined [equator] will thereupon coincide once again with the parecliptic. It will then separate from it after the [centers] depart from the nodes so that the half that had been northern becomes southern and vice versa; Venus comes to be in the half that had been southern but which, upon the arrival of its center there, is now northern,

/14–15/ or tails] β = and tails] α, M. /17/ so that] β = and] α, M.

الفصل العاشر
في عروض الكواكب الخمسة

[١] أما غاية ميل المائل عن المثل : فلزحل جزءان ونصف ، وللمشتري جزء ونصف ، وللمريّخ جزء واحد ، وللزهرة سدس جزء ، ولعطارد نصف وربع جزء . وهي للعلوية ثابتة في الجهتين ، وللسفليين غير ثابتة بل إنما تكون للزهرة أبداً شمالية ولعطارد أبداً جنوبية . وذلك يكون كذلك بسبب حركة منطقة المائل نحو منطقة المثل ، فتقرب منها حتى تنتبق عليها ، ثم تفارقها في الجهة الأخرى إلى أن تبعد عنها غاية بعدها ، ثم ترجع متقاربة إليها إلى أن تنطبق عليها ثانياً ، ثم تفارقها إلى أن تبعد عنها غاية البعد في الجهة الأولى . ويتبادل النصفان في الجهتين بعد كل انطباق بأن يصير الشمالي جنوبياً وبالعكس ، وتتم الأحوال في كل سنة شمسية .

[٢] ومركزا تدويري الزهرة وعطارد يكونان مع رأسيهما / أو ذنبيهما / وقت الانطباق أبداً . فإذا كان مركز تدوير الزهرة مع رأسها ومركز تدوير عطارد مع ذنبه ثم فارقاهما ، فارق المائل المثل / فيصير / مركز الزهرة في النصف الشمالي ومركز عطارد في النصف الجنوبي . ويزداد الميل شيئاً بعد شيء إلى أن ينتهيا إلى منتصف ما بين العقدتين فيبلغ الميل غايته . ثم يتوجه المركزان نحو العقدة الأخرى ويأخذ الميلان في التناقص إلى أن ينتهي مركز الزهرة إلى الذنب ومركز عطارد إلى الرأس فينطبق المائل ثانياً على المثل . ثم يفارقه بعد مفارقتهما العقدة فيصير النصف الذي كان شمالياً جنوبياً وبالعكس ؛ والزهرة تصير إلى النصف الذي كان جنوبياً وصار عند وصول مركزها إليه شمالياً ،

/14–15/ ...] β = وذنبيهما] M , α . /17/ ...] β = ويصير] M , α .

and Mercury comes to be in the half that had been northern but which, upon the arrival of its center there, is now southern. They will then proceed in these halves as the inclination increases until they reach the midpoint between the nodes whereupon the inclination reaches its maximum. They will thereupon head toward the original node as the inclination begins to decrease until the [centers] reach the starting point from which they departed. The result of the above is that Venus's epicycle center will always be either to the north or else on the [parecliptic] equator at the node, and Mercury's epicycle center will always be either to the south or else on the [parecliptic] equator at the node. These two motions require two movers, which have not been referred to by [our] predecessors.

[3] Saturn's head is 140° before its apogee / while its tail is 40° past its apogee. / Jupiter's head is 70° before its apogee / while its tail is 110° past its apogee. / The heads of Mars and Venus are one-quarter revolution before their apogees. Mercury's head is one-quarter revolution past its apogee. In the case of the two lower planets, the head and the tail are indistinguishable except by assumption. Their divergent [values] for the positions of the apogees and nodes, with the stipulation of date, are given in the *zījes*.

[4] As for the epicyclic equators, their diameters that pass through the apices and [epicyclic] perigees are not fixed in the planes of their inclined orbs, and they will not be in them except when, for the upper planets, the epicycle centers are at the nodes and when, for the lower planets, they are at the two distances, *i.e.* the apogee and perigee. Afterwards the apices of the upper planets will always incline in the direction of the ecliptic equator, while their [epicyclic] perigees will incline in the opposite direction. They will reach their maximum [inclinations] at the midpoint between the nodes. At that point, the angle of intersection of the plane of the epicyclic equator and the plane of the inclined equator is: 4½° for Saturn, 2½° for Jupiter, and 2¼° for Mars. Thus the inclination of Saturn when it is at its apex at the northern limit will be seen to be 26′

/10/ while its tail is 40° past its apogee] $\beta = -\alpha$, –M. /11–12/ while its tail is 110° past its apogee] $\beta = -\alpha$, –M.

وعطارد يصير إلى النصف الذي كان شمالياً وصار عند وصول مركزه إليه جنوبياً . فيسيران فيهما والميل متزايد إلى أن ينتهيا إلى منتصف ما بين العقدتين فيبلغ الميل غايته . ثم يتوجهان إلى العقدة الأولى ويأخذ الميل في التناقص إلى أن يبلغا المبدأ الذي فارقاه . ويحصل من ذلك كون مركز تدوير الزهرة دائماً إما في الشمال وإما على المنطقة مع العقدة ، وكون مركز تدوير عطارد دائماً إما في الجنوب وإما على المنطقة مع العقدة . وتحتاج هاتان الحركتان إلى مُحرّكين لم يذكرهما المتقدمون .

[٣] ورأس زحل متقدم على أوجه بمائة وأربعين درجة ، / وذنب زحل متأخر عن أوجه بأربعين درجة / ؛ ورأس المشتري متقدم على أوجه بسبعين درجة ، / وذنب المشتري متأخر عن أوجه بمائة وعشر درجات / ؛ ورأسا المريخ والزهرة متقدمان على أوجيهما بربع دور ؛ ورأس عطارد متأخر عن أوجه بربع دور . والرأس والذنب في السفليين لا يتمايزان إلا بالفرض . ومواضع الأوجات والجوزهرات مذكورة في الزيجات مع قيد التواريخ على اختلافهم فيها .

[٤] وأما مناطق التداوير فأقطارها المارّة بالذُرى والحضيضات لا تثبت في سطوح أفلاكها المائلة ولا تكون فيها إلا عند كون مراكز التدويرات / للعلوية / في العقدتين وللسفليين في البعدين ، أعني الأوج والحضيض . وبَعد ذلك تميل ذُرى العلوية أبداً إلى جهة منطقة البروج وحضيضاتُها إلى خلاف تلك الجهة ، وتنتهي إلى غاياتها في منتصف ما بين العقدتين . وزاوية تقاطُع سطح منطقة التدوير وسطح منطقة المائل حينئذ تكون لزحلَ أربعة أجزاء ونصف ، وللمشتري جزءان ونصف ، وللمرّيخ جزءان وربع . ويُرى لذلك ميل زحل في ذروته في غاية البعد الشمالي ستاً وعشرين

/10/ ...] β = −M , −α . /11–12/ ...] β = −M , −α . /19/ ...] M , β = العلوية] α .

and at the southern 28′; when it is at its [epicyclic] perigee at the northern limit, the inclination will be 33′ and at the southern 35′. The inclination of Jupiter when it is at its apex at the northern limit will be 24′ and at the southern 25′; when it is at its [epicyclic] perigee at the northern limit, the inclination will be 35′ and at the southern 38′. The inclination of Mars when it is at its apex at the northern limit will be 22′ and at the southern 27′; when it is at its [epicyclic] perigee at the northern limit, the inclination will be 3°22′ and at the southern $6\frac{1}{10}$°. As for the two lower planets: as long as Venus's center is descending on the apogee orb, its apex will incline toward the north and its [epicyclic] perigee toward the south, and in the other half the opposite will hold. As for Mercury, as long as its center is descending its apex will incline toward the south and its perigee toward the north, and in the other half the opposite will hold. The angle of intersection of the two planes upon reaching the maximum is $2\frac{1}{2}$° for Venus and $6\frac{1}{4}$° for Mercury. Thus the inclination of Venus's apex at both maximum distances will be seen to be 1°2′ and the inclination of its [epicyclic] perigee will be 6°23′. The inclination of Mercury's apex at both maximum distances will be $1\frac{3}{4}$° and the inclination of its [epicyclic] perigee will be 4°4′. This latter latitude is known as the **deviation** [*mayl*]. The upper planets have only these two latitudes.

[5] As for the lower planets, the diameter that passes through the two mean distances, which intersects the first diameter at right angles, is not fixed in the planes of the inclined orbs nor is it in the parecliptic planes except when the centers of their epicycles are at one of the nodes. After the [centers] depart from the head, the posterior endpoint of that diameter, known as the **evening**, slants toward the north, while the anterior endpoint, known as the **morning**, slants toward the south until the two centers reach the midpoint between the head and the tail. At that point is the apogee for Venus and the point opposite [the apogee] for Mercury; the slants will thereupon reach the maximum. The centers will then cross the midpoint and the slants will decrease

دقيقة وفي الجنوبي ثمانياً وعشرين دقيقة ، وفي حضيضه في غاية البعد الشمالي ثلاثاً وثلاثين دقيقة وفي الجنوبي خمساً وثلاثين دقيقة ؛ وميل المشتري في ذروته في غاية البعد الشمالي أربعاً وعشرين دقيقة وفي الجنوبي خمساً وعشرين دقيقة ، وفي حضيضه في غاية البعد الشمالي خمساً وثلاثين دقيقة وفي الجنوبي ثمانياً وثلاثين دقيقة ؛ وميل المرّيخ في ذروته في غاية البعد الشمالي اثنتين وعشرين دقيقة وفي الجنوبي سبعاً وعشرين دقيقة ، وفي حضيضه في غاية البعد الشمالي ثلاثة أجزاء واثنتين وعشرين دقيقة وفي الجنوبي ستة أجزاء وعُشر جزء . وأما السفليان فالزهرة ما دام مركزها في فلك الأوج هابطاً مالت ذروتها إلى الشمال وحضيضها إلى الجنوب وفي النصف الآخر بالعكس ، وعطارد ما دام مركزه هابطاً مالت ذروته إلى الجنوب وحضيضه إلى الشمال وفي النصف الآخر بالعكس . وزاوية تقاطُع السطحين عند المنتهي إلى الغاية للزهرة جزءان ونصف ولعطارد ستة أجزاء وربع . ولذلك يُرى ميلُ ذروة الزهرة في غايتي البعدين جزءاً ودقيقتين وميل حضيضها ستة أجزاء وثلاثاً وعشرين دقيقة ، وميلُ ذروة عطارد في غايتي البعدين جزءاً وثلاثة أرباع وميل حضيضه أربعة أجزاء وأربع دقائق . وهذا العرض يُعرف بالميل . وليس للعلوية غير هذين العرضين .

[٥] وأما في السفليين فالقطر المارّ بالبعدين الأوسطين المقاطع للقطر الأول على قوائم لا يثبت في سطوح الأفلاك المائلة ولا يكون في سطوح المثلة إلا عند كون مركزي تدويريهما مع إحدى العقدتين . وبعد مفارقتهما الرأسَ فالطرف المتأخر من ذلك القطر – ويُعرف بالمسائي – ينحرف إلى الشمال والطرف المتقدم – ويعرف بالصباحي – إلى الجنوب إلى أن ينتهيا إلى منتصف ما بين الرأس والذنب ، وهناك يكون الأوج للزهرة ومقابلته لعطارد فينتهي الانحرافان إلى الغاية . ثم يجاوز المركزان المنتصف وينتقص

until they disappear when the [centers] arrive at the tail. After leaving the tail, the reverse will occur, namely the evening endpoint will slant toward the south, while the morning endpoint slants toward the north until the [centers] complete their revolutions. The size of the angle of intersection of the plane of the epicycle with a plane passing through its center and parallel to the ecliptic equator is, at the maximum slant, $3\frac{1}{2}^\circ$ for Venus and 7° for Mercury. Therefore, the slant of Venus, due to this [angle], will appear at the apogee and perigee to be $2\frac{1}{2}^\circ$ in either direction; the slant of Mercury at the apogee will be $2\frac{1}{4}^\circ$ in either direction and at the perigee $2\frac{3}{4}^\circ$. This latitude is known as the **slant** [*inḥirāf*], the **slope** [*wirāb*], the **twist** [*iltiwā'*], and the **winding** [*iltifāf*].

[6] Each of these motions requires the establishment of a mover, which was not referred to by the Ancients. We shall state [below] what has reached us concerning this matter from the teachings of recent [astronomers], if God / Most Exalted / so wills. The quantities given in this chapter are derived from observation and calculation based on what is stated in the *Almagest*.

CHAPTER ELEVEN

An Indication of the Solution—of That Which Is Amenable to Solution—of the Difficulties Referred to Previously That Arise from the Aforementioned Motions of the Planets

[1] As for the first difficulty, which was cited [in connection] with the configuration of the moon's orbs, no statement concerning it has reached me from my predecessors. In this matter, I myself have devised what I shall now present.

[2] Let us set forth for that [purpose] a lemma, which is as follows: if two coplanar circles, the diameter of one of which is equal to half the diameter of the other, are taken to be internally tangent at a point, and if a point is taken on the smaller circle—and let it be at the point of tangency—and if the two circles move with simple motions

---

/12/ Most Exalted] β, M = –α.

الانحرافان إلى أن ينعدما عند وصولهما إلى الذنب . وبعد مفارقتهما الذنب بالعكس من ذلك ، أعني ينحرف المسائي إلى الجنوب والصباحي إلى الشمال إلى أن تتم دورتهما . ومقدار الزاوية التي عليها يقاطع سطحُ التدوير سطحاً يمرّ بمركزه ويوازي منطقةَ البروجِ إذا كان الانحرافان في الغاية ثلاثةُ أجزاء ونصف للزهرة وسبعةُ أجزاء لعطارد . فيُرى بحسبها انحراف الزهرة في الجهتين عند الأوج والحضيض جزأين ونصفاً ، وانحراف عطارد في الجهتين عند الأوج جزأين وربعاً وعند الحضيض جزأين وثلاثة أرباع . وهذا العرض يعرف بالانحراف والوراب والالتواء والالتفاف .

[٦] وكل واحد من هذه الحركات محوج إلى إثبات مُحرك لها لم يذكره القدماء ، وسنذكر ما انتهى إلينا من أقوال المتأخرين فيها إن شاء الله / تعالى / . والمقادير المذكورة في هذا الفصل مستخرجة من الرصد والحساب على ما ذكر في المجسطي .

الفصل الحادي عشر
في الإشارة إلى حلّ ما ينحلّ من الإشكالات الواردة علي حركات الكواكب المذكورة التي سبقت الإشارة إليها

[١] أما الإشكال الأول المذكور في هيئة أفلاك القمر فلم يصل إليّ فيه ممن سبقني كلام ، وأنا استنبطتُ فيه ما أذكره ههنا .

[٢] ولنقدم لذلك مقدمة هي هذه : إذا كانت دائرتان في سطح واحد قطرُ إحداهما مساوٍ لنصف قطر /الأخرى/ ، وفُرضتا متماستين من داخل على نقطة ، وفُرضت نقطةٌ على الدائرة الصغيرة ولتكن عند نقطة التماسّ ، ثم تحركت الدائرتان حركتين

/12/ ...] M , β = −α . /21/ ...] M , β = الآخر] α .

in opposite directions in such a way that the motion of the smaller [circle] is twice that of the larger so the smaller completes two rotations for each rotation of the larger, then that point will be seen to move on the diameter of the large circle that initially passes through the point of tangency, oscillating between its endpoints. Let us illustrate this with four drawings so that one may conceive from them how this [may occur], / and they are these: /

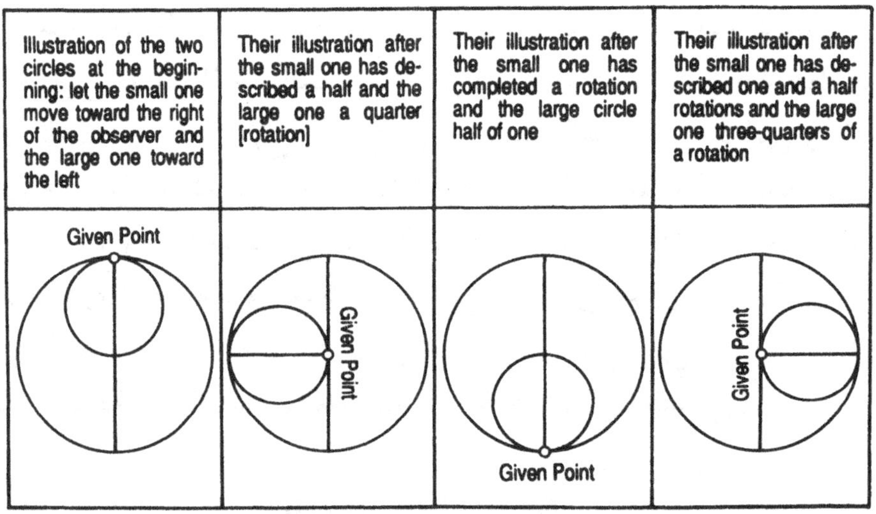

[Fig. T11]

/5/ and they are these] $\beta = -\alpha$, –M.

بسيطتين متخالفتين في الجهة على أن تكون حركة الصغيرة ضعف حركة الكبيرة فتتم للصغيرة دورتان مع دورة واحدة للكبيرة ، رُئيت تلك النقطة متحركة على قطر الدائرة الكبيرة المارّ بنقطة التماسّ أولاً مترددة بين طرفيه . ولنُصوّر لها صوراً أربعة يتوهم منها كيف ذلك ، / وهي هذه / :

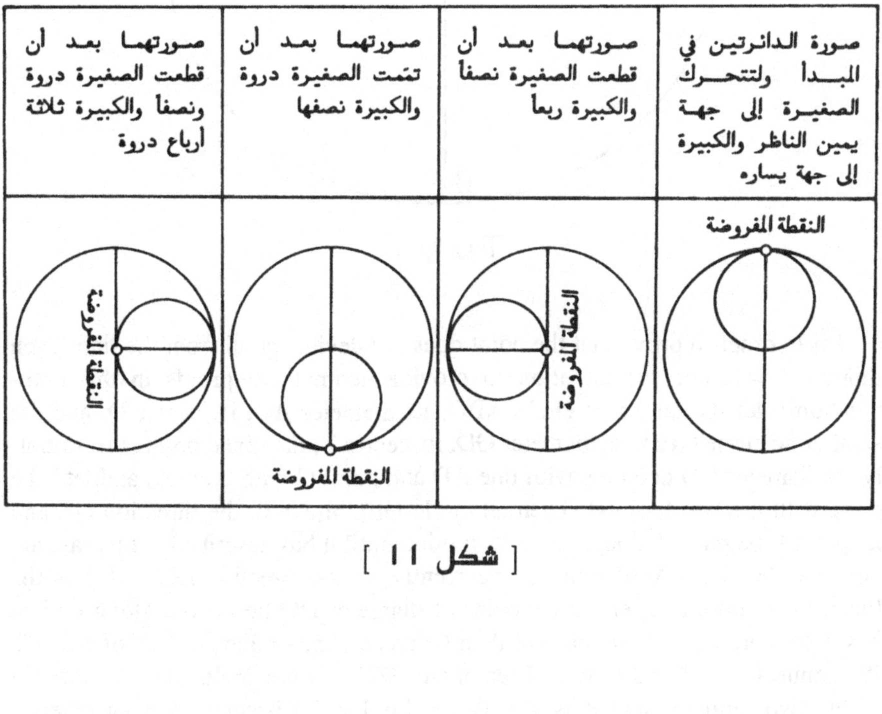

[شكل ١١]

. -α , -M = β [... /5/

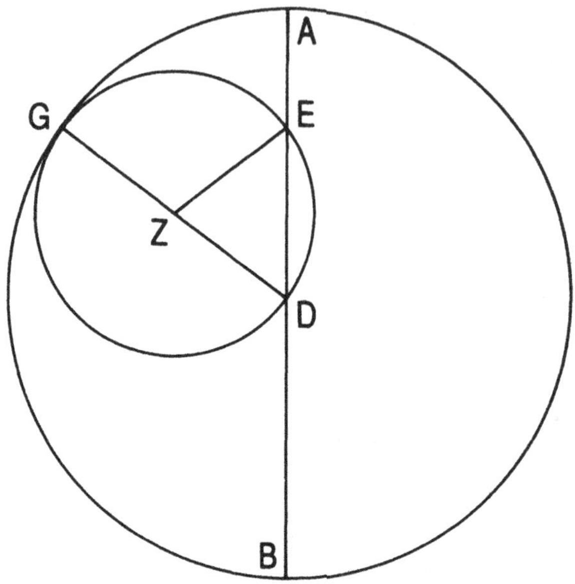

[Fig. T12]

[3] In order to prove that the point does not deviate at all from the line (even though it was not our intention to provide geometrical proofs in this compendium), let the larger be circle ABG, its diameter AB, its center D, and the smaller be circle GED, its diameter GD, its center Z; the given point is E. Initially let diameter GD coincide with line AD and point G with point A, and let E be there with the two [points]. Then let circle GED move in the direction GE and let point E be moved along due to its motion until it has described, let us say, arc GE. And let circle AGB move [concurrently] in the direction AG with half the former's motion, and let the endpoint of diameter DG be moved along until it has described arc AG, which will then be equal [*lit.*, similar] to half of arc GE. We connect E to Z and E to D. Then angle GZE is twice angle GDA on account of the two motions, and it is also twice this [angle] because it is an exterior [angle] of triangle EZD and equal to the two interior [angles] ZED and ZDE, which are equal because the two legs ZE and ZD are equal. Thus the two angles GDE and GDA are equal, and line DE coincides with line DA. Therefore point E is on diameter BA, not deviating from it, and the same will be the case in all other positions. Hence point E will continually oscillate between the endpoints of line AB and will not deviate from it.

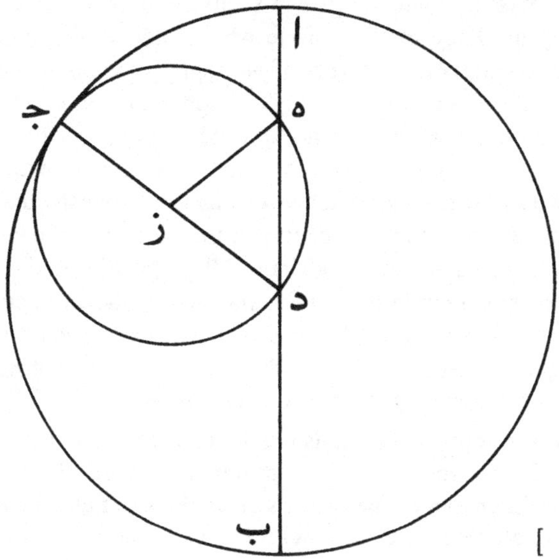

[شكل ١٢]

[٣] ولبيان أنّ النقطة لا تزول عن الخط أصلاً ـ وإن لم نكن نقصد إيراد البراهين الهندسية في هذا المختصر ـ فلتكن الكبيرةُ دائرةَ اب‍ج وقطرها اب ومركزها د ، والصغيرة دائرة ج ه د وقطرها ج د ومركزها ز والنقطة المفروضة ه . ولينطبق أولاً قطر ج د على خط اد ونقطة ج على نقطة ا ولتكن ه هناك معهما . ثم لتتحرك دائرة ج ه د في جهة ج ه ولتنتقل بحركتها نقطة ه إلى أن تقطع قوسَ ج ه مثلاً . ولتتحرك معها دائرةُ اج ب في جهة اج نصف تلك الحركة ولينتقل طرف قطر دج إلى أن يقطع قوسَ اج فهي شبيهة بنصف قوس ج ه . ونصل ه‍ز ه‍د . فزاوية ج ز ه ضعف زاوية ج د ا لأجل الحركتين ، وهي أيضاً ضعفها لكونها خارجة من مثلث ه‍زد ومساوية لداخلتي زه‍د زده المتساويتين لتساوي ساقي زه زد . فإذن زاويتا ج د ه ج د ا متساويتان وخط ده منطبق على خط دا ، فنقطة ه إذن على قطر با غير زائلة عنه وكذلك في سائر الأوضاع . فإذن نقطة ه مترددة دائماً بين طرفي خط اب غير زائلة عنه .

[4] If we wish, we may make the two circles the [inner] equators of two solid orbs. / What must be intended by the small circle is the circuit of the epicycle center and by the large circle a circle whose radius measures the same as the diameter of the small circle. / Then if we replace the point with a given sphere and we wish the diameter of the given sphere to coincide always with the diameter of the large sphere and not deviate from its position, then we assume another sphere enclosing the given one and moving with exactly the same motion and direction as the large [sphere] so as to restore the diameter to its position in the amount by which the excess of the motion of the small [sphere] over the large [sphere] causes it to deviate. / It is [also] stipulated for it that the diameter of the small circle be half the diameter of the large circle and that it always pass through its center. / Thereupon the given sphere is found to move upon a straight line that coincides with its diameter, oscillating between its two endpoints and not deviating from that coincidence.

[5] When this lemma is established, let us then place the moon's epicycle in the position of the given sphere. / Its center is the point E and its circumference has the same dimension as the moon's epicycle. / Let us assume another sphere of any appropriate thickness that encloses it and maintains its position (it should not be large lest it occupy too much space), and two [other] spheres, one of which, the deferent for these two, takes the place of the small sphere and its diameter is the same size as the eccentricity, and the other takes the place of the large [sphere] and contains all the others / and its center is that of a circle that the epicycle center touches at its farthest and nearest distances. / Its diameter is / then / twice the eccentricity. Let us then assume that the large [sphere] is within the thickness of a concentric deferent that is enclosed by the inclined [orb] in such a way that the enclosing [orb] of the epicycle, which is inside it, is tangent to the convex surface of the deferent / at the [epicyclic] apex. / Let the diameter of the deferent that passes through the point of tangency be considered fixed.

/2–4/ What must be intended...small circle] β = What must be intended by the [inner] equator of the small one is the circuit of the epicycle center within it and by the equator of the large one a circle whose radius measures the same as the diameter of the equator of the small one] α, M. /9–10/ It is [also] stipulated...through its center] β = It is [also] stipulated for it that the diameter of the equator of the small one be half the diameter of the equator of the large one and that it always pass through its center] α, M. /14–15/ Its center is the point E...moon's epicycle] β = –α, –M. /19–20/ and its center...nearest distances] β = –α, –M. /20/ then] β = –α, –M. /22/ at the [epicyclic] apex] β = near the [epicyclic] apex] α, M.

[٤] وإن أردنا جعلنا الدائرتين منطقتي فلكين مجسّمين ، وينبغي أن يكون المرادُ من / الدائرة / الصغيرة مدارَ مركز / التدوير / ومن / الدائرة / الكبيرة دائرة نصف قطرها بقدر قطر / الدائرة / الصغيرة . ثم إن جعلنا بدل النقطة كرة مفروضة ، وأردنا أن يكون قطر الكرة المفروضة دائماً منطبقاً على قطر الكرة الكبيرة غير زائل عن وضعها ، فرضنا كرة أخرى محيطة بالمفروضة متحركة مثل حركة الكبيرة بعينها وفي جهتها لترد القطر إلى وضعه بقدر ما يزيله فضل حركة الصغيرة على الكبيرة . ويشترط فيها أن يكون قطر / الدائرة / الصغيرة نصف قطر / الدائرة / الكبيرة ماراً بمركزها أبداً . وحينئذ تُرى الكرة المفروضة متحركة على خط مستقيم منطبق على قطرها ، مترددة بين / طرفيه / غير زائلة عن ذلك الانطباق .

[٥] وإذا تقررت هذه المقدمة فلنُقِم تدوير القمر مكان الكرة المفروضة / مركزه نقطة ةَ ومحيطه بالبعد الذي يكون تدوير القمر / . ولنفرض كرة أخرى محيطة به حافظة لوضعه بأي قدر من الثخن يتّفق وينبغي أن لا تكون عظيمة لئلاّ تشغل مكاناً كثيراً ، وكرتين إحداهما حاملة لهما بدل الكرة الصغيرة قطرها بقدر ما بين المركزين والأخرى بدل الكبيرة متضمنة للجميع / مركزها مركز دائرة يماسّها مركز التدوير في بعديها الأبعد والأقرب فيكون / قطرها بقدر ضعف ما بين المركزين . ثم لنفرض الكبيرة في ثخن حامل موافق المركز يحيط به المائل بحيث يكون المحيط بالتدوير الذي فيه مماساً لمحدب الحامل / عند / الذروة . وليتوهم قطر الحامل ماراً بنقطة التماسّ ثابتاً .

/2/ ...] β = منطقة] M , α . /3/ التدوير] β = التدوير فيها] M , α . الدائرة] β = منطقة] M , α . /4/ ...] β = منطقة] M , α . /9/ ...] β = منطقة] M , α . /10/ ...] β = منطقة] M , α . /12/ ...] = طرفيها] M = طرفها] α . /14–15/ ...] β = –M , –α . /19–20/ ...] β = –M , –α . /22/ ...] β = بقرب من] M , α .

[6] We now assume their motion: the epicycle with its own proper motion, the enclosing and large [spheres] with motions [such that] their rotations are completed with / a rotation / of the deferent, and the small sphere with a motion [such that] its rotation is completed with a half rotation of the deferent. We assume the deferent to move with the motion of the moon's center sequentially and the inclined [orb] with the motion of the moon's apogee countersequentially as is the case with the parecliptic.

[7] If this be the case, the diameter of the epicycle will descend as it adheres to the diameter of the large sphere. The diameter of the large [sphere] departs from its coincidence with the diameter of the deferent that passes through the tangent point cited above; its endpoint, however, will always be tangent to the circumference of the deferent, and the apex of the epicycle will be adjacent to that endpoint. The deferent will, by its motion, turn all these spheres. The epicycle center will then undergo a motion on a circuit resembling the circumference of a circle. Thus if the deferent moves through half a rotation, the epicycle will reach the other endpoint of the large sphere's diameter, which will once more coincide with the diameter of the deferent passing through the tangent point, and the epicycle's enclosing [sphere] touches the concave [surface] of the deferent near the epicyclic perigee. The epicycle will then be at its nearest distance from the center of the World, and that diameter will pass through the farthest and nearest distances. As the orbs continue to move, the epicycle will begin to ascend on the above diameter and move away from the center of the World until it reaches the farthest distance, which is the starting point that it originally departed from. The epicycle [thus] completes its circuit, which takes the place of the eccentric insofar as the inclined [orb] touches one of its points, namely the farthest distance from the center of the World, and [another] point, namely the nearest distance from it, is directly opposite from it. The difference between the farthest and nearest distances is in the amount of twice the eccentricity. Despite this, the motion [of the epicycle center]

/2/ a rotation] β, M = a completion of a rotation] α.

[٦] ثم نفرضها متحركة : أما التدوير فبحركته الخاصة به ، والمحيطة والكبيرة بحركتين يتم دورهما مع / دورة / للحامل ، والصغيرة بحركة يتم دورها مع نصف دورة للحامل . ونفرض الحامل متحركاً بحركة مركز القمر إلى التوالي ، والمائل بحركة أوج القمر إلى خلافه كالممثل .

[٧] وإذا كان كذلك نزل قطر التدوير ملازماً لقطر الكرة الكبيرة ، وزال قطر الكبيرة عن انطباق قطر الحامل المارّ بنقطة التماسّ المذكورة لكن يكون طرفه مماساً لمحيط الحامل أبداً ، وتلي الذروةُ من التدوير ذلك الطرفَ . وأدار الحاملُ بحركته جميعَ تلك الكرات ، فحدث لمركز التدوير حركةٌ على مدار شبيه بمحيط دائرة حتى إذا تحرك الحامل نصف دورة وصل التدوير إلى الطرف الآخر من قطر الكرة الكبيرة ، وانطبق قطرها ثانياً على قطر الحامل المارّ بنقطة التماسّ ، وتماسّ المحيطةُ بالتدوير مقعرَ الحامل بقرب من حضيض التدوير ؛ فكان التدوير في البعد الأقرب من مركز العالم وكان ذلك القطر ماراً بالبعدين الأبعد والأقرب . ثم تتحرك الأفلاك ويأخذ التدوير في التصاعد على القطر المذكور والتباعد عن مركز العالم إلى أن ينتهي إلى البعد الأبعد وهو المبدأ الذي فارقه أولاً . ويتم للتدوير مداره وهو يقوم مقام الخارج المركز من حيث يماسّ المائل نقطة منه هي البعد الأبعد من مركز العالم وتقابلها نقطة هي البعد الأقرب منه ، ويكون الفضل بين البعد والقرب بقدر ضعف ما بين المركزين . وتكون مع ذلك حركتُه

/2/ دورة] M , β = تمام دورة] α .

about the center of the World is uniform. As was the case previously, the apogee will meet it due to the motion of the inclined [orb]. Its illustration is thus:

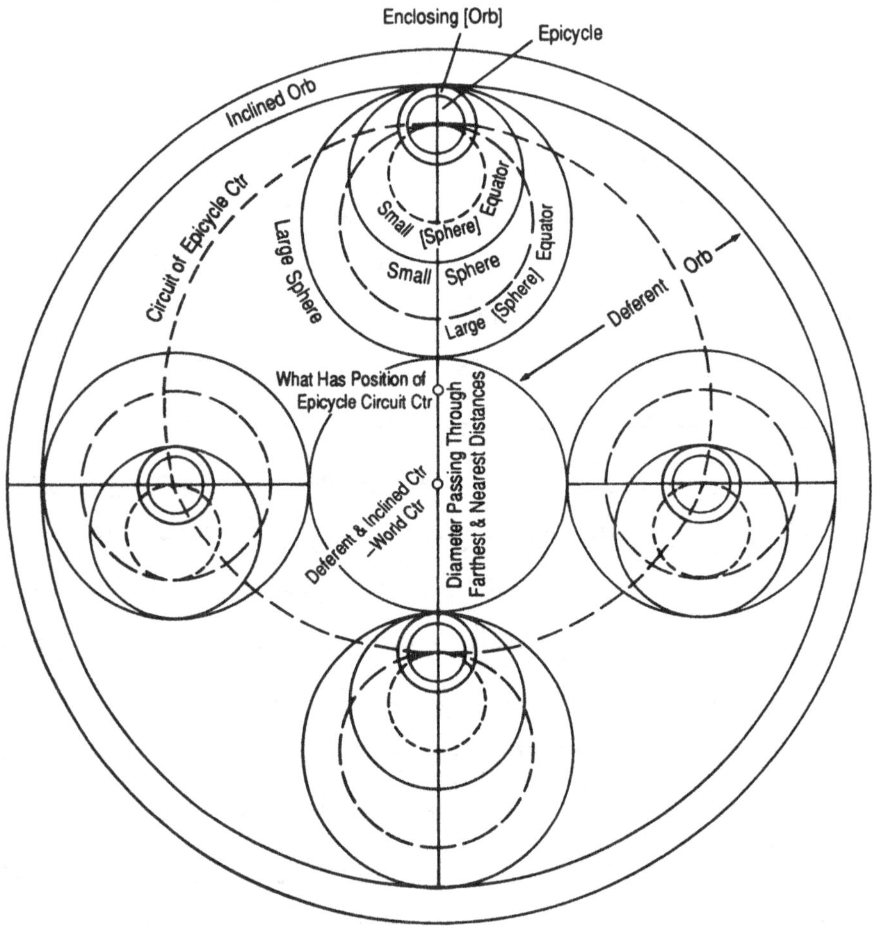

Solid spheres are in red; circles are in black.
[ed. note: solid lines = red, broken lines = black]

[Fig. T13]

[8] This then is my proposal concerning this matter. It is, however, brought about by three additional orbs over and above what has been stated before. The concentric deferent here takes the place of the eccentric orb that was used previously.

حول مركز العالم متشابهة . ويستقبله الأوجُ بحركة المائل كما كان أولاً . وصورته هكذا :

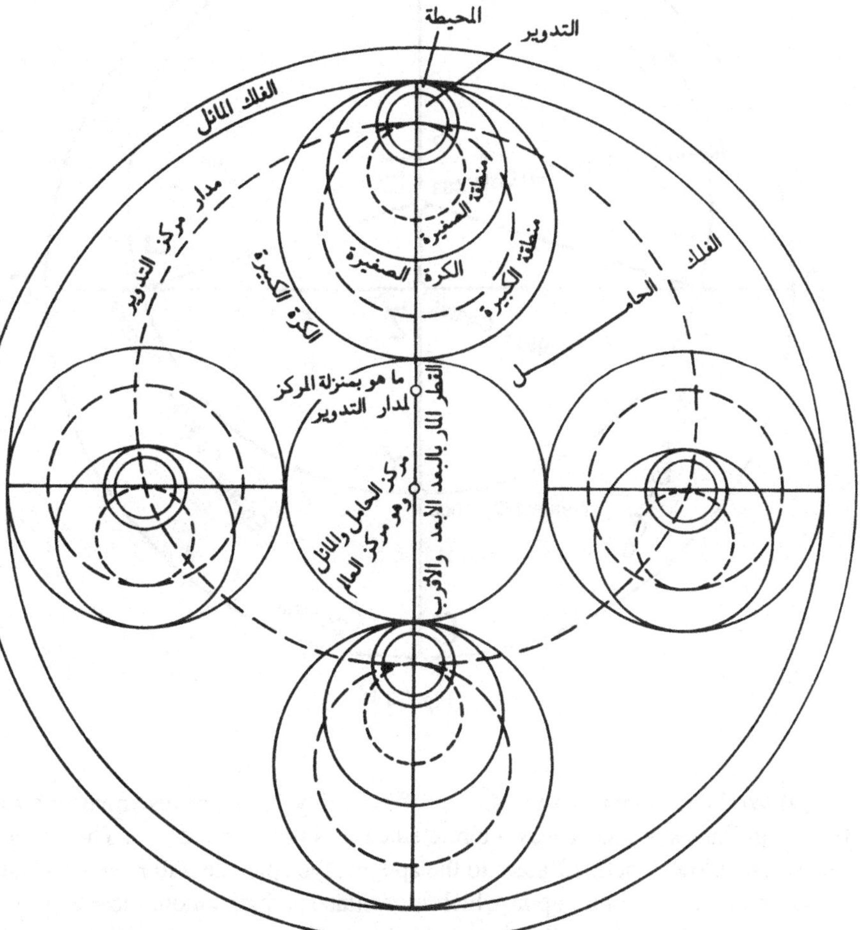

الكرات المجسمة ما هي بالحمرة والدوائر ما هي بالسواد

[شكل ١٢]

[٨] فهذا ما عندي فيه . وإنما يتم ذلك بثلاثة أفلاك زائدة على ما قيل ، ويكون الحامل الموافق المركز بدل الفلك الخارج المركز المذكور .

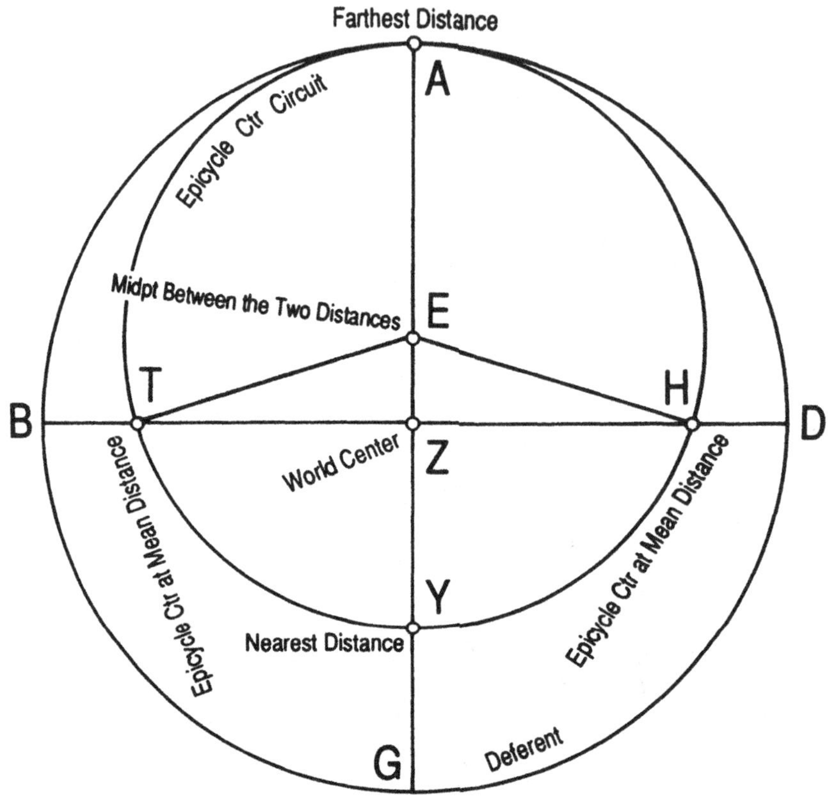

[Fig. T14]

[9] We have stated that the circuit of the epicycle center resembles a circle, but we did not say that it was a circle since it is not a true circle. The proof of this is [as follows]: at quadrature to the apogee, the epicycle will have descended along one-half of the line upon which it oscillates, which amount is equal to the eccentricity. At that point the distance between the center of the World and the center of the epicycle will come to half the distance between the farthest and nearest distances. But for the circuit to be a circle, it would be necessary for the [distance] from the midpoint between the farthest and nearest distances to the epicycle center to be that amount. Thus the aforementioned circuit

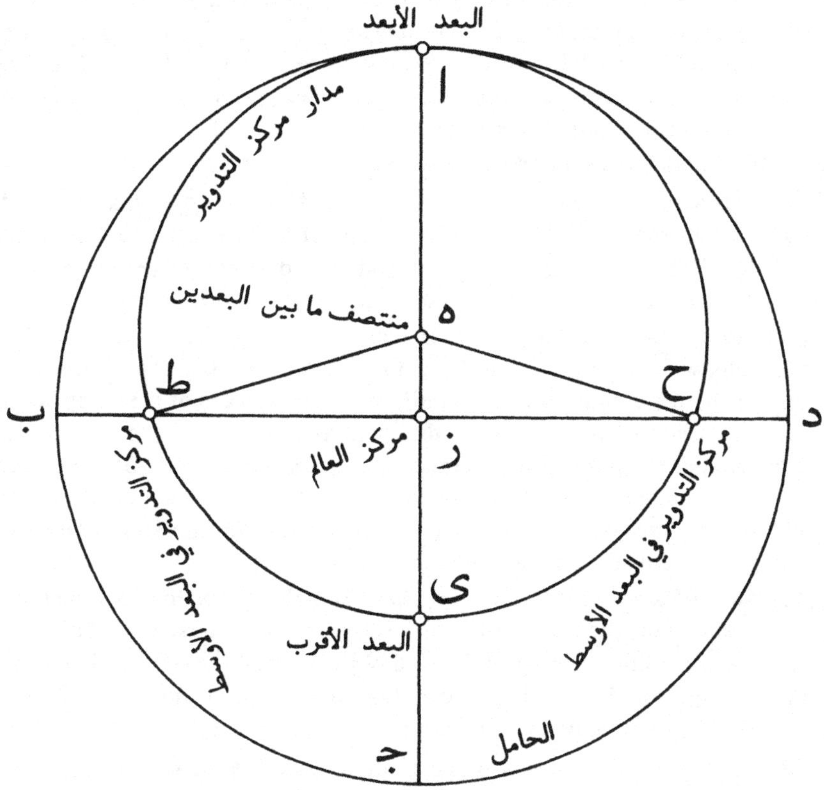

[شكل ١٤]

[٩] وإنما قلنا إنّ مدار مركز التدوير شبيه بدائرة ولم نقُل إنه دائرة لأنه لا يكون دائرة حقيقية . بيان ذلك أنّ التدوير ينزل عند تربيع الأوج نصف الخط الذي يتردد عليه ـ وهو بقدر ما بين المركزين ـ ويبقى البعد بين مركز العالم ومركز التدوير حينئذ بقدر نصف ما بين البعد الأبعد والأقرب ؛ وكان من الواجب أن يكون من منتصف ما بين البعد الأبعد والأقرب إلى مركز التدوير ذلك القدر حتى يكون المدار دائرة . فإذن المدار

is not a circle, and / the [distance] between the two mean distances on it is longer than the [distance] between the other two distances, / *i.e.* the farthest and the nearest. For this reason this method is not completely equivalent to the model they use; however, the difference between the calculations resulting from this method and the model they use does not come to $\frac{1}{6}$ of a degree. The maximum difference will occur at the midpoints between the quarter marks (*i.e.* conjunction, opposition, and quadratures). At those [points] this will be an imperceptible amount in the moon's true position.

[10] This same method may also be postulated for the upper planets and for Venus. We thus make the / diameter of the small circle / equal to the distance between the deferent and equant centers, and the / diameter of the large circle / twice that [amount]. We may then assume an eccentric orb in the thickness of the parecliptic whose center is that of the equant. We also assume the large sphere, along with what is in it, to be in the thickness of this [latter] orb so that the motion about the equant center is uniform. The distances of the epicycle center from the center of the World are the same as resulted from the [Ptolemaic] deferent without there being a difference that might disturb any of the circumstances of these planets. The difficulty concerning them may then be resolved by the addition of three spheres for each one of them, with the solid equant orb replacing the eccentric deferent of the previous [model].

[11] As for Mercury, it has not yet been possible for me to conceive how it should be done. For it is difficult to conceive what could cause the motion to be uniform about a point whereby the moved object in its motion toward and away from it is composed of multiple motions. If God Most High should enable me to accomplish this, I will append it / to that place, / God willing.

[12] Concerning the moon's point of alignment, one of the practitioners of this science has said: Another orb for the moon must be established with this point as its center so that

/1–2/ the [distance]...the other two distances] α, β = the [distance] between the two mean distances on it and the midpoint of the other two distances is longer than half the [distance] between the other two distances] M. /9/ diameter of the small circle] β = diameter of the equator of the small sphere] α, M. /10/ diameter of the large circle] β = diameter of the equator of the large sphere] α, M. /21/ to that place] β = to this place] α, M.

المذكور ليس بدائرة ، / وما بين البعدين الأوسطين فيه أطول مما بين البعدين الآخرين / ، أعني الأبعد والأقرب . ولهذا السبب لا يكون هذا الوجه مطابقاً للأصل الذي يعملون عليه مطابقة تامة لكنَّ التفاوت بين ما يُخرِجه الحساب في هذا الوجه وبين ما يخرجه الحساب على الأصل الذي يعملون عليه لا يبلغ سدس درجة ، وغايته تكون في منتصف الأرباع ، أعني الاجتماع والاستقبال والتربيعين ، وذلك غير محسوس في تقويم القمر هناك .

[١٠] وهذا الوجه بعينه يمكن أن يُفرض في الكواكب العلوية والزهرة . فنجعل قطر / الدائرة / الصغيرة بقدر ما بين مركزي الحامل ومعدل المسير ، وقطر / الدائرة / الكبيرة ضعف ذلك . ثم نفرض في ثخن المثل فلكاً خارج المركز مركزها [!] مركز معدل المسير ، ونفرض الكرة الكبيرة بما فيه[!] في ثخن ذلك الفلك حتى تكون الحركة حول مركز معدل المسير متشابهة . وأبعاد مركز التدوير عن مركز العالم كما كان يقتضيه الحامل من غير تفاوت يختل به شيء من أحوال تلك الكواكب . فينحل الإشكال فيها بزيادة ثلاث أكر في كل واحد منها ، ويكون فلك معدل المسير المجسّم بدل الخارج المركز الحامل المذكور .

[١١] وأما في عطارد فلم يتيسر لي بعدُ توهُّمُ ذلك كما ينبغي ، فإنّ توهُّمَ السبب في تشابه الحركة حول نقطة تتركب حركة المتحرك في القرب إليها والبعد عنها تركباً كثيراً متعذّر . وإن يسّر الله تعالى ذلك ألحقتُه / بذلك / الموضع إن شاء الله تعالى .

[١٢] وأما في نقطة محاذاة القمر فقد قال بعض أهل هذا العلم : ينبغي أن يُثبَت فلكٌ آخر للقمر تكون النقطة مركزه ليحاذي

/1–2/ ...] β , α = وما بين البعدين الأوسطين فيه وبين منتصف البعدين الآخرين أطول من نصف ما بين البعدين الآخرين] M . /9/ ...] β = منطقة الكرة] M , α . /10/ ...] β = منطقة الكرة] M , α . /21/ ...] β = بهذا] M , α .

the diameter of the epicycle passing through the mean apex and perigee will, by means of that orb's motion, stay aligned with its center. He did not, however, show how this motion might occur so as not to disturb the existing motions of the moon.

[13] My own opinion is that just as one may imagine the diameters passing through the apices and the perigees of the epicycles of the five planets to have latitudinal inclinations by which the planes of the equators of their epicycles are displaced from the planes which they were in at the time they had no latitude, one may likewise conceive that diameter of the equator of the lunar epicycle to have a longitudinal inclination. The equator would not, however, become displaced by it from the plane that it is in but instead its parts would undergo a displacement as if they were twisting upon themselves. So as to make this completely clear, let us imagine a line passing through the point of alignment and perpendicular to the diameter passing through the moon's centers and the point of alignment. This [line] will thus divide the deferent into two segments, one of which, the larger, is bisected by the apogee, and the second, the smaller, is bisected by the perigee. When the aforementioned diameter of the epicycle has departed from the diameter passing through the centers, after having coincided with it on the side of the apogee, its apex endpoint will incline in the counter-sequence [of the signs], while the perigee endpoint will incline in the sequence [of the signs]. This inclination will continue to increase until this diameter coincides with the perpendicular passing through the point of alignment. At that point its inclination will be at its maximum. It will thereafter begin to decrease until it disappears when this [diameter of the epicycle] coincides with the diameter passing through the centers on the side of the perigee. When it departs from this [diameter], its apex endpoint will then incline in the sequence [of the signs], whereas its perigee endpoint will incline in the counter-sequence [of the signs] until the [diameter] once more coincides with the perpendicular passing through the point of alignment. At this point its inclination will [again] be at its maximum. It will then begin to decrease until it disappears when the [diameter] reaches the starting point from which it had departed originally, *viz.* when it coincides with the diameter passing through the centers on the side of the apogee. The apex endpoint of the [diameter] will thus move in the counter-sequence [of the signs] in the larger of the two previously mentioned segments, and its maximum speed will occur at the midpoint of the segment at the apogee. In the smaller segment, [its motion will be] in

قطرُ التدوير المارّ بالذروة والحضيض الأوسطين بحركة ذلك الفلك دائماً نحو مركزه . ولم يُبيّن كيفيةَ تلك الحركة على وجه لا يُخِل بالحركات الموجودة للقمر .

[١٣] وأنا أقول : كما توهّم لأقطار تداوير الكواكب الخمسة المارة بالذُرى والحضيضات ميولٌ عرضية تَخرج بها سطوحُ مناطق تداويرها عن السطوح التي كانت فيها وقت انعدام العرض ، فليتوهم لذلك القطر من منطقة تدوير القمر ميلٌ طولي لا تخرج المنطقة به عن / سطحها الذي هي / فيه لكن يحصل لأجزائها زوال عن مواضعها كأنها تلتوي على نفسها . وليتوهم لتمام تقرير ذلك خط يمرّ بنقطة المحاذاة ويكون عموداً على القطر المارّ بمراكز القمر وبنقطة المحاذاة ، فهو يفصل الحامل إلى قطعتين إحداهما أعظم وهي التي ينصّفها الأوجُ والثانية أصغر وهي التي ينصّفها الحضيض . فالقطر المذكور من التدوير إذا فارق القطرَ المارّ بالمراكز بعد انطباقه عليه في جانب الأوج مال طرف الذروة منه إلى خلاف التوالي وطرف الحضيض إلى التوالي . ولا يزال يزيد ذلك الميل إلى أن ينطبق القطر المذكور على العمود المارّ بنقطة المحاذاة فيكون ميله حينئذ في الغاية ، ثم يأخذ في التناقص إلى أن ينعدم عند انطباقه على القطر المارّ بالمراكز من جانب الحضيض ، ثم إذا فارقه مال طرف الذروة منه إلى التوالي وطرف الحضيض إلى خلاف التوالي إلى أن ينطبق على العمود المارّ بنقطة المحاذاة ثانياً ويصير حينئذ ميله في الغاية ، ثم يأخذ في التناقص إلى أن ينعدم عند انتهائه إلى المبدأ الذي فارقه أولاً وهو كونه منطبقاً على القطر المارّ بالمراكز من جانب الأوج . فكان طرف الذروة منه / متحركاً / إلى خلاف التوالي في القطعة العظمى من القطعتين المذكورتين وغاية سرعته في منتصف القطعة عند الأوج ، وفي القطعة الصغرى إلى

/8/ ...] β = سطحه الذي هو] M , α . /23/ ...] β = متحرك] M , α .

the sequence [of the signs], and its maximum speed will occur at the midpoint of this [segment] at the perigee. [The motion of] the perigee [endpoint] in the two [segments] will be the opposite of this.

[14] This diameter will therefore need a mover, and the discussion concerning it is the same as for the movers of the aforementioned diameters of the epicycles. Let us then set forth what has been said about the latter. Ptolemy states in the *Almagest* that the endpoints of the diameters passing through the apices and the perigees of the five [planetary] epicycles revolve along small circles whose planes are perpendicular to the planes of the equators of the epicycles. Their radii are equal to the maximum inclinations of these diameters and their motions are the same as the motions of the epicycle centers upon their deferents. And just as the motions of the epicycle centers are not uniform with respect to the centers of their deferents but rather with respect to other points, so also are these motions nonuniform with respect to the centers of the aforementioned small circles but instead are uniform about other points, the ratio of whose distances from the centers of the small circles to the radii of the small circles is the same as the ratio of the distances from the deferent centers of the points at which there occurs uniform motion of the epicycle centers to the radii of the deferents. This is in order that the arcs described by the endpoints of the diameters of the epicycles on the [small circles] be the same as those described by the centers of the epicycles on the orbs upon which they move. Thus there results a displacement of the endpoints of the epicycle diameters from the planes in which they have no inclination. This will be in each direction in the amount of the radii of the aforementioned small circles, which are equal to the maximum inclinations. He states that a similar situation must also be conceived for the endpoints of the epicycle diameters, known as the morning and evening endpoints, that pass through the mean distances in the case of the two lower planets.

[15] My own opinion is that this explanation is inadequate for our purposes on three counts: (1) it does not take into account the configuration [*hay'a*] of those bodies that are the principles for these motions; (2) it compounds the difficulty that we are expending all this effort to resolve [by making] the motion uniform about a point other than the center of its circuit; (3) just as the aforementioned small circles bring about latitudinal inclinations,

التوالي وغاية سرعته في منتصفها عند الحضيض . والحضيض فيهما بالضد منها .

[١٤] فإذن هذا القطر يحتاج إلى مُحرك . والقول فيه كالقول في المحرّكات التي تُحرك أقطارَ التداوير المذكورة ، فلنورد ما قيل في ذلك . أما بطلميوس فقد ذكر في المجسطي أنّ أطراف أقطار تداوير الخمسة المارة بالذُرى والحضيضات تدور على دوائر صغار سطوحها قائمة على سطوح مناطق التداوير ، وأنصاف أقطارها بقدر غايات ميول تلك الأقطار وحركاتها مساوية لحركات مراكز التداوير على حواملها . وكما أنّ حركات مراكز التداوير لا تتشابه عند مراكز حواملها وإنما تتشابه عند نقط غيرها ، كذلك تلك الحركات لا تتشابه عند مراكز الدوائر الصغار المذكورة وإنما تتشابه حول نقط غيرها ، نسبة أبعادها عن مراكز الدوائر الصغار إلى أنصاف أقطار الدوائر الصغار كنسبة أبعاد النقط التي تتشابه عندها حركات مراكز التداوير عن مراكز الحوامل إلى أنصاف أقطار الحوامل لتكون القسي التي تقطعها أطراف أقطار التداوير منها شبيهة بما تقطعها مراكز التداوير من الأفلاك التي تتحرك عليها . وحينئذ يلزم خروج أطراف أقطار التداوير عن السطوح التي تكون فيها عديمةَ الميول في الجهتين بقدر أنصاف أقطار الدوائر الصغار المذكورة المساوية لغايات الميول . قال : ومثل ذلك ينبغي أن يتوهم في أطراف أقطار التداوير المارّة بالأبعاد الوسطى المعروفة بالصباحية والمسائية للسفليين .

[١٥] أقول : وهذا البيان ليس بمفيد فيما نحن فيه من ثلاثة أوجه : الأول أنه ليس بمشتمل على هيئة الأجسام التي هي مبادئ تلك الحركات ؛ والثاني أنه يُضعّف الإشكال الذي نجهد جميع هذا الجهد في حلّه ، وهو تشابُه الحركة عند نقطة غير مركز مدارها ؛ والثالث أنّ الدوائر الصغار المذكورة كما تُحدِث الميولَ

they also cause inclinations to occur in longitude by which the positions of the apices and the perigees alter from what they should be with respect to the points with which they are aligned.

[16] Ibn al-Haytham has published a treatise in which he discusses the bodies that produce these motions. For each epicycle he adds two spheres to account for the deviation and, in the case of the two lower planets, two other spheres for the slant. He decided to assume a sphere that surrounds the epicycle having poles at a distance from the endpoints of the diameter passing through the apex and the perigee in the two alternate directions equal to the maximum deviation of this diameter for the planet from the plane in which it has no latitude. He [further] assumes this [sphere] to have a motion equal to that assumed for the previously discussed small circle of that planet so that the endpoints of the above diameter will, due to this motion, move upon a circuit identical to the small circle with a motion that is uniform with respect to a point other than its center as was the case for the small circle. However, as a result of this [sphere's] motion, all parts of the epicycle will also move. This includes the mean diameter, which will be displaced because of this motion; thus its morning endpoint will become the evening one and vice versa. The situation will be the same for the other parts of the epicycle. It is therefore necessary to assume another sphere between this sphere and the epicycle sphere whose poles are at the endpoints of the above diameter, *i.e.* the apex and the perigee. Its motion is assumed to be exactly the same as that given for the first sphere except that it is in the opposite direction so as to return to their proper positions all parts of the epicycle that would otherwise be displaced. There will then be no remaining effect of motion from the first sphere except what had resulted from the motion of the above diameter as well as those [parts] of the plane of the epicycle equator connected with it. One may assume in exactly the same way two other spheres for each of the lower planets to account for the slant so that one will slant the epicycle's mean diameter, while the other will preserve the position of the rest of the epicycle lest the apex become the perigee and the perigee the apex. The epicycle of each of the upper planets will then come to comprise

العرضية فهي تُحدِث ميولاً أيضاً في الطول تتغير بها أوضاعُ الذُرى والحضيضات عند النقط التي تحاذيها عما يجب .

[١٦] وقد أورد ابن الهيثم مقالة ذكر فيها الأجسام التي تحرك هذه الحركات ، فزاد في كل تدوير كرتين لأجل الميل وفي السفليين كرتين أخريين لأجل الانحراف . وتقريره أن يفرض كرة تحيط بالتدوير ويكون لها قطبان بعدهما عن طرفي القطر المارّ بالذروة والحضيض في جهتين متبادلتين بقدر غاية ميل ذلك القطر لذلك الكوكب عن السطح الذي هو فيه يكون عديم الميل . ويفرض لها حركة مثل / الذي / فرضت للدائرة الصغيرة المذكورة التي لذلك الكوكب ليتحرك بحركتها طرفا القطر المذكور على مدار مثل الدائرة الصغيرة بعينها حركة متشابهة عند نقطة غير مركزها كما فرضت للدائرة الصغيرة ؛ لكن تلزم من حركتها حركة جميع أجزاء التدوير حتى القطر الأوسط ، فإنه يزول بتلك الحركة عن وضعه فيصير طرفه الصباحي مسائياً وبالعكس ، وكذلك في سائر أجزاء التدوير . فيجب لذلك أن يفرض كرة أخرى بين هذه الكرة وبين كرة التدوير قطباها طرفا القطر المذكور ، أعني نقطتي الذروة والحضيض . ويفرض لها حركة مساوية للحركة المذكورة في الكرة الأولى بعينها لكنها إلى خلاف تلك الجهة لترد جميعَ أجزاء التدوير التي كادت أن تزول عن وضعها إلى وضعها الواجب ، ولا يبقى فيها من الكرة الأولى أثرُ حركة سوى ما كان يلزم بسبب حركة القطر المذكور وما يتّصل به من سطح منطقة التدوير . وتُفرض لكل واحد من السفليين كرتان أخريان لأجل الانحراف بهذه الصفة بعينها لتُحرّف إحداهما القطرَ الأوسط من التدوير وتحفظ الأخرى وضعَ باقي التدوير كي لا تصير الذروة حضيضاً والحضيض ذروة . فيصير تدوير كل واحد من العلوية مشتملاً على

/9/ …] التي = β] M , α .

three spheres, while the epicycle of each of the lower planets will have five spheres. What Ptolemy has described will then be accomplished by establishing solid movers. Ibn al-Haytham states that one could reach the same result by assuming truncated orbs [*manāshīr*] instead of spheres, but to set forth something other than a sphere would not be appropriate for the models of this science.

[17] One should note that his intended result could also have been accomplished had he placed the poles of the first assumed sphere at a distance from the poles of the epicycle equal to the distance that he assumed between them and the endpoints of the diameter of the epicycle.

[18] Furthermore, if one adds another sphere for each of the motions and if one conceives on the surface of the sphere a situation similar to what we have described above, namely the oscillation of a point between two endpoints of a straight line, the third of the three objections I have raised against what was stated by Ptolemy would be eliminated. This was the error in longitude resulting from the longitudinal inclination that follows from [his model].

ثلاث أكر ، وتدوير كل واحد من السفليين خمس أكر ؛ ويتم ما ذكره بطلميوس بحسب إثبات المحرّكات الجسمية . وذكر ابن الهيثم أنه لو فرض بدل الأكر مناشير لتم ذلك لكنّ إثبات غير الكرة لا يصح على أصول هذا العلم .

[١٧] وآعلم أنه إن جعل قطبي الكرة التي فرضها أولاً على بعد من قطبي التدوير مساوٍ للبعد الذي فرضه بينهما وبين طرفي قطر التدوير لتم مقصوده بذلك أيضاً .

[١٨] وأيضاً إن زيد في كل حركة منها كرة أخرى وتوهّم على سطح الكرة مثل ما ذكرنا من قبل في تردد نقطة بين طرفي خط مستقيم ، زال ما ذكرتُ في الوجه الثالث من الوجوه الثلاثة التي أوردتُها على ما ذكر بطلميوس ، وهو الخلل الحادث في الطول بسبب الميل الطولي اللازم منه .

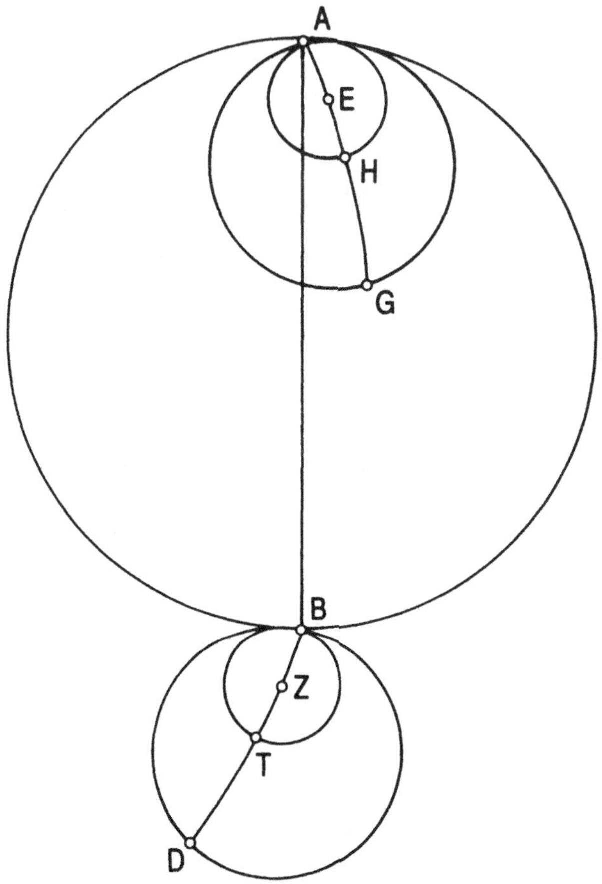

[Fig. T15]

[19] Let us present a lemma to explain this. Let the epicycle be a sphere with diameter AB. We take a great circle occurring on the epicycle that passes through the poles of the epicycle and through points A and B. Let two arcs AG and BD be on this [great circle]; we cut off AE and BZ from these [arcs] equal to half the maximum inclination in one direction in such a way that points E and Z are also endpoints of another diameter of the epicycle. We assume a sphere surrounding the epicycle, which we call the small [sphere], and we take it to move on two poles that are aligned with these two points. Points A and B will then move as a result of this motion; let their circuits [*lit.*, circuit] intersect arcs AG and BD at points H and T, which will also be at the endpoints of another diameter of the epicycle. We now take another sphere, which we shall call the large [sphere], that moves

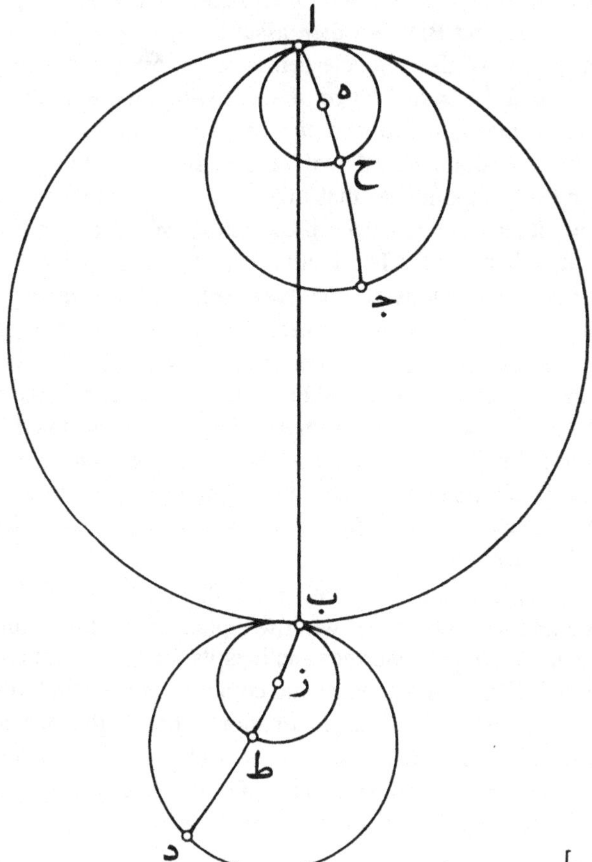

[شكل ١٥]

[١٩] ولنورد لبيان ذلك مقدمة ، فليكن التدوير كرة قطرها اب . ونفرض دائرة من العظام التي تقع على التدوير تمرّ بقطبي التدوير وبنقطتي ا ب . ولتكن قوسا اج بد منها ، ونفصل اه بز منهما مساويتين لنصف غاية الميل في إحدى الجهتين على وجه تكون نقطتا ه ز أيضاً طرفي قطر آخر للتدوير . ونفرض كرة تحيط بالتدوير ونسميها الصغيرة ، ونفرضها متحركة على قطبين محاذيين لهاتين النقطتين . فتتحرك نقطتا ا ب بحركتها ، وليقطع مدارهما قوسي اج بد على نقطتي ح ط وهما أيضاً على طرفي قطر آخر للتدوير . ونفرض كرة أخرى نسميها الكبيرة تتحرك

on two poles aligned with these latter two points. The two circuits AH and BT will then move with its motion. Let the two circuits that are tangent to [AH and BT] be circuits AG and BD. We then assume the large sphere to move with a motion equal to that of the epicycle center on its orb surrounding the Earth on which it moves and the small sphere to move with a motion opposite the former in direction and equal to twice it in amount. The result of these two motions is that the endpoints of diameter AB will continually oscillate on arcs AG and BD between their two endpoints so that they will have no inclination in longitude whatsoever in either direction from these [arcs]. When the endpoint A reaches G, endpoint B will be at D. The inclinations of these [points] are in alternate directions. If one then adds to the two [spheres] the sphere enclosing the epicycle that preserves its position so that the morning endpoint of its diameter does not become the evening endpoint and vice versa, the aforementioned motion will be accomplished, and the above error in the third of the three cited objections will be eliminated. There remains only what was stated concerning the second objection, but I have been unable to conceive a manner in which this difficulty would be eliminated. By the above method, we add three spheres for each of the epicycles of the upper planets and six spheres for each of the epicycles of the lower planets.

[20] In exactly the same way, one may also conceive of how to move the equator of the inclined orb of the two lower planets in latitude until it coincides with the parecliptic equator and then inclines its maximum amount in the opposite direction. It will thereafter return and coincide once again [with the parecliptic] and then return to its original inclination. There will not occur any longitudinal inclination with it that would cause a change to result in what was assumed in the way of longitudinal motion. On this account we would add three spheres surrounding the Earth for each of the lower planets.

[21] Furthermore one may use this method to conceive of how to move the lunar epicycle so as to produce the longitudinal inclination by which its diameter passing through the mean apex and the perigee will remain permanently aligned with the point of alignment. This will occur without that diameter being displaced from the plane of the inclined orb. We also add here three other spheres that surround the epicycle over and above what has previously been presented. This method, however, requires that [the movement of] the inclination in the sequence and in the counter-sequence [of the signs] occur in equal times.

على قطبين محاذيين لهاتين النقطتين فيتحرك مدارا اح بط بحركتها . وليكن المداران اللذان يماسّانهما مداري اج بد . ثم لنفرض الكرة الكبيرة متحركة بحركة مساوية لحركة مركز التدوير على فلكه الذي يتحرك عليه محيطاً بالأرض ، والكرة الصغيرة متحركة بحركة مخالفة لها في الجهة ومساوية لضعفها في المقدار . ويلزم من الحركتين أن لا يزال طرفا قطر اب مترددين على قوسي اج بد بين طرفيهما بحيث لا يميلان في الطول عنهما إلى أحد الجانبين أصلاً _ إذا انتهى طرف ا إلى ج انتهى طرف ب إلى د ، ويكونان بميلهما في الجهتين على التبادل . ثم إذا أضيف إليهما الكرة المحيطة بالتدوير الحافظة لوضعه حتى لا يصير طرف قطره الصباحي مسائياً ولا بالعكس تمت الحركة المذكورة وزال الخلل المذكور في الوجه الثالث من الوجوه الثلاثة المذكورة عنها . وبقي المذكور في الوجه الثاني وحده ، ولم يمكن لي توهُّم وجه يزول به ذلك الإشكال . وعلى هذا الوجه نزيد ثلاث أكر في كل واحد من تدويرات العلوية وست أكر في كل واحد من تدويري السفليين .

[٢٠] وبمثل هذا الوجه بعينه أيضاً يمكن توهُّم تحريك منطقة الفلك المائل للسفليين في العرض إلى أن تنطبق على منطقة الممثل وتميل إلى الجانب الآخر غاية ميلها ، ثم تعود فتنطبق ثانياً وترجع إلى ما كان عليه من الميل أولاً من غير أن يَحدث معه ميل طولي يُحدِث تغيّراً فيما فرض من الحركة الطولية . ونزيد بسببه ثلاث أكر محيطة بالأرض لكل واحد من السفليين .

[٢١] وأيضاً بمثل هذا الوجه يمكن توهُّم تحريك تدوير القمر على وجه يحدث الميل الطولي الذي به يصير قطره المارّ بالذروة والحضيض الوسطيين دائماً محاذياً لنقطة المحاذاة من غير أن يخرج ذلك القطر عن سطح الفلك المائل . ونزيد هناك أيضاً ثلاث أكر أخرى تحيط بالتدوير زائدة على ما مرّ ؛ إلا أنّ هذا الوجه يقتضي أن يكون الميل إلى التوالي وإلى خلافه في زمانين متساويين ،

But the existing state of affairs is different from this, for the inclination is counter-sequential as long as the epicycle center is in the larger of the two previously mentioned segments of the eccentric, while the inclination is sequential as long as it is in the smaller segment. The [epicycle center] will not traverse the two segments in equal times since its motion is uniform, while the two segments are of different sizes.

[22] The same method may also be used for trepidation and for the movement of the obliquity in latitude for the ecliptic orb if the fact of these two motions and their variability is ascertained.

[23] This then is my proposal concerning these difficulties. Perhaps God may grant a reader of this book success in devising a more complete means to solve them all or in eliminating the remaining shortcomings in what we have presented. For He is the One who inspires truth and the Guide to the straightest path.

CHAPTER TWELVE

On Parallax

[1] There may occur for the planets near the Earth, and especially for the moon, a difference between their true positions on the ecliptic orb and their apparent positions. This is because the Earth's radius has an appreciable size in relation to / the diameters of / their orbs. Thus the line extending from the center of the World to the center of the planet and from there to the ecliptic orb will terminate at its true position on it, whereas the line extending from the position of the observer to the center of the planet and from there to the ecliptic orb will terminate at its apparent position on it. The amount occurring between them is the **parallax** [*lit.*, divergence of sight] of the planet, and this is on the altitude circle since the planet's altitude circle passes through the endpoints of the two lines on the ecliptic orb. The apparent position will always be closer to the horizon. The angle occurring from the two lines at the center of the planet is called the **angle of divergence**.

/16/ the diameters of] $\beta = -\alpha$, –M.

والوجود بخلاف ذلك لأنّ الميل إلى خلاف التوالي يكون ما دام مركز التدوير في القطعة الكبرى من قطعتي الخارج المركز المذكورتين والميل إلى التوالي يكون ما دام في القطعة الصغرى ، وهو لا يقطع القطعتين في زمانين متساويين لتشابه حركته واختلافهما بالصغر والكبر .

[٢٢] وبمثل هذا الوجه أيضاً يتم كل واحدة من حركة الإقبال والإدبار وحركة الميل في جهة العرض لفلك البروج إن تحقق وجودهما واختلافهما .

[٢٣] فهذا ما عندي في هذه الإشكالات . ولعل الله يوفّق الناظر في هذا الكتاب أن يستنبط وجهاً تاماً لحلّ جميعها أو يُزيل الخلل الباقي فيما ذكرناه ، إنه ملهم الصواب والهادي إلى سواء الصراط .

الفصل الثاني عشر
في اختلاف المَناظِر

[١] قد يعرض للكواكب القريبة من الأرض وخصوصاً للقمر أن تُخالِف مواضعُها الحقيقية من فلك البروج مواضعَها المرئية ، وذلك لكون نصف قطر الأرض ذا قدر محسوس عند / أقطار / أفلاكها . فإنّ الخط الخارج من مركز العالم إلى مركز الكوكب ومنه إلى فلك البروج ينتهي إلى موضعه الحقيقي منه ، والخط الخارج من موضع الناظر إلى مركز الكوكب ومنه إلى فلك البروج ينتهي إلى موضعه المرئي منه . والقدر الواقع بينهما هو اختلاف مَنْظَر الكوكب في دائرة الارتفاع لأنّ دائرة ارتفاع الكوكب تمرّ بطرفي الخطّين في فلك البروج . ويكون الموضع المرئي إلى الأفق أقرب دائماً . وتسمّى الزاوية الحادثة على مركز الكوكب من الخطّين زاوية الاختلاف .

. −α , −M = β [… /16/

And this is its illustration:

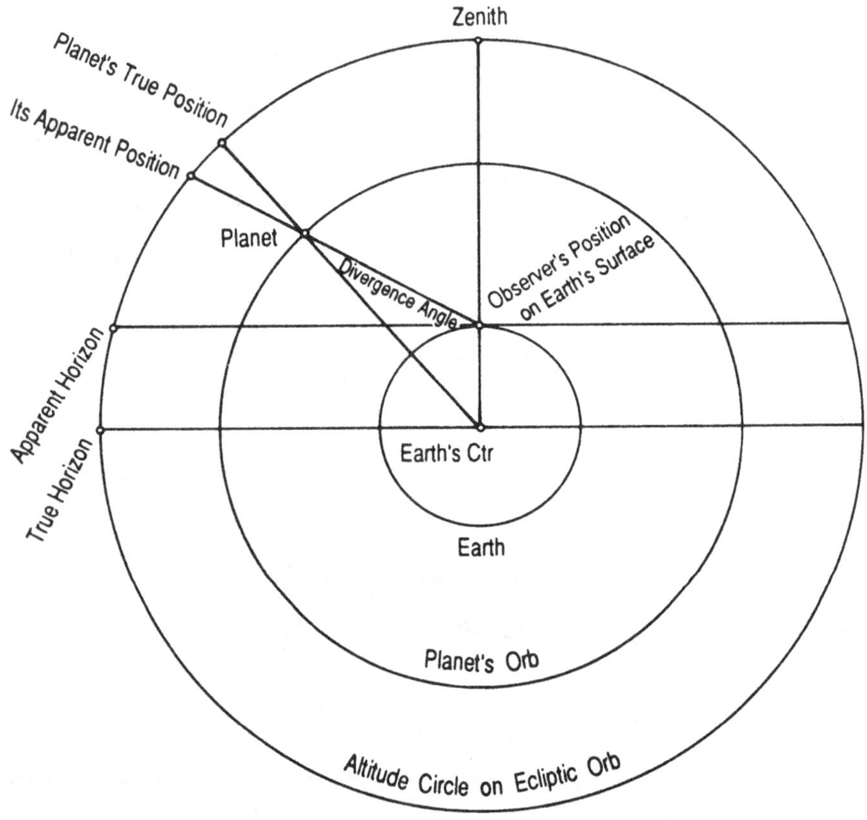

[Fig. T16]

[2] The planet does not have a parallax if it is at the zenith since the two lines are one and the same. Its divergence increases the closer it is to the horizon with the maximum occurring at its rising or setting. The visible part of the planet's orb will be less than half [of the entire orb] by the amount of the difference between the apparent and the true horizon.

[3] As for the planets that are far from the Earth, one does not perceive these divergences. The lines extending from the position of the observer and from the center of the World will be [virtually] the same since the difference with respect to the / diameters of the / orbs of these planets is so small.

/8/ diameters of the] β = [margin of M] = $-\alpha$.

وهذه صورته :

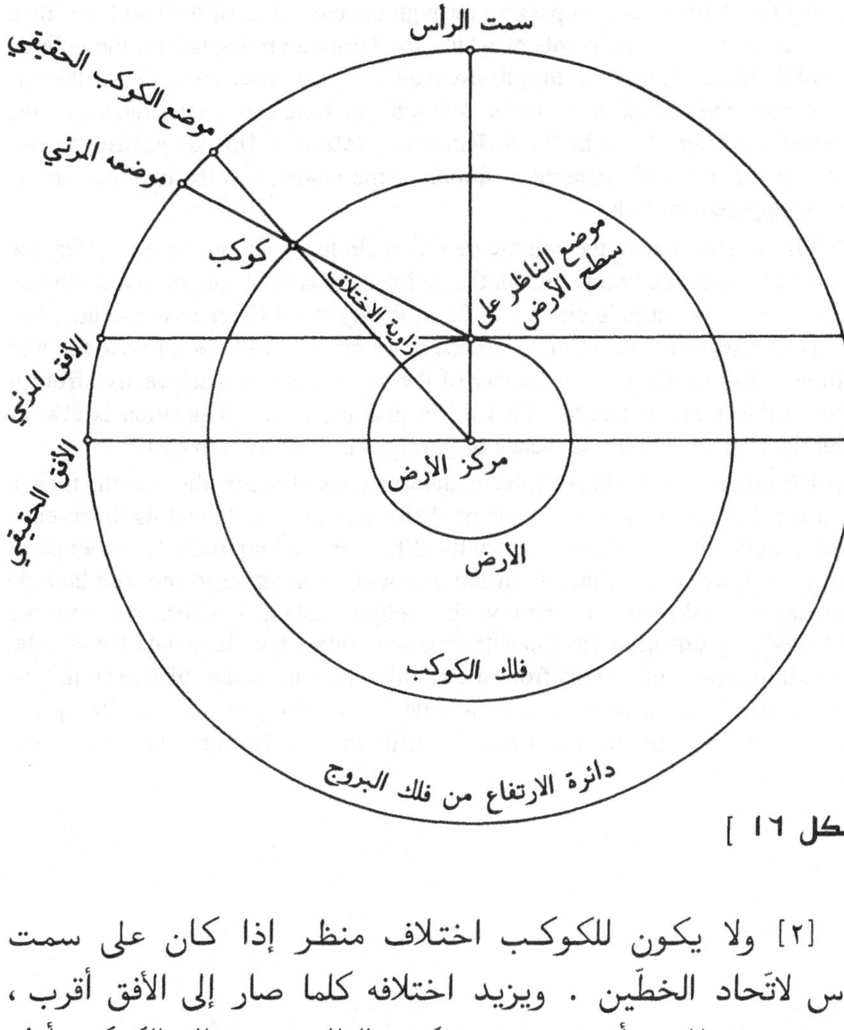

[شكل ١٦]

[٢] ولا يكون للكوكب اختلاف منظر إذا كان على سمت الرأس لاتّحاد الخطّين . ويزيد اختلافه كلما صار إلى الأفق أقرب ، وأكثره عند طلوعه أو غروبه . ويكون الظاهر من فلك الكوكب أقل من نصفه بقدر التفاوت بين الأفق المرئي والأفق الحقيقي .

[٣] وأما في الكواكب البعيدة من الأرض فلا يُحس بهذه الاختلافات . وتكون الخطوط الخارجة من موضع الناظر ومن مركز الأرض متّحدة لقلّة التفاوت بالقياس إلى / أقطار / أفلاك تلك الكواكب .

/8/ ...] β = ها M = α– .

[4] The above divergence requires that the true positions of the planet in longitude and latitude differ from their apparent positions. This is so since when we conceive two latitude circles passing through the endpoints of the two lines, then what is between the two points at which the latitude circles fall on the ecliptic orb will be the **difference in longitude**. And if the two arcs occurring on the circles between the endpoints of the lines and the ecliptic orb are not the same, the difference [in length] will be the **difference in latitude**. This is because the two points are the true and apparent positions of the planet, and the two arcs are its true and apparent latitudes.

[5] If the planet is on the ecliptic meridian circle, it will not have a difference in longitude since the two points on the ecliptic orb are one and the same, and its divergence on the altitude circle will be precisely the difference in latitude. / In other positions, / the [planet's] difference in longitude will be additive to the true position in the visible eastern quarter of the ecliptic orb and subtractive from it in the visible western quarter. This is because the apparent position is always nearer the horizon and the sequence of the signs is from west to east.

[6] Furthermore if the ecliptic equator passes through the zenith, then a planet that has no latitude will have no difference in latitude and its divergence on the altitude circle will be precisely the difference in longitude. / In other positions, / the [planet's] difference in latitude will be additive to the true latitude occurring in the direction of the invisible ecliptic pole and subtractive from the true latitude occurring in the opposite direction unless the planet and the ecliptic orb are in opposite directions from the zenith—thereupon the difference in latitude will also be additive to the true latitude. Now if the planet has no [true] latitude or if its true latitude is less than its difference [in latitude], then the direction

/12/ In other positions] β = Under other circumstances] α, M. /18/ In other positions] β = Under other circumstances] α, M.

[٤] والاختلاف المذكور يقتضي أن يكون موضعا الكوكب في الطول والعرض في الحقيقة مخالفين لموضعيهما المرئيين . وذلك لأنّا إذا توهّمنا دائرتي عرض تمرّان بطرفي الخطّين فهما إن وقعتا على نقطتين من فلك البروج كان ما بينهما اختلاف الطول . وإن اختلف[!] القوسان الواقعتان من الدائرتين بين طرفي الخطّين وبين فلك البروج كان التفاضل اختلاف العرض ، وذلك لأنّ النقطتين هما موضعا الكوكب الحقيقي والمرئي والقوسين عرضاه الحقيقي والمرئي .

[٥] وإذا كان الكوكب على دائرة وسط سماء الرؤية فلا يكون له اختلاف الطول لأنّ نقطتيه تتّحدان على فلك البروج ، ويكون اختلافه في دائرة الارتفاع اختلاف العرض بعينه . وفي غير ذلك / الموضع / يكون له اختلاف في الطول زائد على الموضع الحقيقي في الربع الشرقي الظاهر من فلك البروج ، وناقص عنه في الربع الغربي الظاهر منه . وذلك لكون الموضع المرئي إلى الأفق أقرب دائماً وكون توالي البروج من المغرب إلى المشرق .

[٦] وأيضاً إذا كانت منطقة البروج مارة بسمت الرأس فلا يكون للكوكب الذي لا عرض له اختلاف العرض ، ويكون اختلافه في دائرة الارتفاع اختلاف الطول بعينه . وفي غير ذلك / الموضع / يكون له اختلاف في العرض زائد على العرض الحقيقي الكائن في جهة القطب الخفي من قطبي فلك البروج ، ناقص من العرض الحقيقي الكائن في خلاف تلك الجهة ؛ اللّهم إلاّ أن يكون الكوكب وفلك البروج في جهتين متخالفتين عن سمت الرأس فإنّ اختلاف العرض هناك يكون أيضاً زائداً على العرض الحقيقي . فإن كان الكوكب عديم العرض أو كان عرضه الحقيقي أقل من اختلافه فجهة

/12/ ...] β = الوضع] M , α . /18/ ...] β = الوضع] M , α .

of the difference [in the first case] or the direction of the excess of the difference over the true latitude [in the second case] will be toward the invisible pole, likewise for exactly the same reason as before.

[7] By observation of the moon's parallax, one may find its distances from the Earth, as will later be explained.

[8] As far as the solar parallax is concerned, it is imperceptible even though calculation / yields / a small divergence not exceeding 3 minutes. The divergence for the two lower planets has not been found because of the difficulty in obtaining their true positions in longitude and latitude.

CHAPTER THIRTEEN

On the Variation in the Moon's Illumination and on Lunar and Solar Eclipses

[1] The variation in shape of the moon in accordance with its changing position vis-à-vis the sun indicates that its body is dark, thick and smooth; due to its thickness, it receives light from the sun, which is reflected from it due to its smoothness. Thus approximately half of its spherical body will always be illuminated. A great circle (or one nearly so) on its body separates the illuminated and the dark, and another great circle (or one nearly so) separates that which is apparent to the observer from that which the light of the eye does not reach. The two circles coincide at conjunction when the dark half is discerned, this state being new moon, and at opposition when the illuminated half is seen, which is full moon. At other positions [the two circles] will intersect: at the quadratures at right angles, whereupon that quarter facing the sun of the half that is toward us is illuminated; otherwise at acute and obtuse angles, whereupon in the first and last quarters what is toward the sun will be the part on the side of the acute angle and thus is crescent-shaped, while in the other two quarters it will be the part that is on the side of the obtuse angle and thus is oval-shaped.

/6/ yields] β = yields for it] α, M.

الاختلاف أو جهة فضل الاختلاف على العرض الحقيقي هي جهة القطب الخفي للعلّة المذكورة أيضاً بعينها.

[٧] وبرَصَد اختلاف منظر القمر يُتوصّل إلى معرفة أبعاده من الأرض كما يجيء بيانه.

[٨] وأما اختلاف منظر الشمس فغير محسوس لكنّ الحساب / يُخرج / اختلافاً قليلاً لا يزيد على ثلاث دقائق. والسفليان لا يوقف على اختلافهما لتعذّر الوقوف على مواضعهما الحقيقية في الطول والعرض.

الفصل الثالث عشر
في اختلاف نور القمر وفي الخسوف والكسوف

[١] اختلاف تشكّلات القمر بحسب اختلاف وضعه من الشمس يدلّ على أنّ جرمه مُظلِم كثيف صقيل يقبل من الشمس الضوء لكثافته وينعكس عنه لصقالته. فيكون أبداً المُضيء من جرمه الكري قريباً من نصفه. وتفصل بين المضيء والمظلم دائرة عظيمة أو قريبة من العظيمة على جرمه، وتفصل بين المرئي منه عند الناظر وبين ما لا يصل إليه نور البصر أيضاً عظيمة أو قريبة منها. والدائرتان تتطابقان في الاجتماع، ويكون المُبصَر منه النصف المظلم وتلك الحالة هي المُحاق، وفي الاستقبال، ويكون المبصر منه النصف المضيء وهو البَدْر. وتتقاطعان في سائر الأوضاع: أما في التربيعين فعلى زوايا قائمة، ويكون الربع الذي يلي الشمسَ من النصف الذي يلينا مضيئاً؛ وفي غيرهما على زوايا حادّة ومنفرجة، والذي يلي الشمسَ في الربعين الأوّل والأخير هو القسم الذي يلي الزاوية الحادّة فيكون هِلالي الشكل وفي الربعين الآخرين هو القسم الذي يلي الزاوية المنفرجة فيكون إهْلِيلَجِي الشكل.

/6/ ...] β = يخرج له] α , M .

[2] The Earth is also a thick, dark, spherical body that blocks the light of the sun and so it has a shadow. When it is aligned with the two luminaries at the time of opposition, it blocks the light of the sun from the moon and the moon will fall in its shadow. The moon will thereupon be eclipsed, which can be observed if it is night. The following is an illustration of the **lunar eclipse**:

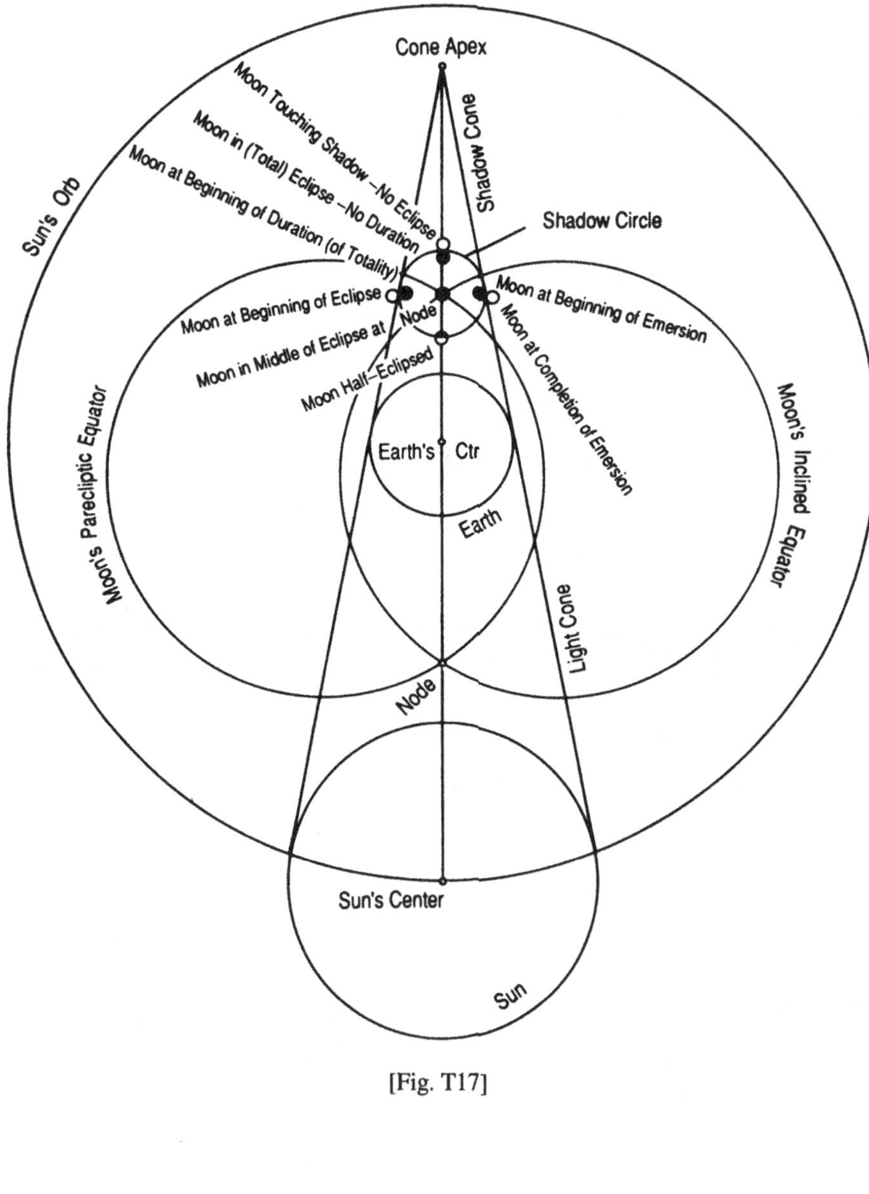

[Fig. T17]

[٢] والأرض أيضاً جسم كثيف مظلم كري يحجب نور الشمس فيقع له ظلّ . وإذا صارت مقاطرة للنيّرين وقت الاستقبال حجبت نور الشمس عن القمر ووقع القمر في ظلّها ، فانخسف القمر ورُئيَ إن كان ذلك ليلاً . وهذه صورة الخُسوف :

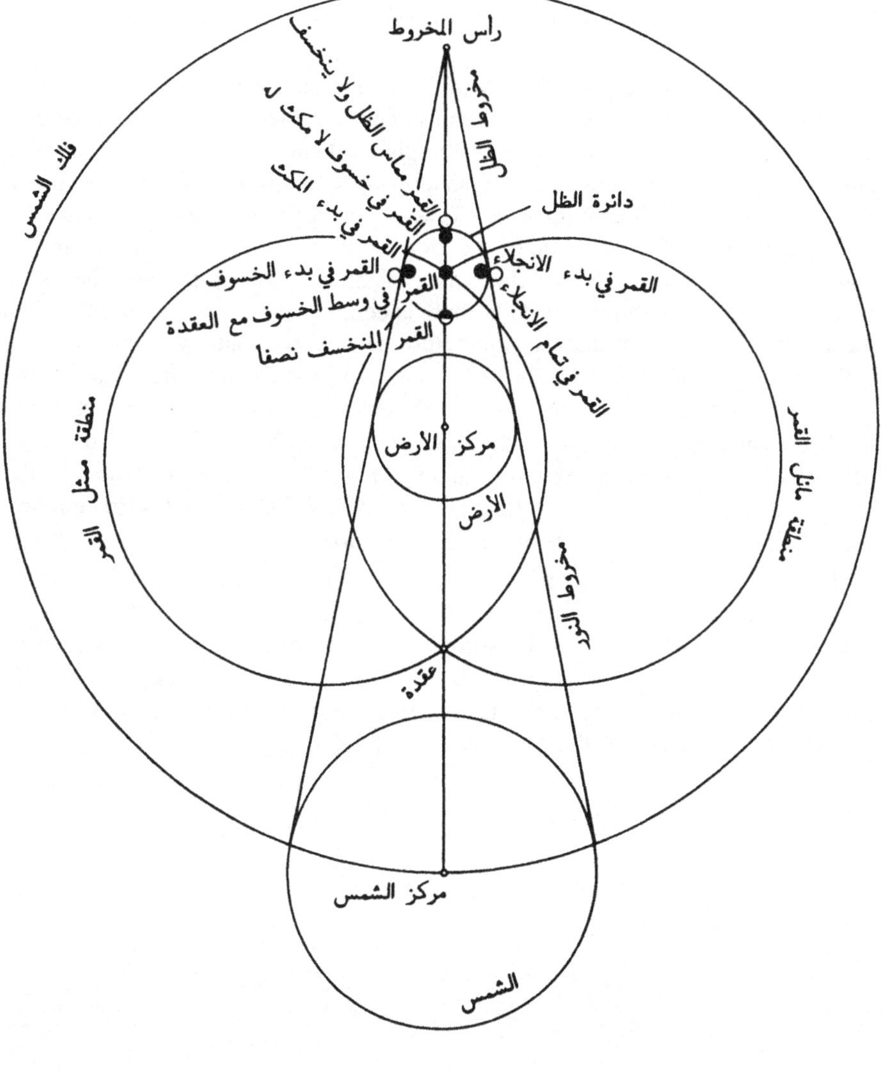

[شكل ١٧]

[3] The farther away the moon is from the Earth, the shorter in duration is its eclipse. It was concluded from this that the shadow narrows with increasing distance from the Earth, which [in turn] indicates that the sun is larger than the Earth. For if the sun were smaller than the Earth, then the shadow would widen with increasing distance from the Earth and thus the moon's duration during a [lunar] eclipse would increase as its distance from the Earth increased, which is contrary to what is found [to be the case]. If the [sun] were the same [size] as the Earth, the shadow would be cylindrical and the duration at all distances would be the same which is also not the case. It is thus clear that the sun is larger than the Earth, that the Earth's shadow is in the shape of a circular cone that vanishes to a point, and that the moon is smaller than the Earth since the latter's shadow, which becomes much smaller than the [Earth] upon [reaching] the moon, conceals it.

[4] The center of the **shadow cone** is always on the [plane of the] ecliptic equator since the sun is always on it and since the center of the Earth is the center [of this equator]. If one imagines the plane of the apparent lunar body as a circle extending until it intersects the shadow cone, it will produce a circle parallel to its base called the **shadow circle** whose center is on [*i.e.* in the plane of] the equator. Now if the latitude of the moon at the time of opposition is greater than the radius of its disk plus the radius of the shadow circle, then the moon will not undergo an eclipse; if its latitude is equal to them, the moon will touch the shadow but will not undergo an eclipse; if it is less than them and equal to the radius of the shadow, the shadow circle will pass through the center of the lunar disk and half its diameter will be eclipsed; if it is equal to the excess of the shadow radius over the radius of the moon, the entire moon will be eclipsed and its surface will touch the shadow circle but it will have no duration in the eclipse; and if it is less than [this excess], it will be eclipsed with duration depending on the extent to which it is in the shadow.

[5] The eclipse limits have been measured as 12° in terms of the distance of the moon from one of the two nodes; for when the moon crosses this limit, its latitude will exceed the two radii. And just as the shadow circle varies according to distance, the circle of the lunar disk will likewise vary according to distance. The two have been compared and the shadow circle's diameter was found to be two and three-fifths times the diameter of the lunar disk

[٣] وكلما كان القمر أكثر بعداً من الأرض كان خسوفه أقلَّ مَكْثاً ، فاستُدلّ بذلك على أنّ الظل يستدق بازدياد بعده من الأرض . ويدلّ ذلك على كون الشمس أكبر من الأرض . وذلك لأنّ الشمس لو كانت أصغر من الأرض لكان الظلّ يستغلظ بازدياد بعده من الأرض ، فكان كلّما زاد بعد القمر من الأرض زاد مكثه في الخسوف على ضدّ ما يوجد ؛ ولو كانت مساوية للأرض لكان الظلّ أُسطوانياً والمكث في جميع الأبعاد متساوياً وليس أيضاً كذلك . فإذن ظهر أنّ الشمس أكبر من الأرض ، وأنّ ظلّ الأرض على هيئة مخروط مستدير ينعدم على نقطة ، وأنّ القمر أصغر من الأرض لستر ظلّها الذي صار أصغر منها كثيراً عند القمر إيّاه .

[٤] ومركز مخروط الظلّ يكون دائماً على منطقة البروج لكون الشمس دائماً عليها وكون مركز الأرض مركزاً لها . وإذا توهّم سطح جرم القمر المرئي كدائرة خارجاً إلى أن يقطع مخروط الظلّ أحدث دائرةً موازية لقاعدته تُسمى دائرة الظلّ ويكون مركزها على المنطقة . فإن كان عرض القمر وقت الاستقبال أكثر من نصفي قطر صفحته وقطر دائرة الظلّ لم يقع للقمر خسوف ؛ وإن كان عرضه مساوياً لهما ماسّ القمرُ الظلّ ولم يقع له خسوف ؛ وإن كان أقلّ منهما وكان مساوياً لنصف قطر الظلّ مرّت دائرة الظلّ بمركز صفحة القمر وانخسف نصف قطره ؛ وإن كان مساوياً لفضل نصف قطر الظلّ على نصف قطر القمر انخسف القمر كلّه وماسّ سطحه دائرةَ الظلّ فلم يكن له مكث في الخسوف ؛ وإن كان أقلّ من ذلك انخسف ومَكَثَ بحسب ما يقع في الظلّ .

[٥] وإنّما قُدّرَ حدود الخسوف باثني عشر جزءاً من بعد القمر عن إحدى العقدتين لأنّ عرضه إذا جاوز هذا الحدّ زاد على نصفي القطرين . وكما أنّ دائره الظلّ تختلف بحسب الأبعاد فدائرة صفحة القمر أيضاً تختلف بحسب الأبعاد . وقد قيس بينهما فوجد قطر دائرة الظلّ مثلي قطر صفحة القمر وثلاثة أخماسه

at all distances. Each of the diameters of the two luminaries and their bodies is divided into 12 equal parts called **digits**, those of the diameters being designated as absolute, while those of the bodies as adjusted.

[6] Since a lunar eclipse is possible at a distance of less than 12° from one of the nodes, then if opposition occurs after the crossing of the node with an eclipse taking place at the edge of the [eclipse] limit, and then opposition occurs five months later at the edge of the eclipse limit before the other node is reached, it will be possible for the moon to be eclipsed once again. This is due to the counter-sequential motion of the node and to its meeting up with the position of the eclipse. If opposition for an eclipse occurs before reaching the first node at the edge of the limit and the other opposition after crossing the second node seven months later, it is not possible [for the moon] to be in the eclipse limit on account of the node, by its counter-sequential motion, having crossed beyond the range necessary for an eclipse. Thus two lunar eclipses cannot be seven months apart. It is most frequently the case that they are six months apart.

[7] Furthermore, if the moon blocks the light of the sun from the observer due to its falling on the line extending from the eye to the sun, the sun will be seen eclipsed, without light; for the obstruction is dark and the [part] of it facing us at that time is not illuminated. This will be during an apparent rather than a true conjunction occurring in the daytime. For this reason parallax is taken into account for solar but not for lunar eclipses, and it is possible for a solar eclipse to occur with respect to one group of people but not another. In order for a solar eclipse to take place, it is necessary that the apparent lunar latitude (that is, adjusted in latitude for parallax) at the time of the apparent conjunction (that is, adjusted in longitude for parallax) be less than the radii of the disks of the two luminaries. For if it is equal to them, the [disks] will touch but the sun will not be eclipsed, and if it is greater than them, then all the more reason [for it not to be eclipsed]; if it is less than them, the eclipse will occur commensurate with that amount.

في كلّ بعد . ويُجزّأ كلّ واحد من قطري النيّرين وجرميهما إلى اثني عشر جزءاً متساوية تسمى الأصابع ، وتقيّد القطرية بالمطلقة والجرمية بالمعدّلة .

[٦] ولمّا كان الخسوف على بعد أقلّ من اثنتي عشرة درجة من إحدى العقدتين ممكناً ، فإن كان الاستقبال بعد التجاوز عن العقدة ووقع خسوف على طرف الحدّ ، ثم وقع استقبال بعد خمسة أشهر قبل الانتهاء إلى العقدة الأخرى على طرف حدّ الخسوف ، أمكن أن ينخسف القمرُ مرة ثانية . وذلك لحركة العقدة إلى خلاف التوالي واستقبالها لموضع الخسوف . وإن كان الاستقبال الخسوفي قبل الوصول إلى العقدة الأولى على طرف الحدّ ، والاستقبال الآخر بعد التجاوز عن العقدة الثانية بعد سبعة أشهر ، لم يمكن أن يقع في حدّ الخسوف لمجاوزة العقدة بحركته[!] إلى خلاف التوالي عن المقدار المقتضي للخسوف ؛ فلا يكون خسوفان بينهما سبعة أشهر . وأمّا بعد ستة أشهر فأكثريّ الوقوع .

[٧] وأيضاً إذا حجب القمرُ نور الشمس عن الناظرين لوقوعه على الخط الخارج من الأبصار إلى الشمس رُئيت الشمس منكسفة عديمة النور ؛ فإنّ الحاجب مظلم والذي يلينا منه غير مضيء في ذلك الوقت . وذلك يكون في الاجتماع الواقع نهاراً المرئي لا الحقيقي ؛ ولذلك يعتبر اختلاف المناظر في الكسوفات دون الخسوفات ، ويمكن أن يقع كسوف بالقياس إلى قوم دون قوم . وينبغي أن يكون العرض المرئي للقمر ، أعني المعدّل باختلاف المنظر في العرض ، وقت الاجتماع المرئي ، أعني المعدّل باختلاف المنظر في الطول ، أقلّ من نصفي قطري صفحتي النيّرين حتى يقع كسوف . فإنّه إن ساواهما تماسّا ولم تنكسف الشمس ، وإن كان أكثر منهما فبالأولى ، وإن كان أقل منهما يقع الكسوف بقدر ذلك .

And this is the illustration of a **solar eclipse**:

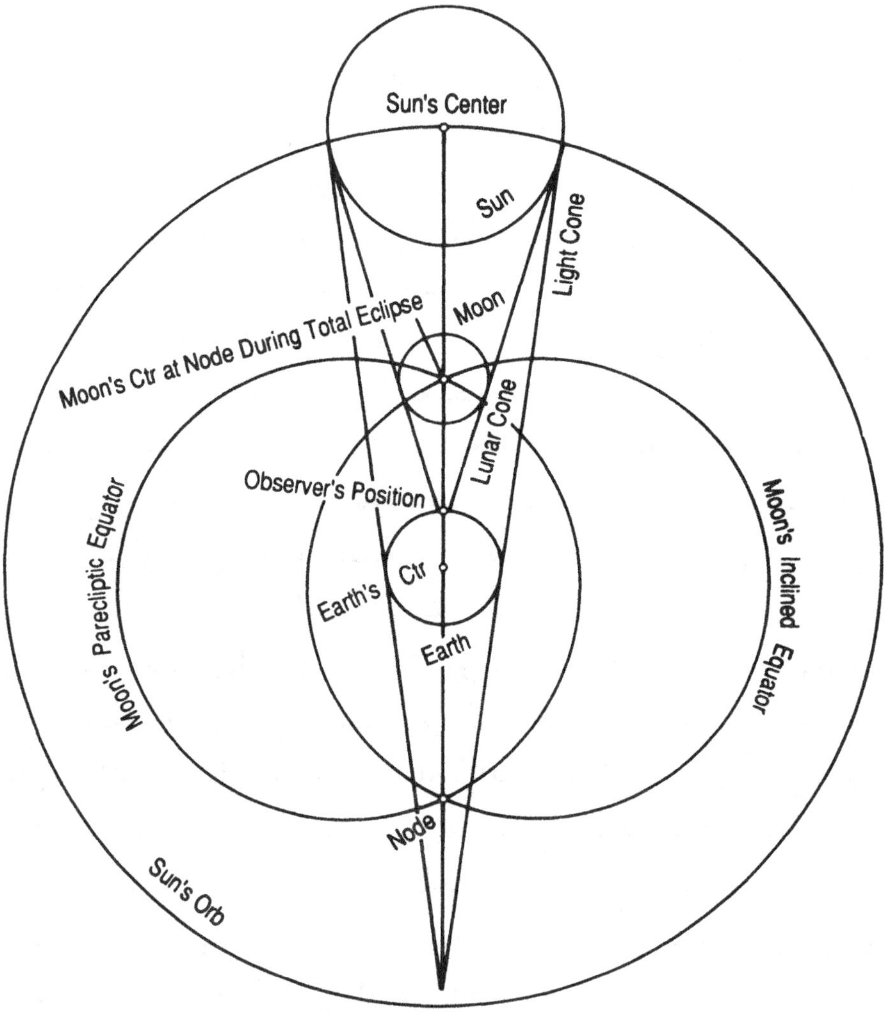

[Fig. T18]

[8] The solar diameter between its two [extremal] distances was found [to vary] from 31 to 34 minutes, while the lunar diameter was found [to vary] from 29 to 36 minutes. Thus if the two centers fall on the line

وهذه صورة الكسوف :

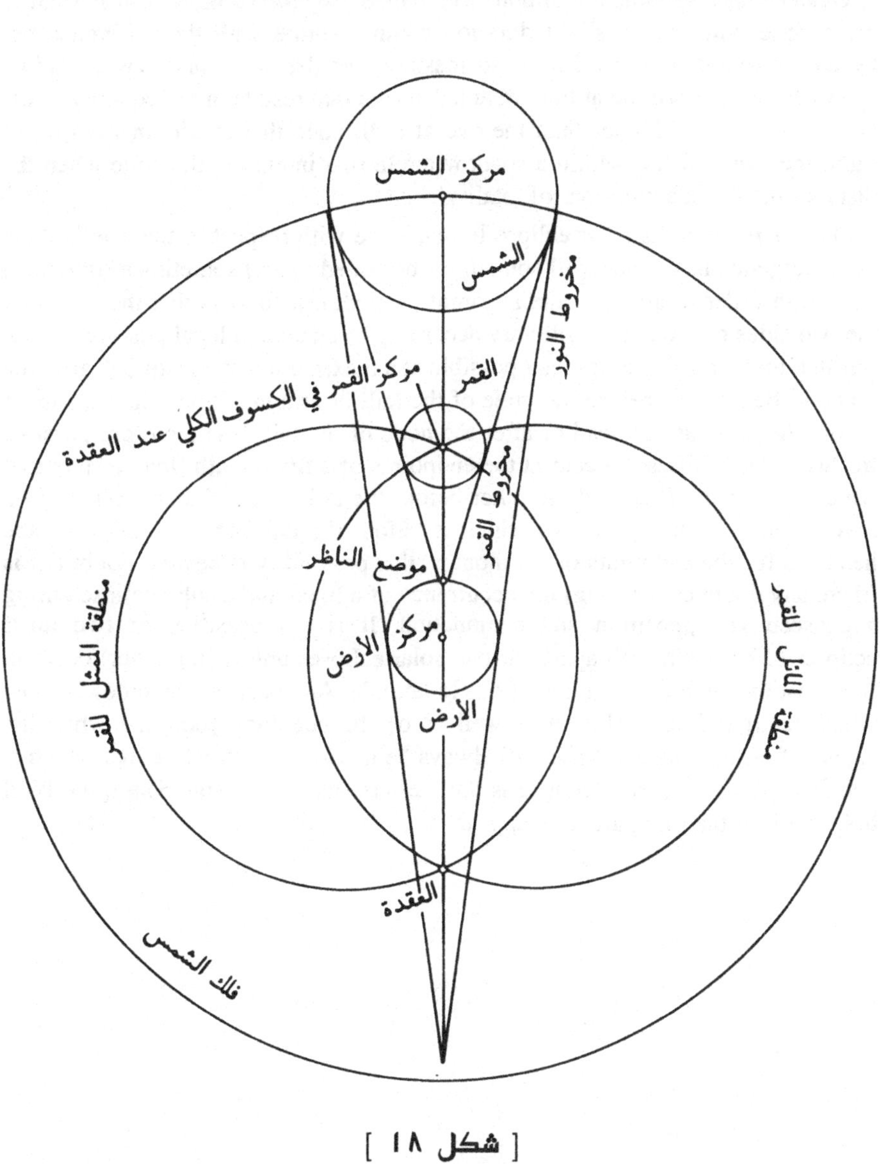

[شكل ١٨]

[٨] وقطر الشمس فيما بين بعديه وُجد من إحدى وثلاثين دقيقة إلى أربعة[!] وثلاثين . وأمّا قطر القمر فقد وُجد من تسع وعشرين دقيقة إلى ست وثلاثين . فإن وقع المركزان على الخط

extending from the eye to the sun, and if the two diameters are equal, the sun will be completely eclipsed and there will be no duration. If the sun's diameter is greater, there remains a luminous ring called the **ring of light**. If it is smaller the eclipse will have a slight duration commensurate with the difference between the two diameters. This is so inasmuch as the moon also has a shadow cone whose apex will be at the eye at a distance that results in the equality of the two diameters and higher than the eye at a distance that results in the ring of light; the eye will fall within a shadow circle that intersects the cone when the distance results in a duration [of totality].

[9] As regards the solar eclipse limits: since with respect to the true latitude the divergence in latitude will sometimes be added to and sometimes subtracted from it in order to arrive at the apparent [latitude], it follows that the limits on the two sides of the nodes will vary according to change in local position. In the fourth clime, / a solar eclipse is / possible at a maximum distance of 18° after the node of the head or before the node of the tail, or at a maximum distance of 7° before the node of the head or after the node of the tail. It is therefore possible for two solar eclipses to occur at the endpoints of a five-month [interval], one of which is after the head and the other before the tail, or at / the endpoints of / a seven-month [interval], one of which is before the tail and the other after the head. As for the endpoints of six months, this possibility is beyond doubt nor is [there any doubt concerning] the occurrence of a lunar and a solar eclipse during a consecutive opposition and conjunction. It is not possible for two lunar eclipses to be one month apart nor two solar eclipses unless [the latter] occur in two locations with different directions in latitude. And because the moon is / that which [can] become eclipsed as well as be the occulting [body], / what disappears first in a lunar eclipse will always be its eastern part while that which is occulted [first] in a solar eclipse is the western part of the sun. The same [will hold true] for the first parts to reappear.

/11/ a solar eclipse is] β = the possibility of a solar eclipse is] α, M. /15/ the endpoints of] β = –α, –M. /19–20/ that which [can] become eclipsed...occulting [body]] β = the occulting [body] as well as that which [can] become eclipsed] α, M.

الخارج من البصر إلى الشمس وكان القطران متساويين انكسفت الشمس كلّها ولم يكن هناك مكث ؛ وإن كان قطر الشمس أكبر بقي حلقة نورانية وتسمى حلقة النور ؛ وإن كان أصغر كان للكسوف مكث قليل بقدر الفضل بين القطرين . وذلك أنّ للقمر أيضاً مخروط ظلّ يكون رأسه عند الأبصار في بعد يقتضي تساوي القطرين ، وأعلى من الأبصار في بعد يقتضي حلقة النور ؛ وتقع الأبصار في دائرة من الظلّ قاطعة للمخروط في بعد يقتضي المكث .

[٩] ولاعتبار حدود الكسوفات إذا اعتُبر العرض الحقيقي وكان اختلاف العرض تارة يزاد عليه وتارة ينقص منه ليصير مرئياً لزم أن تكون الحدود عن جانبي العقدتين مختلفة بحسب اختلاف البقاع . ففي الإقليم الرابع / يكون / الكسوف على بعد غايته بَعد عقدة الرأس أو قبل عقدة الذنب إلى ثماني عشرة درجة أو على بعد غايته قبل عقدة الرأس أو بَعد عقدة الذنب إلى سبع درجات ممكناً . ولذلك يمكن كسوفان على طرفي خمسة أشهر أحدهما بعد الرأس والآخر قبل الذنب أو على / طرفي / سبعة أشهر أحدهما قبل الذنب والآخر بعد الرأس . وأمّا على طرفي ستة أشهر فلا اشتباه في إمكانه ولا في وقوع خسوف وكسوف في استقبال واجتماع متواليين . ولا يمكن خسوفان بينهما شهر ولا كسوفان إلاّ في بقعتين مختلفتي جهة العرض . ولكون القمر هو / الداخل في الخسوف والكاسف / يكون المنخسف أولاً أبداً شرقيَهُ والمنكسف غربي الشمس ، وكذلك المنجلي أولاً .

/11/ يكون] β = يكون إمكان] M , α . /15/ …] β = M– , α– . /19–20/ …] β = الكاسف والداخل في الخسوف] M , α .

CHAPTER FOURTEEN

/ On Sectors and Conjunctions and the Situation of Visibility and Invisibility /

[1] The first and third initial points of the [planetary] **sectors** are the apogee and apex and the two perigees. These are the farthest and nearest distances from the center of the World, and the positions at which the fastest and slowest motions occur. The initial points of the remaining two [sectors] on the two sides are either according to distance, so that for the apogee orb it is where the two lines produced to it from the center of the World and the [center] of that orb are equal, while for the epicyclic orb [the initial points] are where the circumferences of the epicycle and the deferent intersect; or else, they are according to speed, so that for the apogee orb it is where a line passing through the center of the World perpendicular to the diameter passing through the centers reaches the [orb], while for the epicyclic orb it is where a line produced from the center of the World to it is tangent to its circumference. Something moving in the two orbs is ascending in the third and fourth [sectors], descending in the first and second; elevated in the fourth and first [sectors], depressed in the others. [Sector] measurements are provided in the practical handbooks.

[2] With respect to visibility and invisibility, for the stars these vary first, according to their differing sizes; second, according to differences in [both] their measure and direction in latitude; and third, according to different horizons. Thus some stars do not become invisible at all, whereas some of them become invisible for a long period. Venus does not become invisible in the fourth clime when it is in Pisces; during its day of combust, it may be seen in the evening and following morning while retrograding. But it will be invisible for a considerable period when it combusts in Virgo / while retrograding. / Mercury is not visible in the evenings round about the autumnal point in the vicinity of its apogee; nor [is it visible] in the mornings round about the vernal point in the vicinity of [the point] opposite its apogee. When the upper planets

---

/2–3/ On Sectors...Invisibility] β = On Sectors and Conjunctions and the Situation of Visibility and Invisibility and Conjunctions] M = On Sectors, the Situation of Visibility and Invisibility, and Conjunctions] α. /21/ while retrograding] β = while undergoing direct motion] α, M.

الفصل الرابع عشر
في النطاقات / والاقترانات
وأحوال الظهور والاختفاء /

[١] مبادئ الاوّل والثالث من النطاقات هي الأوج والذروة والحضيضان ، وهي الأبعاد البعيدة والقريبة من مركز العالم والمواضع التي يكون هناك أسرع الحركات وأبطؤها . ومبادئ الباقيين في الجانبين إمّا بحسب البعد ، ففي فلك الأوج حيث يتساوى الخطّان الخارجان من مركزي العالم وذلك الفلك إليه ، وفي فلك التدوير حيث يتقاطع محيطا التدوير والحامل ؛ وإمّا بحسب السير ، ففي فلك الأوج حيث ينتهي إليه العمود المارّ بمركز العالم القائم على القطر المارّ بالمراكز ، وفي فلك التدوير حيث يماسّ محيطه الخطّ الخارج إليه من مركز العالم . والسائر في الفلكين صاعد في الثالث والرابع ، هابط في الأوّل والثاني ؛ مستعلٍ في الرابع والأوّل ، منخفض في الباقيين . ومقاديرها تُورَد في كتب العمل .

[٢] وأما الظهور والاختفاء فيختلف في الكواكب أوّلاً بحسب كبرها وصغرها ، وثانياً بحسب اختلاف مقادير عروضها وجهاتها ، وثالثاً بحسب اختلاف الآفاق . ولذلك لا يختفي بعض الكواكب أصلاً ويختفي بعضها مدّة طويلة . والزهرة لا تختفي في الإقليم الرابع في الحوت ، تُرى يوم احتراقها راجعة بُكرة وعشياً ؛ وتختفي إذا احترقت في السُنبلة / راجعة / مدّة كثيرة . وعطارد لا يظهر بالعشيات حوالى النقطة الخريفية وحدود أوجه ، ولا بالغدوات حوالى النقطة الربيعية وحدود مقابلة أوجه . والكواكب

/2–3/ ...] β = والاقترانات وأحوال الظهور والاختفاء والقرانات] M = وأحوال الظهور والاختفاء والاقترانات] α . /21/ ...] β = مستقيمة] M , α .

become visible after the sun departs from them, they may be seen rising in the mornings in the east until the sun crosses its quadratures with them. They will then be seen rising in the evenings until the sun is in opposition to them, after which they may be seen setting in the mornings until the second quadratures. They are then seen setting in the evenings in the west, and thereafter they become invisible. When the lower planets go ahead of the sun, they will be visible in the evenings in the west, thus setting in the evenings, until they retrograde and become invisible in the evenings. They then become visible, rising in the mornings in the east, until they become invisible in the mornings. As for the moon, one may add to the previously cited reasons for variability [the moon's] parallax and its variable distance from the sun that leads to increases and decreases in the light[ed part] of its body. The minimum that it is invisible is two nights and the maximum is three nights. [The matter] has been subjected to testing and the limits of visibility and invisibility of the six wandering [planets] have been found whereby the altitude at the [time of the] rising or setting of the sun is: 11° for Saturn; 10° for Jupiter; 11½° for Mars; 5° for Venus; 10° for Mercury; and 8° for the moon, which, in its case only, is apparent.

[3] As for the conjunction of two planets, [it occurs when] they fall on the same latitude circle in the same direction from one of the two poles. A **true latitudinal conjunction** is when a single line produced from the center of the World passes through them. An **apparent latitudinal conjunction** is when a single line produced to the two [planets] from the position of the / observer / passes through them.

/18/ observer] β, M = observers] α.

العلوية إذا فارقتها الشمسُ وظهرت فهي تُرى تطلع بالغدوات مُشرّقة إلى أن تجاوز الشمس تربيعاتها ؛ ثم تُرى تطلع بالعشيّات إلى أن تقابلها الشمسُ ؛ وبعد ذلك تُرى تغرب بالغدوات إلى التربيعات الثانية ؛ ثم تُرى تغرب بالعشيات مُغرّبة ؛ ثم تختفي . والسفليان إذا سبقا الشمسَ ظهرا بالعشيات مُغرّبين فيغربان بالعشيات إلى أن يرجعا ويختفيا بالعشيّات ؛ ثم يظهران ويطلعان بالغدوات مشرّقين إلى أن يختفيا بالغدوات . وأمّا القمر فينضاف فيه إلى أسباب الاختلافات المذكورة اختلاف مَنظره واختلاف بعده من الشمس المقتضي لزيادة نور جرمه ونقصانه . وأقلّ ما يختفي ليلتان وأكثره ثلاث ليالٍ . وقد امتُحِنَ فوُجد حدود ظهور السيّارات الستّة وخفائها حيث يكون الارتفاع عند طلوع الشمس أو غروبها : لزحل أحد عشر جزءاً ، وللمشتري عشرة أجزاء ، وللمرّيخ أحد عشر جزءاً ونصفاً ، وللزهرة خمسة أجزاء ، ولعطارد عشرة أجزاء ، وللقمر ثمانية أجزاء مرئية له فقط .

[٣] وأمّا اقتران الكوكبين فهو وقوعهما على دائرة عرض واحدة في جهة واحدة من أحد القطبين . والاقتران العرضي الحقيقي هو أن يمرّ بهما خط واحد خارج من مركز العالم ، والاقتران العرضي المرئي أن يمرّ بهما خط واحد خارج من موضع / الناظر / إليهما .

. α [الناظرين = β , M [... /18/

BOOK III

On the Configuration of the Earth and the [Consequences] Accruing to It Due to the Changing Positions of the Celestial Bodies

Twelve Chapters

CHAPTER ONE

A General Summary of the Configuration and Circumstances of the Earth

[1] It was shown in the first part of the book that the Earth is as a whole spherical and that in all directions the head of someone standing upon it will be toward the circumference, which is up, while his feet will be toward the center, which is down, and that the surface of the Earth, which is its convexity, is parallel to the concavity of the orb surrounding [the surface]. The direction of the head of someone traveling upon the Earth is necessarily at each instant [toward] another part of the [celestial] orb. If it were possible to travel around the entire Earth and then three individuals were assumed to become separated at some location, one of them traveling toward the west, the second traveling toward the east, and the third staying in place until the two travelers had circled the Earth—the traveler who went west returning to him from the east and the traveler who went east returning to him from the west—then the first [traveler] will have one fewer than the total [number] of days that have been generally counted because he has lengthened [the period for each of] the revolutions of the orb due to his travel so that he distributes a revolution among their total [number]. The second will have one more because he has shortened [the period for each of] the revolutions due to his motion so that a revolution accumulates for him from the decreases. This is also something that is asked about and found to be strange.

[2] The great circle / that is / on the surface of the Earth occurring in the plane of the equinoctial is called the **equator**. If another great circle is imagined

/19/ that is] β, M = $-\alpha$.

الباب الثالث
في هيئة الأرض وما يلزمها بحسب اختلاف أوضاع العلويات
اثنا عشر فصلاً

الفصل الأول
في جمل من هيئة الأرض وأحوالها

[١] قد تبيّن في أول الكتاب أنّ الأرض بجملتها مستديرة ، وأنّ الواقف عليها من جميع الجوانب رأسه إلى ما يلي المحيط وهو الفوق ورجله إلى ما يلي المركز وهو التحت ، وأنّ سطح الأرض وهو محدّبه موازٍ لمقعر الفلك المحيط به . والسائر على الأرض يجب أن يصير سمت رأسه في كل وقت جزءاً آخر من الفلك . ولو كان السير على جميع الأرض ممكناً ، ثم فُرض تفرّق ثلاثة أشخاص عن موضع ، فسار أحدهم نحو المغرب والثاني نحو المشرق وأقام الثالث حتى دار السائران دوراً من الأرض ورجع السائر في المغرب إليه من الشرق والسائر في المشرق من الغرب ، نقص من الأيّام التي عدّوها جميعاً للأول واحد لأنّه زاد بسيره في أدوار الفلك فوزّع دوراً على جملتها ، وزاد للثاني واحد لأنّه نقص بسيره عن الأدوار فاجتمع له من النقصانات دور . وهذا أيضاً مما يُسأل عنه ويُستغرب .

[٢] والدائرة العظيمة / التي / على سطح الأرض الكائنة في سطح معدّل النهار تسمى خط الاستواء . وإذا توهّمت عظيمة

. –α = β , M [… /19/

that passes through its two poles, the Earth will be divided by them into fourths. One of the two northern [fourths] is the **populated quarter**, and the others are either submerged in the seas and not populated or else their circumstances are unknown. One should then conceive the division of the Earth's surface to be in longitude according to the division of the equinoctial and in latitude to the poles according to the divisions of the circles of declination. One should also conceive on the [surface] circles that are exactly aligned with the day-circles in order to allow one location to be differentiated from another. Distances and quantities are measured just as they are on the orb.

[3] The **inhabited world** has been determined to be a quarter because observations of celestial phenomena, such as lunar eclipses, have not been found [to occur] for those living in the farthest eastern [regions] earlier than 12 hours ahead [of their occurrence] for those in the farthest western regions. From this they discerned that the length of the populated area does not exceed one-half revolution of the orb. It was [further] determined that [this] quarter is in the north because the shadows at noon of the equinoxes are not southerly in any part of [the inhabited world] except that it has been reported that they are southerly in a few populated regions at the edge of Zanzibar [*al-Zanj*], Abyssinia [*al-Ḥabasha*], and some other [areas]; in any case their latitudes do not exceed a few degrees. In the northern region as well, it is not possible to live beyond the latitude that is the complement of the obliquity due to the intensity of the cold.

[4] The sea surrounds most sides of the aforementioned area of the Earth. This is well-established for the western side, the north, and most of the south, especially the eastern part of it. As for the southwest, it has been stated that persons traveling in the direction of the sources of Egypt's Nile have reached locations whose southern latitude is not more than 10° to 20°. They saw mountains at a distance to their south white with snow, which were named for the moon, from which [arise] the headwaters of the Nile. They did not reach a body of water. Furthermore we do not have definitive knowledge about the sea in the northeast.

[5] In the area that has been uncovered to allow for habitation, there are also numerous bodies of water some of which are connected to the encompassing [ocean] such as: that between the Maghrib and Andalusia and that between Andalusia and Syria; the southern sea, which is connected to the eastern region, from which extends four gulfs into the middle of the inhabited world: the Gulf of Barbary [Gulf of Aden], which is the nearest of them to

أخرى تمرّ بقطبيها انقسمت الأرض بهما أرباعاً : أحد الشماليين هو الربع المسكون ، والباقية إمّا غامرة في البحار غير مسكونة وإما غير معلومة الأحوال . فينبغي أن يتوهّم تجزئة سطح الأرض طولاً بحسب تجزئة معدّل النهار وعرضاً إلى القطبين بحسب تجزئة دوائر الميول ، وتتوهّم عليه مدارات محاذية للمدارات اليومية بعينها لتمكّن امتياز بعض المواضع عن بعض . وتُقدّر المسافات والمقادير كما على الفلك .

[٣] وإنّما حُكم بأنّ المعمور ربع لأنّه لم يوجد في أرصاد الحوادث الفلكية كالخسوفات تقدّم ساعات الواغلين في المشرق لها على ساعات الواغلين في المغرب زائداً على اثنتي عشرة ساعة ؛ فعلموا من ذلك أنّ طول المسكونة لا يزيد على نصف دور الفلك . وإنّما حُكم بأنّ الربع شمالي لأنّه لم توجد أظلال أنصاف نهار الاعتدالين في شيء منها جنوبية إلاّ قليل من مساكن على أطراف الزنج والحبشة وغيرها حُكي أنّها جنوبية لكنّ لا تزيد عروضها على نيف درجات . وفي جانب الشمال أيضاً لا يمكن أن يُسكن فيما جاوز عرضه تمام الميل الكلي لشدّة البرد .

[٤] والبحر محيط بأكثر جوانب القدر المذكور من الأرض . أمّا من جانب المغرب والشمال وأكثر الجنوب لا سيّما الشرقي منه فمعلوم . وأمّا جنوب المغرب فقد ذُكر أنّ السائرين على سمت منابع نيل مصر انتهوا إلى مواضع زاد عرضها الجنوبي على بضع عشرة درجة ، وشاهدوا الجبال البيض من الثلج المنسوبة إلى القمر التي منها منابع النيل في جنوبهم من بعيد ؛ ولم يصلوا إلى بحر . وأيضاً ليس لنا على البحر الذي في شمال المشرق وقوف يقيني .

[٥] وفي القدر المكشوف للعمارة أيضاً بحار كثيرة بعضها متّصل بالمحيط كالذي بين المغرب وأندلس والذي بين أندلس والشام ؛ والبحر الجنوبي المتّصل بالجانب الشرقي الذي خرج منه أربع خليجات إلى وسط العمارة : الخليج البربري وهو أقربها إلى

the Maghrib, the Red Gulf [Red Sea], the Persian Gulf, and the Green Gulf, for each of which the longitude and latitude are well-established; and such as the Sea of *Warank* [Baltic Sea] in the northern region. Some [of the bodies of water] are unconnected [to the encompassing ocean] such as the Sea of Ṭabaristān [Caspian Sea], Lake Khwārazm [Aral Sea], and other channels and basins. In addition to the seas, there are numerous other obstacles to habitation such as deserts, mountains, hills, sands, jungles, and so on, which the learned have imparted in the geographical placebooks [*al-masālik*] and which travelers and others [have also made known].

[6] Some practitioners of this science have stated that the reason for the lack of habitation in the southern region is that on account of its proximity to the day-circle of the sun's perigee it is warmer since the sun will there be larger in size and more intense in its rays and its effects because it is nearer. This is unconvincing because the difference between the smallest size of the sun from its being at the apogee and its largest size from being at the perigee is imperceptible to the senses. It is thus far-fetched that its effect would reach an extent whereby one of two similarly positioned locations would be populated, while the other would not be populated. Furthermore, if this were the reason, then what is beyond [this day-circle] in the south, namely the [potential] populated regions whose latitude is greater than the maximum declination [*i.e.* obliquity], would be inhabited. In addition some have stated that the southern region is generally warmer than the northern region during the period that the perigee is in the southern signs; the heat draws the moisture and so the seas are drawn to the southern hemisphere and the uncovered part of the Earth is in the northern hemisphere. The inhabited world would then move as the apogee moved. This is likewise unconvincing because the existence of seas north of the inhabited world contradicts this judgment. Also some have said that the region that is beneath those southern day-circles occurring between the "falls" of the two luminaries is not populated; it is called the combust way whence they named what on the orb is between the two "falls" by this name as well. This is one of the fairy tales of the astrologers. In general there is no known cause for the above-mentioned amount of uncovering of the Earth except divine providence; otherwise, why would one of the northern quarters be characterized by [habitation] and not the other despite their situation with respect to the heavens being the same?

المغرب ، والخليج الأحمر ، وخليج فارس ، والخليج الأخضر ، ولكلّ واحد منها طول وعرض صالحان ؛ وكبحر ورنك من جانب الشمال . وبعضها غير متّصل كبحر طبرستان ، وبحيرة خوارزم ، وغيرهما من البطائح والمغائض . وغير البحار من موانع العمارة كالبراري والجبال والتلال والرمال والآجام وغيرها أيضاً كثير ، يعرّفها أهل العلم بالمسالك والسيّاح وغيرهم.

[٦] وقد قال بعض أهل هذا العلم في علّة عدم العمارة في الناحية الجنوبية أنّها لقربها من مدار حضيض الشمس تكون أحرّ إذ الشمس توجد هناك لقربها أعظم جرماً وأشدّ شعاعاً وأثراً . وهذا ليس بيقيني لأنّ التفاوت بين صغر الشمس من جهة كونها في الأوج وكبرها من جهة كونها في الحضيض ليس ببيّن عند الحسّ ، فمن البعيد أن يبلغ تأثيرها إلى حدّ يصير أحد موضعين متساويين في الوضع مسكوناً والآخر غير مسكون . وأيضاً لو كان السبب ذلك لكان ما جاوزه في الجنوب من المساكن التي يزيد عرضها على غاية الميل معموراً . وذكر أيضاً بعضهم أنّ ناحية الجنوب بالجملة أحرّ من ناحية الشمال مدّة كون الحضيض في البروج الجنوبية . والحرارة تجذب الرطوبة فلذلك انجذبت البحار إلى النصف الجنوبي ، وصار المنكشف من الأرض في النصف الشمالي . وتنتقل العمارة بانتقال الأوج . وهذا أيضاً ليس بيقيني لأنّ وجود البحار في شمال العمارة ينافي ذلك الحكم . وقال بعضهم أيضاً أنّ المواضع التي تكون تحت المدارات الجنوبية التي تقع بين هبوطي النيّرين غير مسكونة ، وتُسمّى بالطريقة المحترقة ولذلك سمّوا ما بين الهبوطين من الفلك بهذا الاسم أيضاً . وهذا من خرافات الأحكاميين . وبالجملة ليس لانكشاف القدر المذكور من الأرض سبب معلوم غير العناية الإلهية ؛ وإلاّ لِما اختُصّ أحد الربعين الشماليين بها دون الآخر مع تساوي أوضاعهما بالقياس إلى السماويّات ؟

[7] Most of the inhabited world on the northern side falls between the [area] beyond 10 degrees in latitude up to 50 [degrees]. The practitioners of the science have divided it into **seven climes** [stretching] lengthwise so that each clime is beneath a day-circle, the conditions of the places in it then being similar. Thus each clime extends from east to west in longitude, while its [extent in] latitude is a small amount, which results in a difference of half an hour in lengths of longest daylight. The majority have made the initial point for the longitudes on the western side so that the longitude increases in magnitude in the direction of the sequence of the signs; the starting point for the latitudes is the equator since it, rather than something else, is determined by nature. They have stated that the beginning of the inhabited world in the west is in the islands named the Eternals (which at the present time are uninhabited) so some have made them the initial point for longitude. Another group made the coast of the western ocean the initial point. Between them is 10 degrees of revolution of the equinoctial. Among their learned the end of the inhabited world on the eastern side is Kangdezh, which is the initial point for those who make it be from the eastern side. What is on the equator between the two ends they named the **cupola** of the Earth, and it is at a distance of one-quarter revolution from the initial western point. There will thus be a divergence attendant upon [its location] because of the difference concerning [the initial point].

[8] Turning to the initial and midpoints of the climes according to latitudes and the longest periods of daylight, they are as follows: the initial point of the first is where the longest day is (12 + ½ + ¼) hours and its latitude is 12⅔ degrees. Its midpoint is where the day is 13 hours and its latitude is (16 + ½ + ⅛) degrees. The initial point of the second is where the day is 13¼ and the latitude is (20 + ¼ + ⅕); its midpoint is where the day is 13½ and the latitude is (24 + (½ of ⅙)) [*sic*]. The initial point of the third is where the day is (13 + ½ + ¼) and the latitude is 27½; its midpoint is where the day is 14 and the latitude is 30⅔. The initial point of the fourth is where the day is 14¼ and the latitude is (33 + ½ + ⅛); its midpoint is

[٧] ومعظم العمارة في طرف الشمال يقع بين ما يجاوز عشر درجات في العرض إلى حدود الخمسين . فقسمها أهل الصناعة بالأقاليم السبعة طولاً ليكون كل إقليم تحت مدار ، فتتشابه أحوال البقاع التي فيه . فإذن كل إقليم يمتدّ ما بين الخافقين طولاً ، ويكون عرضه قدراً قليلاً وهو ما يوجب تفاضل نصف ساعة في مقادير النهار الأطول . والجمهور جعلوا مبدأ الأطوال من جانب المغرب ليكون ازدياد عدد الطول في جهة توالي البروج ، ومبدأ العروض خط الاستواء لأنّه بالطبع متعيّن دون ما عداه . وقد ذكروا أنّ بداية العمارة في المغرب كانت في جزائر منسوبة إلى الخالدات ـ وهي الآن غير معمورة ـ فجعلها بعضهم مبدأ الطول . وقوم آخر جعلوا ساحل البحر الغربي مبدأه ، وبينهما عشر درجات من دور معدّل النهار . ونهاية العمارة من الجانب الشرقي عند علمائهم كَنْكْدِز ، وهي المبدأ عند من يجعله من جانب المشرق . وسمّوا ما بين النهايتين على خط الاستواء قبّة الأرض ، وهي على بعد ربع الدور من المبدأ الغربي ، فيلزمها الاختلاف بسبب الاختلاف فيه .

[٨] وأمّا مبادئ الأقاليم وأوساطها بحسب العروض وساعات النهار الأطول فهي هذه : أمّا الأول فمبدؤه حيث النهار الأطول اثنتا عشرة ساعة ونصف وربع ، وعرضه اثنتا عشرة درجة وثُلثا درجة ؛ ووسطه حيث النهار ثلاث عشرة ساعة ، وعرضه ستّ عشرة درجة ونصف وثُمن . وأمّا الثاني فمبدؤه حيث النهار ثلاث عشرة وربع ، والعرض عشرون ورُبع وخُمس ؛ ووسطه حيث النهار ثلاث عشرة ونصف ، والعرض أربع وعشرون ونصف سُدس . وأمّا الثالث فمبدؤه حيث النهار ثلاث عشرة ونصف ورُبع ، والعرض سبع وعشرون ونصف ؛ ووسطه حيث النهار أربع عشرة ، والعرض ثلاثون وثُلثان . وأمّا الرابع فمبدؤه حيث النهار أربع عشرة وربع ، والعرض ثلاث وثلاثون ونصف وثُمن ؛ ووسطه

where the day is $14\frac{1}{2}$ and the latitude is $(36 + \frac{1}{5} + \frac{1}{6})$. The initial point of the fifth is where the day is $(14 + \frac{1}{2} + \frac{1}{4})$ and the latitude is 39 less $\frac{1}{10}$; its midpoint is where the day is 15 and the latitude is $41\frac{1}{4}$. The initial point of the sixth is where the day is $15\frac{1}{4}$ and the latitude is $(43 + \frac{1}{4} + \frac{1}{8})$; the midpoint is where the day is $15\frac{1}{2}$ and the latitude is $(45 + \frac{1}{4} + \frac{1}{10})$. The initial point of the seventh is where the day is $(15 + \frac{1}{2} + \frac{1}{4})$ and the latitude is $47\frac{1}{5}$; its midpoint is where the day is 16 and the latitude is $(48 + \frac{1}{2} + \frac{1}{4} + \frac{1}{8})$; its endpoint is where the day is $16\frac{1}{4}$ and the latitude is $50\frac{1}{3}$. The end of each clime other than the [last] is the beginning of the following one. One group has made the initial point of the first clime the equator. The end of the seventh is the termination of the inhabited world.

[9] The longest [period] of daylight reaches 17 hours where the latitude is 54 degrees plus a fraction; it reaches 18 where the latitude is 58; it reaches 19 where the latitude is 61; it reaches 20 where the latitude is 63; it reaches 21 where the latitude is $64\frac{1}{2}$; it reaches 22 where the latitude is 65 plus a fraction; it reaches 23 where the latitude is 66; and it reaches 24 where the latitude is equal to the colatitude of the obliquity. It reaches 1 month where the latitude is $67\frac{1}{4}$, 2 months where the latitude is 70 less $\frac{1}{4}$, 3 months where the latitude is $73\frac{1}{2}$, 4 months where the latitude is $78\frac{1}{2}$, 5 months where the latitude is 84, and half a year where the latitude is a quarter revolution.

[10] Let us now go into the characteristics of the day-circles.

حيث النهار أربع عشرة ونصف ، والعرض ستّ وثلاثون وخُمس وسُدس . وأمّا الخامس فمبدؤه حيث النهار أربع عشرة ونصف وربُع ، والعرض تسع وثلاثون إلاّ عُشر ؛ ووسطه حيث النهار خمس عشرة ، والعرض إحدى وأربعون وربع . وأمّا السادس فمبدؤه حيث النهار خمس عشرة وربع ، والعرض ثلاث وأربعون وربع وثُمن ؛ ووسطه حيث النهار خمس عشرة ونصف ، والعرض خمس وأربعون وربع وعُشر . وأمّا السابع فمبدؤه حيث النهار خمس عشرة ونصف وربع ، والعرض سبع وأربعون وخُمس ؛ ووسطه حيث النهار ستّ عشرة ، والعرض ثمان وأربعون ونصف وربع وثُمن ؛ وآخره حيث النهار ستّ عشرة وربع ، والعرض خمسون وثُلث . وآخر كل إقليم ما عداه أوّل الذي يليه . وقوم جعلوا مبدأ الإقليم الأول خط الاستواء . وآخر السابع منتهى العمارة .

[٩] والنهار الأطول يبلغ سبع عشرة ساعة حيث العرض أربع وخمسون درجة وكسر ؛ ويبلغ ثماني عشرة حيث العرض ثمان وخمسون ؛ ويبلغ تسع عشرة حيث العرض إحدى وستّون ؛ ويبلغ عشرين حيث العرض ثلاث وستّون ؛ ويبلغ إحدى وعشرين حيث العرض أربع وستّون ونصف ؛ ويبلغ اثنتين وعشرين حيث العرض خمس وستّون وكسر ؛ ويبلغ ثلاثاً وعشرين حيث العرض ستّ وستّون ؛ ويبلغ أربعاً وعشرين حيث العرض مثل تمام الميل كلّه . ويبلغ شهراً حيث العرض سبع وستّون وربع ، وشهرين حيث العرض سبعون إلاّ ربع ، وثلاثة أشهر حيث العرض ثلاث وسبعون ونصف ، وأربعة أشهر حيث العرض ثمان وسبعون ونصف ، وخمسة أشهر حيث العرض أربع وثمانون ، ونصف السنة حيث العرض ربع الدور .

[١٠] ولنشرع الآن في خواصّ المدارات .

CHAPTER TWO

On the Characteristics of the Equator

[1] Horizon circles of localities that are on the equator bisect all the day-circles since they pass through the poles of the equinoctial. Therefore / night and day / during the entire year are equal. Also the period of visibility of each point on the orb is equal to the period of its invisibility. If there is a difference, it is because of variability in speed in the two halves due to the second[ary] motion; this will not be perceptible. The sun will in a year pass twice over their zenith, this being when it is at the / point / of the two equinoxes. It moves away from their zeniths only to the extent of the ecliptic orb's maximum declination from the equinoctial; thus [the sun's] maximum altitude is never less than the complement of the obliquity. The sun is in each direction for half the year, the noon shadow being in the opposite direction. The poles of the ecliptic are at the horizon when one of the two equinox points are at the zenith; thereupon the intersection of the ecliptic equator and the horizon is at right angles. During the passage of the northern half of the [ecliptic] equator across the meridian, the southern of the two ecliptic poles is visible; during the passage of the southern half, the northern is visible. The altitude [of each pole] does not exceed the magnitude of the obliquity. Since [by definition] the beginning of summer is the time in which the sun is nearest the zenith and the beginning of winter is the time in which it is farthest from it, the time at which it is at the two points / of the two equinoxes / is the beginning of their summer and the time at which it is at the solstitial points is the beginning of their winter. The beginnings of the other two seasons are [at] the midpoints of the quarter [divisions of the ecliptic]. It follows from this that they have eight seasons in a year. The turning of the orb there is wheel-like because the planes

---

/5/ night and day] β = day and night] α, M. /9/ point] β = two points] α, M. /20/ of the two equinoxes] β = of the equinox] α, M.

الفصل الثاني
في خواصّ خط الاستواء

[١] دوائر آفاق البقاع التي تكون على خط الاستواء تنصّف جميع المدارات اليومية لكونها مارّة بقطبي معدل النهار ؛ فلذلك يكون / الليل والنهار / في جميع السنة متساويين . وأيضاً يكون زمان ظهور كل نقطة على الفلك مساوياً لزمان خفائه ؛ فإن كان تفاوت كان بسبب اختلاف السير بالحركة الثانية في النصفين ، وذلك لا يكون محسوساً . وتمرّ الشمس في السنة مرّتين بسمت رؤوسهم ، وذلك عند كونها في / نقطة / الاعتدالين . ولا تبعد عن سمت رؤوسهم إلاّ بقدر غاية ميل فلك البروج عن معدل النهار ، فلا تنقص غاية ارتفاعها عن تمام الميل كلّه . وتكون الشمس نصف السنة في كل جهة وظل نصف النهار إلى خلاف تلك الجهة . وقطبا البروج يكونان على الأفق عند كون إحدى نقطتي الاعتدالين على سمت الرأس ، وهناك يكون قطع منطقة البروج للأفق على قوائم . وفي مدّة مرور النصف الشمالي من المنطقة على نصف النهار يكون الظاهر من قطبي البروج جنوبيّهما ، وفي مدّة مرور النصف الجنوبي يكون الظاهر شماليّهما ؛ ولا يزيد ارتفاعهما على قدر الميل الكلي . ولكون مبدأ الصيف الوقت الذي تكون الشمس فيه إلى سمت الرأس أقرب ومبدأ الشتاء الوقت الذي تكون فيه منه أبعد ، يكون وقت كونها في نقطتي / الاعتدالين / مبدأ صيفهم ووقت كونها في نقطتي الانقلاب مبدأ شتائهم . وتكون مبادئ الفصلين الآخرين أوساط الأرباع . فيلزم على ذلك أن يكون لهم في سنة ثمانية فصول . ويكون دور الفلك هناك دولابياً لأنّ سطوح

/5/ ...] β = النهار والليل] M , α . /9/ ...] β = نقطتي] M , α . /20/ ...] β = الاعتدال] M , α .

of all the day-circles intersect the horizon at right angles. Their horizons are therefore called the **horizons of the right orb**. And since the horizon circle [at the equator] is one of the circles of declination, the **ortive amplitude** of each point, which is the arc along the horizon between its rising place and the rising place of the equinoctial, is in the amount of its declination; the same [holds] for the **occasive** [setting] **amplitude**.

[2] The Grand Master Abū ᶜAlī ibn Sīnā judged the [equator] to be the most temperate locality. He stated: Because the sun does not linger there long at the zenith, but rather it passes by it at the times of its crossing from one of the directions to the other and its motion in declination will there be at its fastest, then the heat of their summer will therefore not be intense. This is because even though being directly overhead leads to heating, nevertheless duration of this [state] is more effective for that than the [state] itself. It is because of this that summer is warmer than spring and the afternoon is warmer than before [noon] despite the equality of alignments in each case. Furthermore since the periods of their day and their night are equal, the severity of each of the weather conditions arising from them will be quickly broken by the other; thus [each] period will be temperate. He also argued that the hottest localities in summer are those whose latitudes are equal to the obliquity. For the sun will be directly overhead and will linger near this alignment for nearly two months. Their days at that time will be long and their nights short.

[3] The eminent Imām Fakhr al-Dīn al-Rāzī rejected the first argument, saying: Even though the sun lingers on the equator only briefly, it nonetheless is never too far from being directly overhead; it is thus virtually overhead for the length of the year. We see localities whose maximum solar altitudes do not greatly exceed the minimum [solar] altitude at the equator, and the heat of their summers is extremely intense. / One learns / from this that the heat of the winter at the equator is many times the heat of the summer of those localities. He judged the most temperate clime to be the fourth.

/23/ One learns] β = Let one learn] α, M.

جميع المدارات تقطع سطح الأفق على قوائم ، وتسمى لذلك آفاقها بآفاق الفلك المستقيم . ولكون دائرة الأفق إحدى دوائر الميول تكون سعة مشرق كل نقطة ، وهي القوس التي تكون من الأفق بين مطلعها ومطلع معدّل النهار ، بقدر ميلها ؛ وكذلك سعة المغرب .

[٢] والشيخ الرئيس أبو علي بن سينا حكم بأنّها أعدل البقاع . قال : لأنّ الشمس لا تلبث على سمت الرؤوس هناك كثيراً بل إنّما تمرّ به وقتي اجتيازها عن إحدى الجهتين إلى الأخرى وتكون هناك حركتها في الميل أسرع ما تكون ، فلا تكون لذلك حرارة صيفهم شديدة . وذلك لأنّ المسامتة ـ وإن كانت مقتضية للتسخين ـ لكنّ المكث عليها أبلغ في ذلك من نفسها ؛ ولذلك يكون الصيف أحرّ من الربيع وبعد الزوال أحرّ من قبله مع تساوي المسامتة فيهما . وأيضاً لتساوي زماني نهارهم وليلهم تنكسر سَورتا كل واحدة من الكيفيتين الحادثتين منهما بالآخر[!] سريعاً ، فيعتدل الزمان . وحكم أيضاً بأنّ أحرّ البقاع صيفاً هي التي تكون عروضها مساوية للميل الكلي : فإنّ الشمس تسامتها وتلبث في قرب مسامتتها قريباً من شهرين ، ونهارها حينئذ يطول وليلها يقصر .

[٣] وردّ الإمام الفاضل فخر الدين الرازي عليه الحكم الأول بأن قال : لبثُ الشمس في خط الاستواء ـ وإن كان قليلاً ـ لكنّها لا تبعد كثيراً عن المسامتة ، فهي طول السنة في حكم المسامتة . ونحن نرى بقاعاً أكثر ارتفاعات الشمس / بها / لا يزيد كثيراً على أقلّ ارتفاعاتها بخط الاستواء وحرارة صيفها في غاية الشدّة . / فيُعلم / من ذلك أنّ حرارة شتاء خط الاستواء تكون أضعاف حرارة صيف تلك البقاع . وحكم بأنّ أعدل البقاع الإقليم الرابع .

/21/ بها] β = بها ما] M , α . /23/ ...] β = فليعلم] M , α .

[4] The truth of the matter is that if one means by temperate a uniformity in the conditions, then there is no doubt that it is most so at the equator as stated by the Master. But if one means by it a balancing of the two [extreme] weather conditions, then there is no doubt that the equator is not this way; this is indicated by the extreme blackness in color of its inhabitants among the peoples of Zanzibar and Abyssinia, the extreme frizziness of their hair, and other things that are brought about by the heat of the air. The opposite of this among the people of the fourth clime indicates that the state of its air is more temperate; indeed, the general reason for the profusion of habitations and the magnitude of propagation and reproduction in the seven climes, but not in the rest of the uncovered parts of the Earth, indicates that they are more temperate than other [places]. And what is nearer to the middle [of the climes] is most certainly nearer [to being temperate] than what is at their fringes; for becoming seared and remaining immature, which result from the two [extreme] weather conditions, are clearly evident at the two fringes.

CHAPTER THREE

On the Characteristics of Locations Having Latitude Which Are Called the Oblique Horizons

[1] For every location that is beneath one of the day-circles between the equator and one of the poles of the / first / motion, the turning of the orb there is slanted. The altitude of the pole that is in the direction to which the location is inclined is in the amount of the local latitude. The distance from the equinoctial of the day-circles that are permanently visible or permanently invisible is greater than the local colatitude; the distance of the largest of them, which touches the horizon, is equal to it. The remaining day-circles are divided by the horizon into two unequal parts—the largest visible ones being closest to the visible pole, the [largest] invisible ones being farthest away. The two parts are conversely equal for any two day-circles equidistant from and on opposite sides of the equinoctial. There is an increase in daylight up to the apex of the solstice that is adjacent to the visible pole, and a decrease in it until

/15/ first] $\beta = -\alpha$, –M.

[٤] والحق في ذلك أنّه إن عُني بالاعتدال تشابه الأحوال فلا شكّ أنّه في خط الاستواء أبلغ كما ذكره الشيخ . وإن عُني به تكافؤ الكيفيتين فلا شكّ أنّ خط الاستواء ليس كذلك ــ يدلّ عليه شدّة سواد لون سكّانه من أهل الزنج والحبشة وشدّة جُعودة شعورهم وغير ذلك مما تقتضيه حرارة الهواء . وأضداد ذلك في أهل الإقليم الرابع تدل على كون هوائه أعدل ؛ بل السبب الكلّي في توفّر العمارات وكثرة التوالُد والتناسُل في الأقاليم السبعة دون سائر المواضع المنكشفة من الأرض يدلّ على كونها أعدل من غيرها . وما يقرب من وسطها يكون لا محالة أقرب مما يكون على أطرافها ، فإنّ الاحتراق والفَجاجة اللازمين من الكيفيتين ظاهران في الطرفين .

الفصل الثالث
في خواصّ المواضع التي يكون لها عرض وتُسمّى بالآفاق المائلة

[١] كل موضع يكون تحت أحد المدارات اليومية بين خط الاستواء وأحد قطبي الحركة / الأولى / يكون دور الفلك هناك حمائلياً . ويكون ارتفاع القطب الذي يكون في الجهة التي مال الموضع إليها بقدر عرض البلد . وكان بُعد المدارات الأبدية الظهور والأبدية الخفاء عن معدّل النهار أكثر من تمام عرض البلد ؛ وبُعد أعظمها الذي يماسّ الأفق مساوياً له . وسائر المدارات ينقسم بالأفق إلى مختلفين : أعظمهما الظاهر فيما هو إلى القطب الظاهر أقرب ، والخفي فيما هو أبعد . ويتساوى القسمان على التبادل في كل مدارين متساويي البعد عن معدّل النهار على جنبتيه . وتزايد النهار يكون إلى رأس المنقلب الذي يلي القطب الظاهر وتناقصه إلى

/15/ ...] β = M– , α– .

the apex of the other solstice. Daylight will only be equal to night when the sun is at the equinox points.

[2] If two circles of declination are assumed to pass through the two points at which the day-circle of the sun or of some star and the horizon intersect, there will occur two triangles between / those two circles, / the horizon, and the equinoctial; one is easterly and the other westerly. One of the sides of each [triangle] is the declination of the sun or the distance of the star from the equinoctial, which is along the declination circle; the second of them is the ortive amplitude of the sun or of the star, which is along the horizon circle; and the third [of the sides] is the **equation of daylight** of the sun or of the star, which is along the equinoctial, and it is half the difference between the daylight of the sun or of the star and the daylight of the equator. [When] this triangle is in the direction of the visible pole, it is below the Earth [*i.e.* horizon], while in the direction of the invisible pole it is above it.

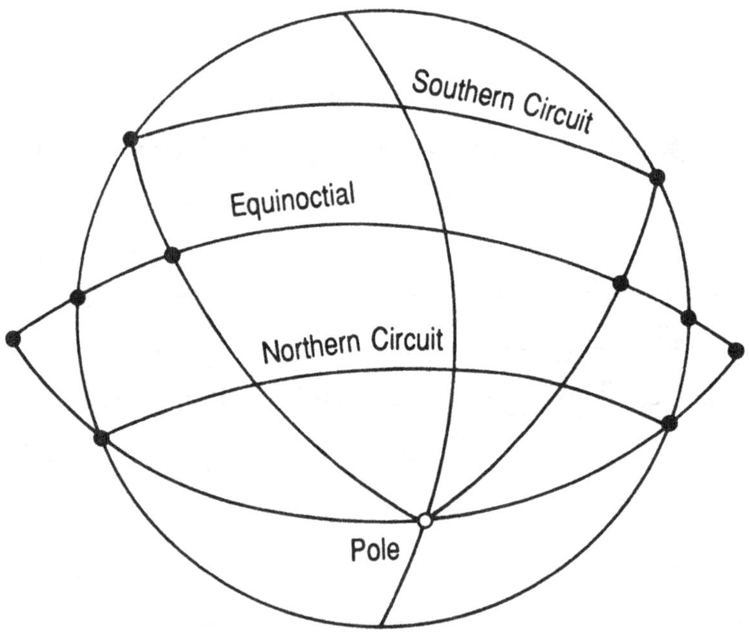

[Fig. T19*]

/5/ those two circles] β = that circle] α, M. [* Figure T19 occurs only in MS L.]

رأس المنقلب الآخر . ولا يكون النهار مساوياً للّيل إلاّ عند كون الشمس في نقطتي الاعتدالين .

[٢] وإذا فُرضت دائرتا ميل تمرّان بالنقطتين اللتين عليهما يتقاطع مدار الشمس أو كوكب من الكواكب والأفق حدث مثلّثان بين / تينك الدائرتين / والأفق ومعدّل النهار ، أحدهما شرقي والآخر غربي : أحد أضلاع كل واحد منهما ميل الشمس أو بُعد الكوكب عن معدّل النهار وهو الذي يكون من دائرة الميل ، وثانيها سعة مشرق الشمس أو الكوكب وهو الذي يكون من دائرة الأفق ، وثالثها تعديل نهار الشمس أو الكوكب وهو الذي يكون من معدّل النهار وهو نصف الفضل بين نهار الشمس أو الكوكب وبين نهار خط الاستواء . ويكون ذلك المثلّث في جانب القطب الظاهر تحت الأرض وفي جانب القطب الخفي فوقها .

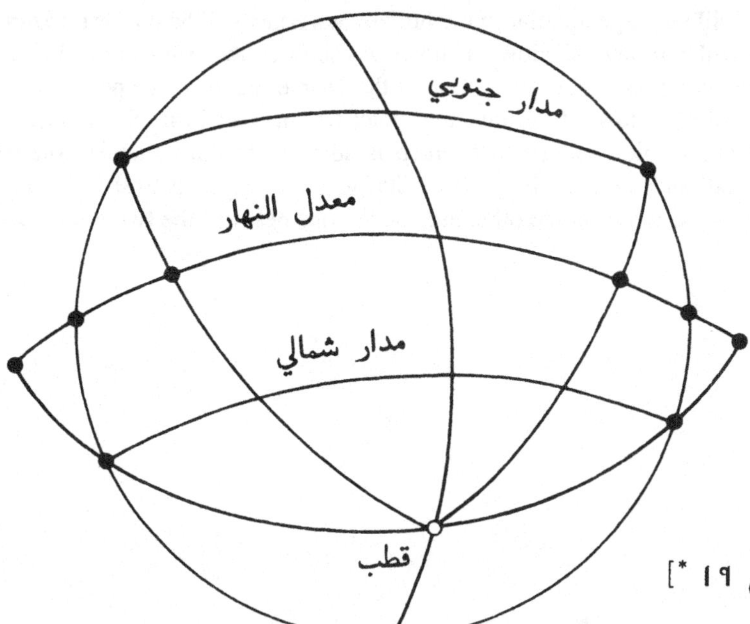

[شكل ١٩ *]

/5/ ...] β = تلك الدائرة] M , α . [* لا يوجد الشكل إلاّ في مخطوط ل.]

[3] For every day-circle / whose elongation / from the equinoctial is on the side of the invisible pole, what revolves on [the part of the circle] above the Earth will not reach the prime vertical circle. Every day-circle whose elongation from the equinoctial in the direction of the visible pole is equal to the local latitude will pass through the zenith and be tangent to the prime vertical circle above the Earth; every [day-circle] whose elongation is greater than that will pass beyond the zenith in the direction of the visible pole, and it will not meet the prime vertical circle; and every one whose elongation is less than that will intersect the prime vertical at two points, one of which is east and the other west. The star, as long as it is between the two points, is beyond the prime vertical circle in the direction of the invisible pole.

CHAPTER FOUR

On the Characteristics of Locations Whose Latitude Does Not Exceed the Complement of the Obliquity

[1] They are divided into four divisions:

[2] The first is that whose latitude is less than the obliquity. In those locations the sun will pass over the zenith at two points whose declination is equal to the local latitude in the direction of the visible pole. At that time the ecliptic equator will be perpendicular to the horizon, its poles will be on the horizon, and objects will not have shadows at noon. As long as the [sun] is on the arc between the two points in the direction of the visible [equinoctial] pole, the [noon] shadow will fall toward the invisible pole; [in this case] the visible one of the ecliptic orb's poles will be that which is adjacent to the invisible pole of the equinoctial, and the invisible will be that which is adjacent to the visible pole. As long as the sun is on the other arc, *i.e.* the one between the two points in

---

/1/ whose elongation] $\beta = -\alpha$, –M.

[٣] وكل مدار يكون / بُعده / من معدّل النهار في جانب القطب الخفي فلا يصل ما يدور عليه فوق الأرض إلى دائرة أوّل السموت . وكل مدار يكون بعده عن معدّل النهار في جهة القطب الظاهر مثل عرض البلد فهو يمرّ بسمت الرأس ويماسّ دائرة أول السموت فوق الأرض ؛ وكل ما يكون بُعده أكثر من ذلك فهو يمرّ عن سمت الرأس في جهة القطب الظاهر ولا يلاقي دائرة أول السموت ؛ وكل ما يكون بُعده أقلّ من ذلك فهو يقطع أول السموت على نقطتين ، إحداهما شرقية والأخرى غربية ، ويكون الكوكب ما دام بين النقطتين عن دائرة أول السموت في جهة القطب الخفي .

الفصل الرابع
في خواصّ المواضع التي عرضها لا يجاوز تمام الميل الكلّي

[١] وهي تنقسم أربعة أقسام :

[٢] الأول ما يكون عرضها أقلّ من الميل الكلّي . وفي تلك المواضع تمرّ الشمس بسمت الرأس في نقطتين ميلهما يساوي عرض البلد في جهة القطب الظاهر ؛ وحينئذ تقوم منطقة البروج على الأفق على قوائم ، ويكون قطباها على الأفق ، ولا يكون للأشخاص في انتصاف النهار ظل . وما دامت في القوس التي بين النقطتين في جهة القطب الظاهر يقع الظل إلى جهة القطب الخفي ؛ ويكون القطب الظاهر من قطبي فلك البروج هو الذي يلي القطب الخفي من معدل النهار ، والخفي هو الذي يلي القطب الظاهر . وما دامت الشمس في القوس الأخرى ، أعني التي تكون بين النقطتين في

/1/ ...] β = −M , −α .

the direction of the invisible [equinoctial] pole, the [noon] shadow will fall toward the visible pole; [in this case] the visible one of the ecliptic orb's poles will be that which is adjacent to the visible pole of the equinoctial, and the invisible will be that which is adjacent to the invisible. The seasons of the year in those regions [*lit.*, horizons] are not equal; even if [their number] were increased beyond the four, they would still not be uniform.

[3] The second division is that whose latitude is equal to the obliquity. In those locations the sun will pass once a year over the zenith. One of the ecliptic orb's poles will be permanently visible, while the second will be permanently invisible; in their revolution, they will not touch the horizon except once, this being when the solstice point that is in the direction of the visible [equinoctial] pole reaches the zenith. At that time only, the ecliptic equator will be perpendicular to the horizon. Throughout the year, shadows will be toward the visible pole. The altitude of the sun increases from one of the solstices to the other; it then returns, [the altitude] decreasing until [the sun] comes back to [the original solstice]. The seasons of the year will be four—no more, no less.

[4] The third division is that whose latitude is greater than the obliquity and less than its complement. There the sun will not reach the zenith; it will have a highest altitude, which is equal to the sum of the obliquity and the local colatitude, and a lowest, which is equal to the excess of the local colatitude over the obliquity. The remaining conditions are as we have [already] explained. Where the local latitude does not exceed the obliquity by [more than] the amount of the latitudes of the other wandering [planets], those whose latitude is greater than the excess of the local latitude over the obliquity will pass over the zenith twice, while those whose latitude is equal to the excess [will pass over] once. In these latitudes, the equation of daylight and the ortive and occasive amplitudes increase with increasing latitude.

جهة القطب الخفي ، يقع الظل إلى جهة القطب الظاهر ؛ ويكون القطب الظاهر من قطبي فلك البروج هو الذي يلي القطب الظاهر من معدل النهار ، والخفي هو الذي يلي الخفي . ولا تكون فصول السنة في تلك الآفاق متساوية ؛ وإن زادت على الأربعة لم تكن متشابهة .

[٣] القسم الثاني ما يكون عرضها مساوياً للميل الكلي . وفي تلك المواضع تمرّ الشمس في السنة مرة واحدة بسمت الرأس . ويصير أحد قطبي فلك البروج أبدي الظهور والثاني أبدي الخفاء ، لا يماسّان الأفق في دورتهما إلاّ مرة واحدة وذلك عند انتهاء نقطة المنقلب الذي يكون في جهة القطب الظاهر إلى سمت الرأس ؛ وحينئذ تقطع منطقة البروج الأفق على قوائم فقط . وتصير الأظلال في جميع السنة إلى جهة القطب الظاهر . وارتفاعات الشمس تتزايد من أحد الانقلابين إلى الآخر ، ثم ترجع وتتناقص إلى أن تعود إليه . وتصير فصول السنة أربعة لا غير .

[٤] القسم الثالث ما يكون عرضها زائداً على الميل الكلي وناقصاً من تمامه . وهناك لا تنتهي الشمس إلى سمت الرأس ، ويكون / لها / ارتفاعان : أعلى وهو يكون بقدر مجموع الميل الكلي وتمام عرض البلد ، وأسفل وهو يكون بقدر فضل تمام عرض البلد على الميل الكلي . ويكون سائر الأحوال كما بيّنّا . فإن كان عرض البلد لا يزيد على الميل الكلي بقدر عروض سائر السيّارة مرّ منها بسمت الرأس مرّتين ما زاد عرضه على فضل عرض البلد على الميل الكلي ، ومرة ما ساوى عرضه الفضل . وفي هذه العروض يزداد تعديل النهار وسعة المشرق والمغرب بازدياد العرض .

/17/ ...] β = لها [M , α .

[5] The fourth division is that whose latitude is equal to the complement of the obliquity. The day-circle of the solstice that is in the direction of the visible pole becomes permanently visible there, while the day-circle of the other solstice is permanently invisible. The day-circle of the visible pole of the ecliptic orb passes over the zenith, and the day-circle of the other pole over the nadir [*lit.*, that opposite it]. Then when the visible solstice comes to touch the horizon, it will do so at the point of the pole of the prime vertical that is in the direction of the visible pole, while the invisible solstice will touch it at the other pole; the two poles [of the ecliptic] will thereupon be at the zenith and the nadir [*lit.*, that opposite it], and the ecliptic equator will coincide with the horizon. Then when the pole departs from the zenith, and the visible solstice rises from [the horizon], the eastern half of the [ecliptic] equator rises in one stroke from the horizon. The point subsequent to the invisible solstice will then be upon the pole of the prime vertical, being about to set, and the point subsequent to the visible solstice will be at its other pole, being about to rise. The visible half is what is between them, *i.e.* the half that the vernal equinox is in the middle of if the visible pole is northerly, or the autumnal if it is southerly; the invisible half is the other half. Then the invisible half will rise point by point in all parts of the eastern half of the horizon, and the visible half will similarly set point by point, during the period of a nychthemeron [*lit.*, the day with its night] until the position of the orb returns to its original condition. [The maximum for] each of the ortive amplitude and the equation of daylight will there be a quarter revolution. Daylight will increase until the measure of the nychthemeron becomes entirely daylight; night will thereafter occur, increasing until the measure of the nychthemeron becomes entirely night. The altitude of the sun increases until it reaches twice the obliquity; it will then begin to decrease, decreasing until it becomes zero [whereupon] the sun will touch the horizon. The rising of a half revolution of the ecliptic equator occurs with a revolution of the equinoctial; the rising of the other half of the ecliptic equator does not [require] time.

[٥] القسم الرابع ما يكون / عرضها / مساوياً لتمام الميل الكلي . وهناك يصير مدار المنقلب الذي يكون في جهة القطب الظاهر أبدي الظهور ومدار المنقلب الآخر أبدي الخفاء . ويمرّ مدار قطب فلك البروج الظاهر بسمت الرأس ومدار القطب الآخر بمقابله . فإذا وافى المنقلبُ الظاهرُ مماسةَ الأفق ، ماسّهُ على نقطة قطب أول السموت التي في جهة القطب الظاهر ، وماسّه المنقلب الخفي على القطب الآخر ، وصار القطبان على سمت الرأس ومقابله ، وانطبقت منطقة البروج على الأفق . ثم إذا زال القطب عن سمت الرأس وارتفع المنقلب الظاهر عنه ، ارتفع النصف الشرقي من المنطقة دفعة عن الأفق ؛ فيكون الجزء التالي للمنقلب الخفي على قطب أول السموت يريد الغروب ، والجزء التالي للمنقلب الظاهر على قطبه الآخر يريد الطلوع . ويكون النصف الظاهر ما بينهما ، أعني النصف الذي يتوسّطه الاعتدال الربيعي إن كان القطب الظاهر شمالياً أو الخريفي إن كان جنوبياً ؛ والنصف الخفي هو النصف الآخر . ثم يطلع النصف الخفي جزءاً بعد جزء في جميع أجزاء نصف الأفق الشرقي ، ويغيب النصف الظاهر جزءاً بعد جزء كذلك ، في مدة اليوم بليلته إلى أن يعود وضع الفلك إلى حاله الأولى . ويكون هناك كل واحد من سعة المشرق وتعديل النهار ربعاً من الدور . وزيادة النهار إلى أن يصير مقدار يوم بليلته نهاراً كله ؛ ثم يحدث ليل ويزيد إلى أن يصير مقدار يوم بليلته ليلة كله . ويزيد ارتفاع الشمس إلى أن يبلغ ضعف الميل الكلي ؛ ثم يأخذ في التناقص ويتناقص إلى أن يفنى ، وتماسّ الشمس الأفق . ويكون طلوع نصف دور من منطقة البروج مع دور من معدل النهار ؛ وطلوع النصف الآخر من منطقة البروج لا في زمان .

/1/ ...] β = عرضه] M, α .

CHAPTER FIVE

On the Characteristics of Locations Whose Latitude Exceeds the Complement of the Obliquity But Does Not Reach One-Quarter Revolution

[1] In these locations, the largest permanently visible day-circle will intersect the ecliptic equator at two points whose declinations in the direction of the visible pole are equal. The largest permanently invisible day-circle will intersect it at two points opposite these two in the direction of the invisible pole. The ecliptic equator is divided into four arcs: one of them is permanently visible at the middle of which is the solstice that is in the direction of the visible pole; the second is permanently invisible at the middle of which is the other solstice. The endpoints of the first arc touch the horizon but do not disappear, and the end-points of the second arc touch it but do not rise. As for the two remaining arcs, the one at the middle of which is the first of Aries rises in reverse order and sets in regular order if the visible pole is northerly, and it rises in regular order and sets in reverse order if the visible pole is southerly; and the one at the middle of which is the first of Libra is the opposite of this. The visible solstice has a highest altitude, which is equal to the sum of the obliquity and the local colatitude along the meridian circle in the direction of the invisible pole, and a lowest [altitude], which is equal to the excess of the local latitude over the complement of the obliquity along the meridian circle in the direction of the visible pole. The visible pole of the ecliptic orb also has a highest altitude, which is equal to the sum of the local colatitude and the complement of the obliquity, and a lowest, which is equal to the excess of the local latitude over the obliquity; the pole will be simultaneously with the solstice on the meridian, but [they are] in opposite directions from the zenith and their altitudes are at opposite [extremes]. One may draw analogous conclusions from this for the situation of the invisible solstice and the invisible pole.

[2] In order to conceive positions in these latitudes, we shall take an example. Let the latitude in the north be 70: the arc that is permanently visible will be Gemini

الفصل الخامس
في خواصّ المواضع التي يجاوز عرضها تمام الميل الكلي ولا يبلغ ربع الدور

[١] في هذه المواضع يكون أعظم المدارات الأبدية الظهور قاطعاً لمنطقة البروج على نقطتين يتساوى ميلهما في جهة القطب الظاهر ؛ وأعظم المدارات الأبدية الخفاء قاطعاً لها على نقطتين متقابلتين لهما في جهة القطب الخفي . وتنقسم منطقة البروج إلى أربع قسي : إحداها أبدية الظهور ، وهي التي يتوسطها المنقلب الذي يكون في جهة القطب الظاهر ؛ والثانية أبدية الخفاء ، وهي التي يتوسطها المنقلب الآخر . وطرفا القوس الأولى يماسّان الأفق ولا يغيبان ، وطرفا القوس الثانية يماسّانه ولا يطلعان . وأما القوسان الباقيتان فالتي يتوسطها أول الحمل تطلع معكوسة وتغرب مستوية إن كان القطب الظاهر شمالياً ، وتطلع مستوية وتغرب معكوسة إن كان القطب الظاهر جنوبياً ؛ والتي يتوسطها أول الميزان تكون بالضد من ذلك . ويكون للمنقلب الظاهر ارتفاعان : أعلى ، وهو يكون بقدر مجموع الميل الكلي وتمام عرض البلد على دائرة نصف النهار في جهة القطب الخفي ؛ وأسفل ، وهو يكون بقدر فضل عرض البلد على تمام الميل الكلي على دائرة نصف النهار في جهة القطب الظاهر . ويكون لقطب فلك البروج الظاهر أيضاً ارتفاعان : أعلى ، وهو يكون بقدر مجموع تمام عرض البلد وتمام الميل الكلي ؛ وأسفل ، وهو يكون بقدر فضل عرض البلد على الميل الكلي . ويكون القطب مع المنقلب على نصف النهار معاً ولكن في الجهتين المتقابلتين عن سمت الرأس والارتفاعين المتبادلين . وقِس عليه حال المنقلب الخفي والقطب الخفي .

[٢] ولكي نتصور الأوضاع في هذه العروض نمثّل له مثالاً . وليكن العرض في الشمال سبعين : والقوس الأبدية الظهور الجوزاء

and Cancer, and the arc that is permanently invisible will be Sagittarius and Capricornus; the arc that rises in reverse order and sets in regular order is from the first of Aquarius to the last of Taurus, and that which rises in regular order and sets in reverse order is from the first of Leo to the last of Scorpius. Then when the first of Cancer is at the meridian on the southern side, its altitude being at its maximum [value], namely (43 + ⅓ + ¼)°, the visible pole of the ecliptic orb is on the northern side on the meridian as well and its altitude is at its minimum [value], which is (46 + ¼ + ⅙)°. At the rising place of the equinox is the first of Libra, which is on the verge of rising, and at its setting place is the first of Aries, which is on the verge of setting. The visible half of the ecliptic orb [extends] from west to east in the south as in this illustration:

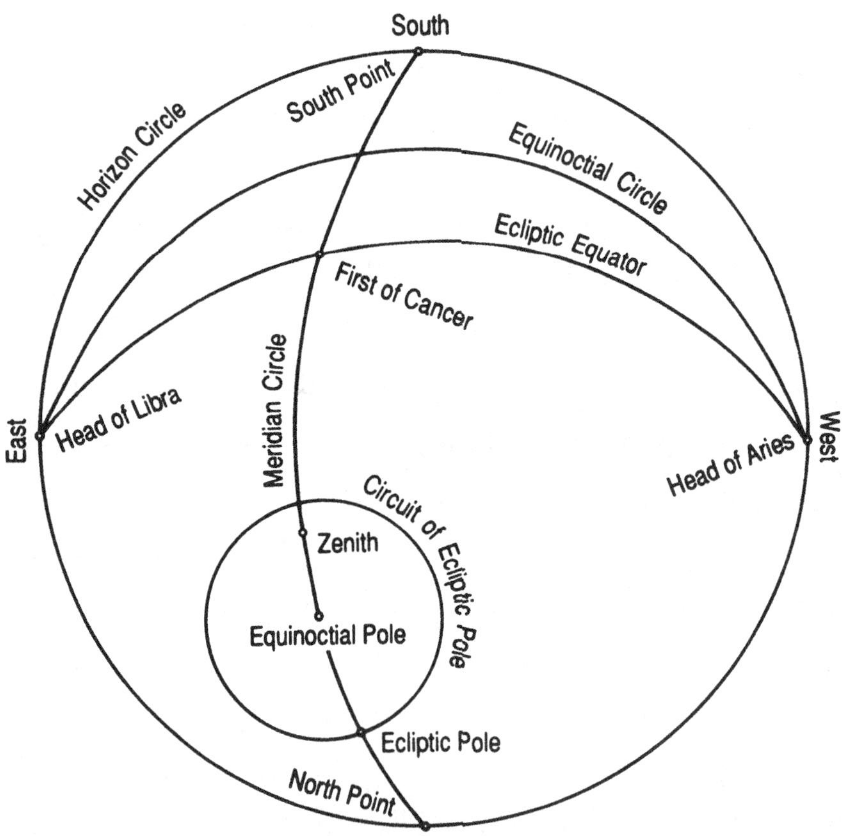

[Fig. T20]

والسرطان ، والقوس الأبدية الخفاء القوس والجدي ؛ والقوس التي تطلع معكوسة وتغرب مستوية من أول الدلو إلى آخر الثور ، والتي تطلع مستوية وتغرب معكوسة من أول الأسد إلى آخر العقرب . فإذا كان أول السرطان على نصف النهار من جانب الجنوب وارتفاعه في غاية الزيادة وهو ثلاث وأربعون درجة وثُلث وربع ، كان قطب فلك البروج الظاهر من جانب الشمال أيضاً على نصف النهار وارتفاعه في غاية النقصان وهو ست وأربعون درجة وربع وسدس . ويكون على مطلع الاعتدال أول الميزان يريد الطلوع ، وعلى مغيبه أول الحمل يريد الغروب . ونصف فلك البروج الظاهر من المغرب إلى المشرق في الجنوب على هذه الصورة :

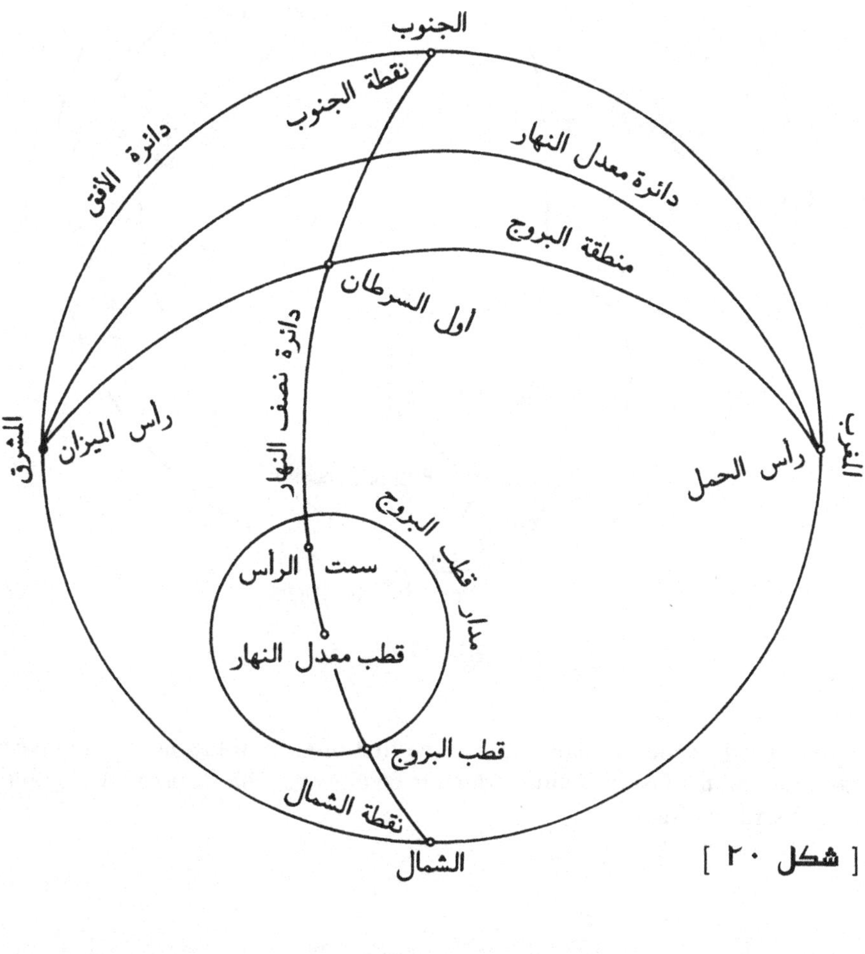

[شكل ٢٠]

[3] Then let the orb move with the first motion so that Libra and Scorpius will rise in regular order, and their ortive amplitude will span the southeastern quarter [of the horizon]. Aries and Taurus will set likewise in regular order, and their occasive amplitude will span the northwestern quarter [of the horizon]. The first of Cancer will begin its descent toward the west and the pole of the ecliptic orb its ascent toward the east until the first of Sagittarius comes to touch the horizon at the south point and the first of Gemini [comes] to touch the horizon at the north point. The visible half of the ecliptic equator comes to be on the western side [extending] from south to north as in this illustration:

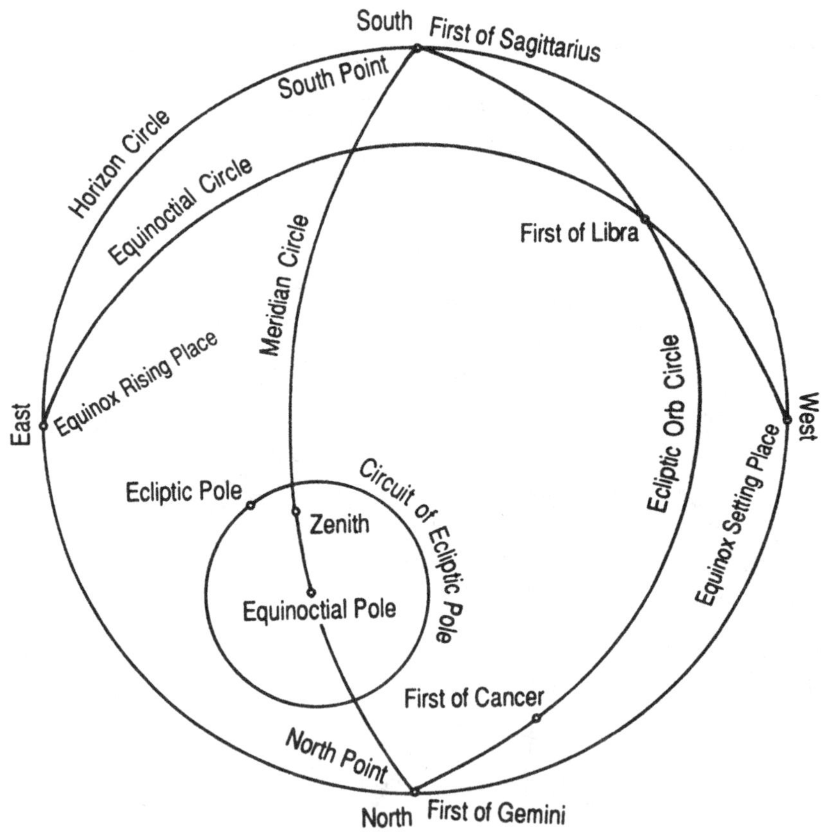

[Fig. T21]

[4] Then let the orb move so that the first of Gemini begins its rise toward the east, and the last of Taurus, which is contiguous with it, rises little by little until Taurus has risen.

[٣] ثم ليتحرك الفلك بالحركة الأولى : فيطلع الميزان والعقرب مستويين وليستغرق الربع الشرقي الجنوبي سعة مشرقهما ؛ ويغرب الحمل والثور أيضاً مستويين وليستغرق الربع الغربي الشمالي سعة مغربهما . وليأخذ أول السرطان في / الانحطاط / نحو المغرب وقطب فلك البروج في / الارتفاع / نحو المشرق إلى أن ينتهي أول القوس إلى مماسّة الأفق على نقطة الجنوب وأول الجوزاء إلى مماسّة الأفق على نقطة الشمال . ويصير النصف الظاهر من منطقة البروج في الجانب الغربي من الجنوب إلى الشمال على هذه الصورة :

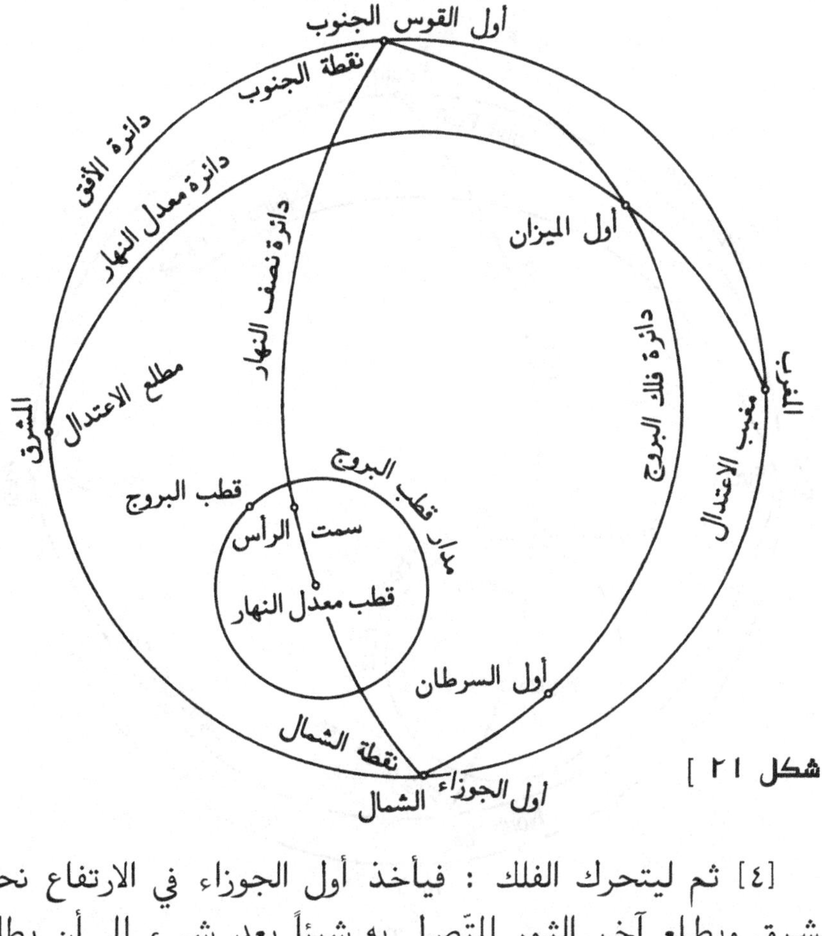

[شكل ٢١]

[٤] ثم ليتحرك الفلك : فيأخذ أول الجوزاء في الارتفاع نحو المشرق ويطلع آخر الثور المتّصل به شيئاً بعد شيء إلى أن يطلع

/4/ ...] انحطاط = β [M , α . /5/ ...] ارتفاع = β [M , α .

Then Aries rises from its last to its first. The ortive amplitude of these two signs spans the northeastern quarter [of the horizon], and the first of Aries reaches its rising place. Directly opposite this, the first of Sagittarius begins its descent below the horizon, and the last of Scorpius, which is contiguous with it, sets little by little until Scorpius has disappeared; then Libra sets from its last to its first. Their occasive amplitude spans the southwestern quarter [of the horizon], and the first of Libra reaches its setting place. The first of Cancer reaches the meridian circle on the northern side; it is at its lowest altitude, which is $(3 + \frac{1}{3} + \frac{1}{4})^\circ$. The pole of the ecliptic orb [reaches] its highest altitude [which is] on the southern side, and this is $(86 + \frac{1}{4} + \frac{1}{6})^\circ$. The visible half of the ecliptic orb is on the northern side between the rising and setting places of the equinox, the directional sequence [of the signs here being] opposite the conventional one, as in this illustration:

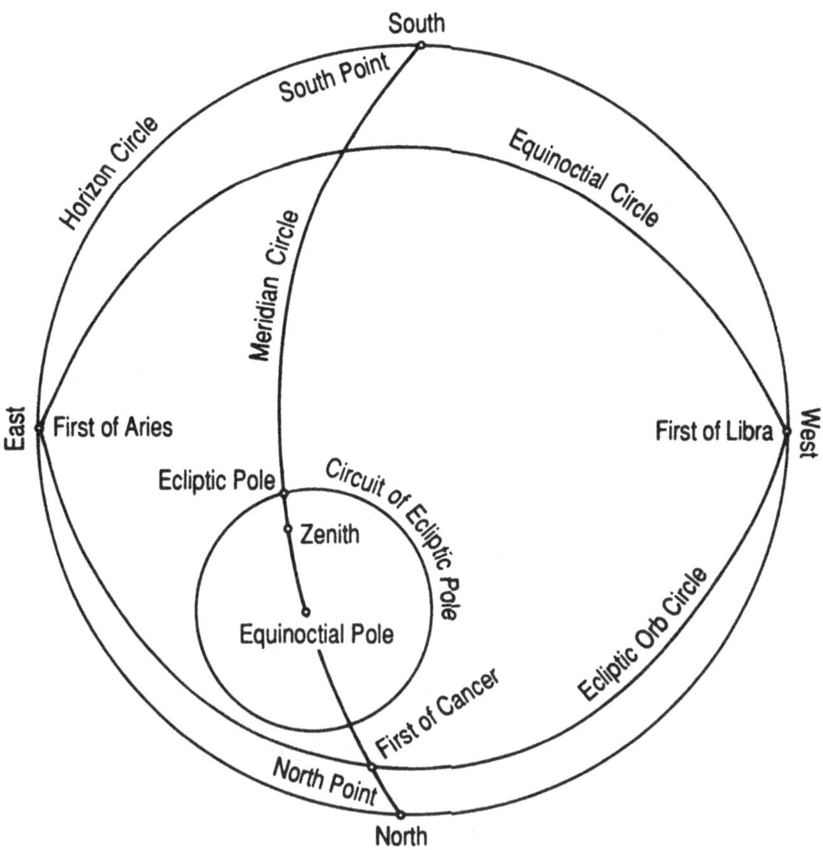

[Fig. T22]

الثور ، ثم يطلع آخر الحمل إلى أوله ؛ ويستغرق الربع الشمالي الشرقي سعة مشرق هذين البرجين ، وينتهي أول الحمل إلى مطلعه . ويأخذ بإزاء ذلك أول القوس في الانحطاط تحت الأفق ويغرب آخر العقرب المتّصل به شيئاً بعد شيء إلى أن يغيب العقرب ، ثم يغرب آخر الميزان إلى أوله ؛ ويستغرق الربع الجنوبي الغربي سعة مغربهما ، وينتهي أول الميزان إلى مغيبه . وينتهي أول السرطان إلى دائرة نصف النهار في جانب الشمال ويكون في ارتفاعه الأسفل وهو ثلاث درجات وثُلث وربع ، وقطب فلك البروج إلى ارتفاعه الأعلى في جانب الجنوب وهو ست وثمانون درجة وربع وسدس . ويكون النصف من فلك البروج الظاهر في جانب الشمال بين مطلع الاعتدال ومغيبه على توالي مخالف للمعهود على هذه الصورة :

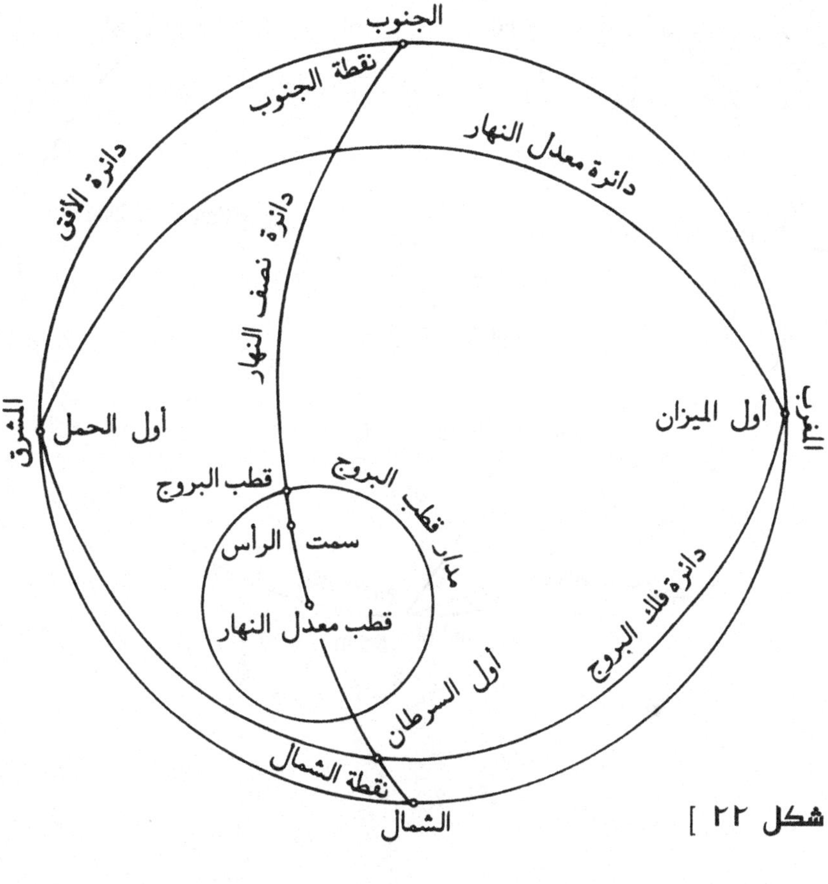

[شكل ٢٢]

[5] Then let the orb move so that Pisces rises from its last to its first, then Aquarius from its last to its first. Their ortive amplitude spans the southeastern quarter [of the horizon]. Directly opposite them, Virgo disappears from its last to its first, then Leo from its last to its first. Their occasive amplitude spans the northwestern quarter [of the horizon]. The first of Aquarius comes to touch the horizon at the south point, and the first of Leo comes to touch the horizon at the north point; the visible half of the ecliptic circle is in between them in the eastern direction. The first of Cancer has risen higher on the eastern side, and the pole has begun its descent on the western side as in this illustration:

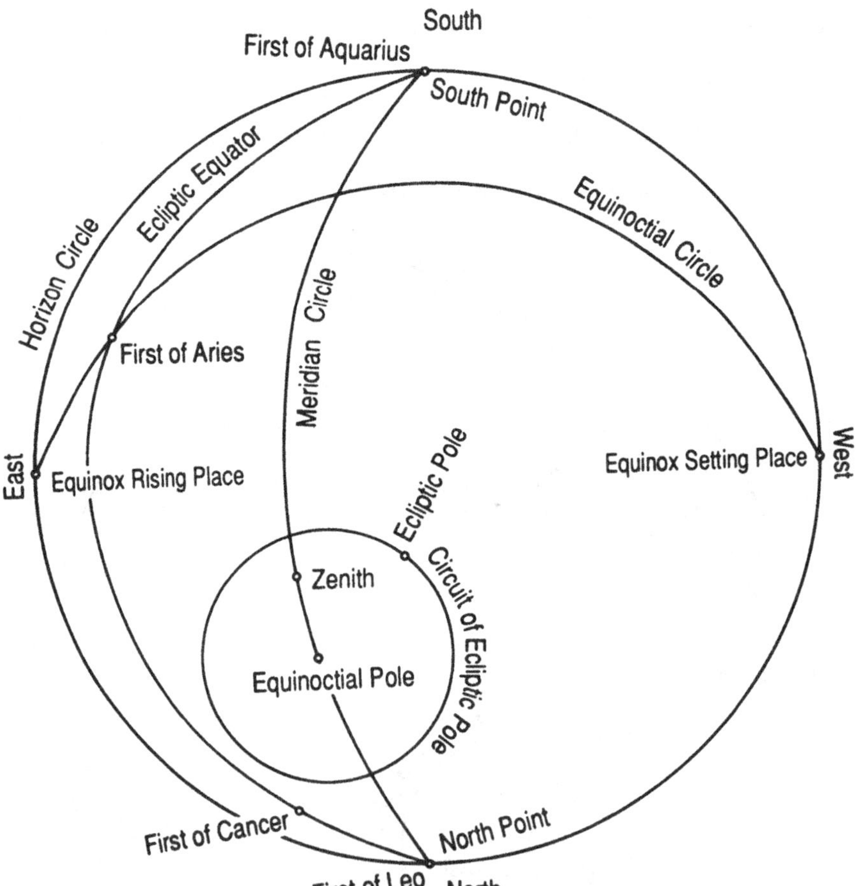

[Fig. T23]

[٥] ثم ليتحرك الفلك : فيطلع آخر الحوت إلى أوله ، ثم آخر الدلو إلى أوله ؛ ويستغرق الربع الشرقي الجنوبي سعة مشرقهما . ويغيب بإزائهما آخر السنبلة إلى أولها ، ثم آخر الأسد إلى أوله ؛ ويستغرق الربع الغربي الشمالي سعة مغربهما . ويصير أول الدلو على نقطة الجنوب مماساً للأفق ، وأول الأسد على نقطة الشمال مماساً للأفق ، ونصف دائرة البروج الظاهر فيما بينهما من جهة المشرق . وأول السرطان قد ارتفع في جانب المشرق ، والقطب قد أخذ في الانحطاط في جانب المغرب على هذه الصورة :

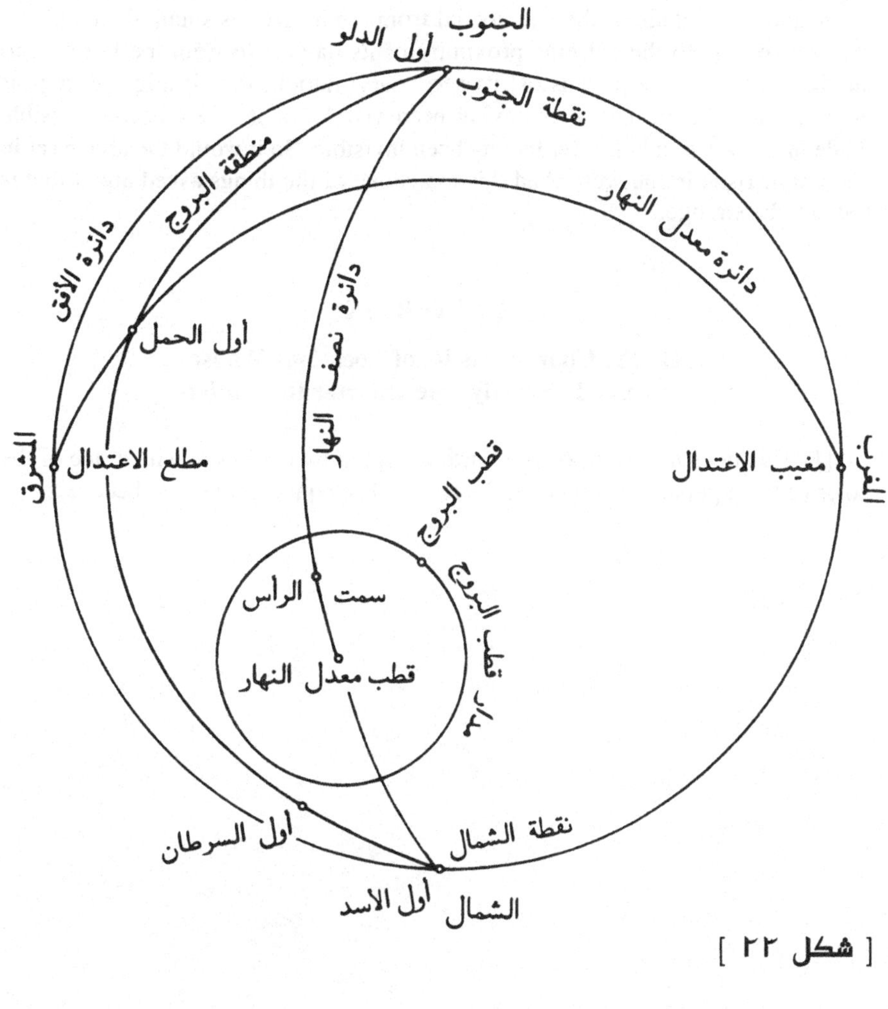

[شكل ٢٢]

[6] Then let the orb move so that the first of Leo rises from the horizon moving toward the eastern half; the points of Leo will then rise sequentially until its last, then the points of Virgo. Their ortive amplitude spans the northeastern quarter [of the horizon]. Directly opposite this, the first of Aquarius will drop below the horizon to beneath the Earth. Thus Aquarius will set followed by Pisces sequentially. Their occasive amplitude spans the southwestern quarter [of the horizon]. The [action of] rising will then have reached the first of Libra and that of setting the first of Aries. At that time the first of Cancer will have come to be at the meridian circle at its highest altitude, and the visible pole of the ecliptic orb will be at its lowest altitude on the meridian. The visible half of the ecliptic orb comes to be on the southern side, and the situation returns to what we assumed originally. So the revolution is completed and what we have [earlier] described should become clear. We have been excessively verbose in this chapter only because of the difficulty of imagining these positions.

[7] In these regions [*lit.*, horizons] when the local latitude approaches the extreme and the altitude of the equinoctial from the horizon is small, then a planet may move, due to the extreme proximity of its day-circle from the horizon, to another day-circle by [means of] its secondary motion; thus it might disappear while in the eastern half after having been visible, or it might become visible while in the western half after having been invisible. So it would then have set in the east or risen in the west. And this too is one of the things asked about that is found to be strange.

CHAPTER SIX

On the Characteristics of Locations Whose Latitude Is Exactly One-Quarter Revolution

[1] This does not occur on the Earth except at two points at which one of the poles of the equinoctial is there at the zenith. The equinoctial circle becomes

[٦] ثم ليتحرك الفلك : وارتفع أول الأسد عن الأفق آخذاً نحو النصف الشرقي ، فتطلع أجزاء الأسد على التوالي إلى آخره ، ثم أجزاء السنبلة ؛ ويستغرق الربع الشمالي الشرقي سعة مشرقهما. وبإزاء ذلك ينخفض أول الدلو عن الأفق إلى تحت الأرض ، فيغرب الدلو ثم الحوت على التوالي ؛ ويستغرق الربع الجنوبي الغربي سعة مغربهما . ثم ينتهي الطلوع إلى أول الميزان والغروب إلى أول الحمل . ويصير حينئذ أول السرطان إلى دائرة نصف النهار وارتفاعه الأعلى ، والقطب الظاهر من فلك البروج إلى ارتفاعه الأسفل من نصف النهار . ويصير النصف الظاهر من فلك البروج في جانب الجنوب ، ويعود الوضع إلى ما فرضناه مبدأً ؛ فيتم الدور ويتَضح ما وصفناه . وإنما أطنبنا القول في هذا الفصل لعسر تصور هذه الأوضاع .

[٧] وفي هذه الآفاق إذا قرب عرض البلد من النهاية وصار ارتفاع معدل النهار من الأفق قليلاً ، فربما ينتقل كوكب /لقرب/ مداره من الأفق جداً إلى مدار آخر بحركته الثانية ؛ فيغيب بعد ما كان ظاهراً وهو في النصف الشرقي أو يظهر بعد ما كان خفياً وهو في النصف الغربي . فيكون قد غرب في المشرق أو طلع من المغرب . وهذا أيضاً من الأسئلة المستغربة .

الفصل السادس
في خواصّ المواضع التي يكون عرضها ربعاً من الدور سواءً

[١] وذلك لا يكون على الأرض إلا عند نقطتين يكون أحد قطبي معدل النهار على سمت الرأس هناك . وتصير دائرة معدل

/14/ ...] M , β = بقرب] α .

coincident with the horizon, and the orb turns with the first motion in a spinning manner; there is no longer on the horizon an east or a west. Thus the half of the orb that is in the direction of the visible pole from the equinoctial is permanently visible, and the other half is permanently invisible.

[2] As long as the sun is in the visible half of the ecliptic orb, it will be daytime; and as long as it is in the invisible half of it, it will be night. Thus its [the sun's?] entire year will be a nychthemeron [*lit.*, a day with its night], with one [*i.e.* daytime or night] exceeding the other due to the variability in speed [*lit.*, the slowness and fastness] of the [sun's] motion. Beneath the northern pole at the present time, their daytime is greater than their night by seven of our nychthemerons. This is because the apogee of the sun is at the end part of Gemini and its perigee is at the end part of Sagittarius. The period for the setting of dusk or the rising of dawn occurs for them in 50 of our days as we shall explain when describing these two [terms] later on. The maximum altitude of the sun and its maximum depression is in the amount of the maximum declination [*i.e.* the obliquity]. Neither the rising nor the setting of the sun and planets that is due to the secondary motion will occur at the same place on the horizon.

[3] Stars whose latitude is less than the obliquity will have a rising and a setting; the periods of visibility and invisibility will vary depending on how far or how near their day-circle is from the ecliptic orb [*sic*]. Stars whose latitude is equal to the obliquity will touch the horizon once during one revolution of the second motion; neither they nor those [stars] whose latitude exceeds the obliquity will have a rising or a setting, but instead they will be either permanently visible or invisible. Let it be recalled what we said regarding the positions of the orb due to the first two motions, and let [the situation] be determined here on that basis.

[4] This is the last of the descriptions of the localities that are beneath the day-circles and what is comparable to them.

النهار منطبقة على الأفق ، ويدور الفلك بالحركة الأولى رَحَوية ولا يبقى في الأفق مشرق ولا مغرب . فيكون النصف من الفلك الذي يكون من معدل النهار في جهة القطب الظاهر أبدي الظهور والنصف الآخر أبدي الخفاء .

[٢] والشمس ما دامت في النصف الظاهر من فلك البروج يكون نهاراً ، وما دامت في النصف الخفي منه يكون ليلاً ؛ فتكون سنتها كلها يوماً بليلته ويفضل أحدهما على الآخر من جهة بطء حركتها وسرعتها . فيكون تحت القطب الشمالي في هذا التأريخ نهارهم أكبر من ليلهم بسبعة أيام بلياليها من أيامنا ، وذلك لكون أوج الشمس في أواخر الجوزاء وحضيضها في أواخر القوس . وتكون مدة غروب الشفق أو طلوع الصبح لهم في خمسين يوماً من أيامنا على ما تتبين عند وصفهما فيما بعد . وتكون غاية ارتفاع الشمس وغاية انحطاطها بقدر غاية الميل . ويكون طلوع الشمس والكواكب بالحركة الثانية وغروبها لا في موضع بعينه من الأفق .

[٣] ويكون للكواكب التي عرضها ينقص من الميل كله طلوع وغروب ؛ وتختلف مدّتا الظهور والخفاء بحسب بُعد مدارها عن فلك البروج وقربها إليه . والكواكب التي عرضها مساوٍ للميل كله تماس الأفق في دور واحد من الحركة الثانية مرة واحدة ؛ ولا يكون لها ولا للتي يزيد عرضها على الميل الكلي طلوع ولا غروب بل تكون إما ظاهرة وإما خفية أبداً . وليتذكر ما قُلنا في أوضاع الفلك بسبب الحركتين الأوليين ، وليحكم هاهنا بحسب ذلك .

[٤] وهذا آخر أوصاف البقاع التي تحت المدارات اليومية وما يجري مجراها .

CHAPTER SEVEN

On the Co-ascensions of the Ecliptic

[1] The arc on the equinoctial that rises with a given arc on the ecliptic orb [*sic*] is called the **co-ascension** of that arc; the arc on the ecliptic orb is referred to as the **equal degrees**.

[2] The co-ascension varies according to the changing horizons. At the [Earth's] equator, each quarter that is bounded by two of the four [cardinal] points will rise with a quarter because when the equinox point, which is simultaneously one of the two boundaries for two quarters for [each of] the two equators, reaches the zenith, the solstitial colure will coincide with the horizon; thus the solstice point will be on the horizon and the two other boundaries [*i.e.* other than the equinox] of the two quarters will both be on the horizon. One should draw analogous conclusions from this for the remaining quarters. But 30 time units, *i.e.* (½ of ⅙) of the equinoctial, will not rise with a zodiacal sign, for example one that adjoins one of the four [cardinal] points, which is / (⅙ of ½) / of the ecliptic equator. This is so because if the zodiacal sign is one that adjoins the equinox point, then one of the boundaries, namely this point, is common to both [the sign and its co-ascension]. When the zodiacal sign's other boundary reaches the horizon, there occurs a triangle [formed] from the sign, the arc that rises with it along the equinoctial, *i.e.* the co-ascension, and what is between them along the horizon; its angle that is bounded by the equinoctial and the horizon is right, and the remaining two are acute. Since the sign is the subtense of a right [angle], and its co-ascension is the subtense of an acute angle, the sign is larger than its co-ascension; the same will hold for two signs that adjoin the equinox point and their co-ascension. Now if the zodiacal sign is one that adjoins the solstitial point, its co-ascension will be larger than it. This is because what remains [after taking] the co-ascension of the [above] two signs, which is less than ⅙ of a revolution, from a full quarter will be greater than (½ of ⅙) of a revolution, and [this remainder] will rise with the remaining sign.

/12/ ⅙ of ½] β, M = ½ of ⅙] α.

الفصل السابع
في مَطالِع البروج

[١] القوس من معدل النهار التي تطلع مع قوس مفروضة من فلك البروجِ يقال لها مَطالِـع تلك القوس ، ويقال للقوس من فلك البروجِ الدرَج السَوَاء .

[٢] والمطالع تختلف بحسب اختلاف الآفاق . أما في خط الاستواء فكل ربع يتحدّد بنقطتين من النقط الأربع يطلع مع ربع لأنّ نقطة الاعتدال التي هي أحد حدّي الربعين من المنطقتين معاً إذا انتهت إلى سمت الرأس انطبقت الدائرة المارة بالأقطاب الأربعة على الأفق ؛ فتكون على الأفق نقطة الانقلاب ويكون الحداَن الآخران للربعين معاً على الأفق . وقِس عليه سائر الأرباع . ولا يطلع مع برج مثلاً يلي إحدى نقط الأرباع ، وهو / سدس نصف / منطقة البروجِ ، ثلاثون زماناً أعني نصف سُدس معدل النهار ؛ وذلك لأنّ البرج إن كان مما يلي نقطة الاعتدال كان أحد حدّيهما مشتركاً وهو تلك النقطة . وإذا انتهى الحدّ الآخر للبرج إلى الأفق حدث من البرج والقوس الطالعة معه من معدل النهار – أعني مطالعه – ومما يقع بينهما من الأفق مثلث زاويته التي يحيط بها معدل النهار والأفق قائمة والباقيتان حادّتان . فلكون البرج وترَ قائمةٍ ومطالعه وترَ حادةٍ ، يكون البرجِ أعظم من مطالعه ؛ وكذلك القول في برجين يليان نقطة الاعتدال ومطالعهما . أما إن كان البرج مما يلي نقطة الانقلاب فيكون مطالعه أعظم منه ، وذلك لأنّ الباقية من مطالع البرجين التي هي أصغر من سدس الدور إلى تمام الربع تكون أعظم من نصف سدس الدور ، وهي تطلع مع البرج الباقي .

/12/ ...] M , β = نصف سدس] α .

It is apparent from the above that for any two equal arcs that are equidistant from one of the four [cardinal] points, *i.e.* the two equinoxes and the two solstices, their co-ascensions at the equator will be equal. The ecliptic equator may be divided into four segments whose initial [points] are at the midparts of the quarters [of the ecliptic]. Each segment at the midpart of which falls one of the equinoxes will be greater than its co-ascension; each segment at the midpart of which falls one of the solstices is smaller than its co-ascension.

[3] The passage of the equinoctial and the ecliptic equator across the meridian circles in all localities is as their rising for the equator since each one of [these meridian circles] is one of the horizons for the equator; the same conclusion holds for all the declination circles. The co-descensions are as the co-ascensions for these horizons [*i.e.* those of the equator].

[4] As for the oblique horizons, a quarter will not rise with a quarter since the plane of the equinoctial is not perpendicular to the plane of the horizon. A half will rise with a half when they are bounded by the two equinox points. When an arc adjoining the equinox point that is away from the equinoctial in the direction of the visible pole rises, it will be larger than its co-ascension because in the above-mentioned triangle it will be the subtense of an obtuse [angle] and its co-ascension will be the subtense of an acute [angle]. And if it is away from the equinoctial in the direction of the invisible pole, then its co-ascension is greater than [the arc] because the disposition becomes the opposite of what it was [before]. It is apparent from the above that for equal arcs that are equidistant from one of the equinox points their co-ascensions are equal. The orb [*i.e.* the ecliptic equator] may be divided into two segments: one of them is bisected by that equinox which is such that a star, when crossing it, comes to be in the direction of the visible pole; the other is that bisected by the other equinox. The first is larger than its co-ascension, while the other is smaller. The co-ascensions of northern arcs for northern horizons are like [*i.e.* equal to] the co-ascensions of the corresponding [equivalent arcs] opposite to them in the south for the [equivalent] southern horizons; the same holds for the southern [arcs]. The co-descension of any arc for any horizon is like [*i.e.* equal to] the co-ascension of the arc directly opposite.

[5] As for those regions [*lit.*, horizons] in which the day-circles of the two solstice points are the largest of the permanently visible and invisible day-circles, we have shown that a half of the ecliptic orb

وقد ظهر من ذلك أنّ كل قوسين متساويتين متساويتي البعد عن إحدى النقط الأربع ــ أعني الاعتدالين والانقلابين ــ فمطالعهما في خط الاستواء متساوية . ومنطقة البروج تنفصل إلى أربع قطع تكون مبادئها أواسط الأرباع . ويكون كل قطعة يقع في وسطها أحد الاعتدالين أعظم من مطالعها ، وكل قطعة يقع في وسطها أحد الانقلابين أصغر من مطالعها .

[٣] ومرور معدل النهار ومنطقة البروج على دوائر أنصاف النهار في جميع البقاع يكون كطلوعها في خط الاستواء لأنّ كل واحدة منها أفق من آفاق خط الاستواء ؛ وكذلك الحكم في جميع دوائر الميول . والمغارب كالمطالع في تلك الآفاق .

[٤] وأما في الآفاق المائلة فلا يطلع ربع مع ربع لكون سطح معدل النهار غير قائم على سطح الأفق ؛ ويطلع نصف مع نصف إذا كانا متحدّدين بنقطتي الاعتدالين . وإذا طلعت قوس تلي نقطة الاعتدال وكانت من معدل النهار في جهة القطب الظاهر ، فهي أعظم من مطالعها لأنها في المثلث المذكور تكون وترَ منفرجةٍ ومطالعها وترَ حادةٍ . وإن كانت من معدل النهار في جهة القطب الخفي فمطالعها أعظم منها لأنّ الحكم يصير بضدّ ما كان . ويظهر من ذلك أنّ القسي المتساوية التي تتساوى أبعادها عن إحدى نقطتي الاعتدال تكون مطالعها متساوية . والفلك ينقسم إلى قطعتين : إحداهما التي يتوسطها الاعتدال الذي إذا جاوزه الكوكب صار في جهة القطب الظاهر ، والأخرى التي يتوسطها الاعتدال الآخر . والأولى تكون أعظم من مطالعها والأخرى تكون أصغر . ومطالع القسي الشمالية في الآفاق الشمالية كمطالع نظائرها من الجنوبية في الآفاق الجنوبية ، وكذلك في الجنوبية . ومغارب كل قوس في كل أفق تكون كمطالع نظير تلك القوس .

[٥] وأما في الآفاق التي يكون فيها مدارا نقطتي الانقلابين أعظم المدارات الأبدية الظهور والخفاء ، فقد بيّنا أنّ نصفاً من فلك

rises with the entire equinoctial and the other half rises without [requiring] time; in setting, the two halves exchange [properties].

[6] As for the regions [*lit.*, horizons] in which there are arcs of the ecliptic orb that are permanently visible and invisible: let the example be the one in the northern region we have [previously] used, namely the region whose latitude is 70 where Gemini and Cancer are permanently visible and Sagittarius and Capricornus are permanently invisible. When the vernal equinox point rises, Pisces will rise after it in reverse order from last to first, then Aquarius in reverse order from last to first; then will begin the rising of Leo from its first in regular order, then Virgo, then Libra, then Scorpius, likewise [in regular order]. Now when the first of Sagittarius is reached, the last of Taurus will begin to rise reversed, and Taurus and Aries will rise in reverse order. Thereupon the vernal equinox point will return to the horizon. One should draw analogous conclusions from this for the remaining regions [*lit.*, horizons], and from rising for setting.

CHAPTER EIGHT

On the Lengths of the Nychthemerons

[1] The **nychthemeron** [*lit.*, a day with its night] is the time that falls from either the sun's occurrence on the horizon—whether rising or setting—or else [from its occurrence] on the meridian until its return there after one complete revolution by the first motion. Its length is one revolution of the equinoctial increased by the amount of it that rises with the arc traversed by the sun during that nychthemeron.

[2] Since what is traversed by the sun is variable, the lengths of the nychthemerons will [also] be variable; for [the sun] will traverse in the far half from the Earth smaller arcs, while in the near half larger arcs, and furthermore [those arcs] that rise along the equinoctial with the arcs from the ecliptic orb are variable, [the former] being sometimes smaller, sometimes larger than [the latter]. However, the variation [in the nychthemerons] is imperceptible in one or two days due to the smallness of the difference; one does perceive it over many days.

البروج يطلع مع جميع معدل النهار والنصف الآخر يطلع لا في زمان ؛ وفي الغروب يتبادل النصفان .

[٦] وأما في الآفاق التي تكون فيها قسي من فلك البروج أبدية الظهور والخفاء – وليكن الأفق ما تمثّلنا به من الآفاق الشمالية ، وهو أفق عرضه سبعون ، والجوزاء والسرطان فيه أبديا الظهور ، والقوس والجدي أبديا الخفاء – فإذا طلعت نقطة الاعتدال الربيعي طلع بَعدها الحوت معكوساً من الآخر إلى الأول ، ثم الدلو معكوساً من الآخر إلى الأول ؛ ثم يبتدئ طلوع الأسد من أوله مستوياً ، ثم السنبلة ، ثم الميزان ، ثم العقرب كذلك . فإذا انتهى إلى أول القوس ابتدأ آخر الثور بالطلوع المعكوس ويطلع الثور والحمل معكوسين ، فتعود نقطة الاعتدال الربيعي إلى الأفق . وقِس عليه في سائر الآفاق ، والغروب على الطلوع .

الفصل الثامن
في مقادير الأيّام بلياليها

[١] اليوم بليلته هو الزمان الذي يقع بين كون الشمس إمّا على الأفق طالعة أو غاربة وإمّا على نصف النهار وبين عودها إلى هناك بعد دورة واحدة تامة بالحركة الأولى . ومقداره دورة من أدوار معدّل النهار مع زيادة تطلع منه مع القوس التي تقطعها الشمس في ذلك اليوم بليلته .

[٢] ولكون ما تقطعها الشمس مختلفاً – فإنّها تقطع في النصف البعيد من الأرض قسياً أصغر وفي النصف القريب قسياً أكبر ، وأيضاً ما يطلع من معدّل النهار مع القسي من فلك البروج مختلف فإنّه تارة يكون أصغر منها وتارة يكون أكبر – تكون مقادير الأيّام بلياليها مختلفة ؛ لكنّ اختلافها غير محسوس في يوم أو يومين لصغر التفاوت ، ويحسّ به في أيّام كثيرة . وأهل

The calculators, being obliged to use nychthemerons of equal size in order to find the mean and other motions, have taken the above increase to be in the amount of the mean motion of the sun during a nychthemeron. They named those days that were taken to be equal the **mean days**, each day of which is the measure of a revolution of the equinoctial plus the amount traveled by the mean sun in a day. As for the true state of affairs, one needs to find out all about each of the two differences.

[3] As for the difference that is due to the sun's variable speed: in the period that the sun moves from the apogee to the mean distance that follows it, it will be the increase of the sun's mean over its true position, [which is] in the amount of the maximum anomaly; in the period that it moves from the other mean distance to the apogee, it will be the same as this. Thus the increase of the mean over the true position in the far segment from the Earth of the sun's orb is in the amount of twice the anomaly; in the near segment, the increase of the true position over the mean is the same as this as well. The difference between the two segments is four times the anomaly.

[4] As for the difference due to the co-ascension: if the beginning of the day is made to be when the sun reaches the horizon, this difference will vary according to the changing horizons, and it will not be the exact same thing for all localities. This will be the case whether the beginning is [upon] its reaching the eastern horizon, at that place being according to the difference between the equal degrees and their co-ascension, or whether [the beginning] is upon its reaching the western horizon, at that place being according to the difference between the equal degrees and the co-ascension of the [arc] directly opposite them. But if the beginning of the day is made to be upon its reaching the meridian, the difference agrees for all horizons, and that is according to the co-ascension for the equator. They thus chose this rather than the first alternative.

[5] It has been mentioned above that the ecliptic orb may be divided into four segments. The two that are bisected by the equinoxes are greater than their co-ascensions, and they are from the middle part of Aquarius to the middle part of Taurus, and from the middle part of Leo to the middle part of Scorpius. The amount of excess of each one of them over its co-ascension is five degrees for the [Earth's] equator.

الحساب ، لمّا اضطرّوا إلى استعمال أيّام بلياليها متساوية الأقدار لمعرفة حركات الأوساط وغيرها ، أخذوا تلك الزيادة مقدار حركة الشمس الوسطى في يوم بليلته . وسمّوا تلك الأيّام المأخوذة بالتساوي الأيّام الوسطى ، كل يوم منها يكون مقدار دور من معدّل النهار مع سير وسط الشمس ليوم . وأمّا التحقيق فمحوج إلى معرفة جملة كل واحد من التفاوتين .

[٣] أمّا التفاوت الذي يكون بسبب اختلاف سير الشمس فيكون في المدّة التي تسير الشمس من الأوج إلى البعد الأوسط الذي يليه زيادةَ وسط الشمس على تقويمها بقدر غاية الاختلاف ، وفي المدة التي تسير من البعد الأوسط الآخر إلى الأوج مثل تلك . فتكون زيادة الوسط على التقويم في القطعة البعيدة من الأرض من فلك الشمس بقدر ضعف الاختلاف ؛ وتكون في القطعة القريبة زيادة التقويم على الوسط أيضاً بمثل ذلك . ويكون الفضل بين القطعتين بأربعة أمثال الاختلاف .

[٤] وأمّا التفاوت الذي يكون بسبب المطالع : فإن جعل مبادئ الأيّام انتهاءَ الشمس إلى الأفق اختلف ذلك التفاوت بحسب اختلاف الآفاق ، ولم يكن في جميع البقاع شيئاً واحداً بعينه ؛ ويكون ذلك إن كان المبدأ انتهاءها إلى الأفق الشرقي بحسب التفاوت بين درج السواء ومطالعها في ذلك الموضع ، وإن كان انتهاءها إلى الأفق الغربي بحسب التفاوت بين درج السواء ومطالع نظيرها في ذلك الموضع . وإن جعل مبادئ الأيّام انتهاءها إلى نصف النهار اتّفق التفاوت في جميع الآفاق ، ويكون ذلك بحسب مطالع خطّ الاستواء ؛ فاختاروا ذلك دون الوجه الأول .

[٥] وقد مرّ أنّ فلك البروج ينقسم إلى أربع قطع : اثنتان منها اللتان يتوسّطهما الاعتدالان تزيدان على مطالعيهما ، وهما من أواسط الدلو إلى أواسط الثور ومن أواسط الأسد إلى أواسط العقرب ؛ ومقدار زيادة كل واحدة منهما على مطالعها بخطّ

The other two segments, which are bisected by the solstices, are less than their co-ascensions, and they are from the middle part of Taurus to the middle part of Leo and from the middle part of Scorpius to the middle part of Aquarius. The amount of deficiency of each one of them from its co-ascension is likewise five degrees for the [Earth's] equator.

[6] When one additively combines the two differences when they are both additive or both subtractive, or takes the difference when they differ, there results the total amount of difference between the mean days and the true days during the year.

[7] One day must necessarily be taken as the initial one from which the other days are measured. Noon of that day is then the initial [time] for both mean and true days. For any day of the year that is taken to be the initial one, the difference between elapsed mean and elapsed true days [measured] from that day is sometimes additive and sometimes subtractive, except for [initial days at] the end part of Aquarius and the beginning part of Scorpius; for if the initial [point] is made the end part of Aquarius, the true days will always be shorter than the mean ones, and if it is made the beginning part of Scorpius, the true days will always be longer than the mean ones. / The practitioners / of the profession have agreed on making [the initial point] the end part of Aquarius.

/16/ The practitioners] β = Therefore the practitioners] α, M.

الاستواء خمس درجات . والقطعتان الأخريان وهما اللتان يتوسّطهما الانقلابان تنقصان عن مطالعيهما ، وهما من أواسط الثور إلى أواسط الأسد ومن أواسط العقرب إلى أواسط الدلو ؛ ومقدار نقصان كل واحدة منهما من مطالعها بخط الاستواء أيضاً خمس درجات .

[٦] وإذا تركّب التفاوتان بالجمع إذا كانا زائدين معاً أو ناقصين معاً أو بالتفريق إذا اختلفا ، حصل مقدار التفاوت بين الأيّام الوسطى والأيّام الحقيقية جملة في السنة .

[٧] ولا بدّ من يوم يُفرض مبدأً ويقاس سائر الأيّام إليه ؛ فيكون نصف نهار ذلك اليوم مبدأً للأيّام الوسطى والحقيقية جميعاً . وكل يوم من السنة يفرض مبدأً يكون التفاوت بين الأيّام الماضية الوسطى والحقيقية الماضية من ذلك اليوم تارة زائداً وتارة ناقصاً إلاّ أواخر الدلو وأوائل العقرب ؛ فإنّ المبدأ إذا جعل أواخر الدلو كانت الأيّام الحقيقية دائماً ناقصة من الوسطى ، وإذا جعل أوائل العقرب كانت الأيّام الحقيقية دائماً زائدة على الوسطى . / واتفق / أهل الصناعة على جعلة أواخر الدلو .

/16/ ... [β = فاتفق] M , α .

This is an illustration of the segments, the apogee being /1/ in the end part /1/ of Gemini.

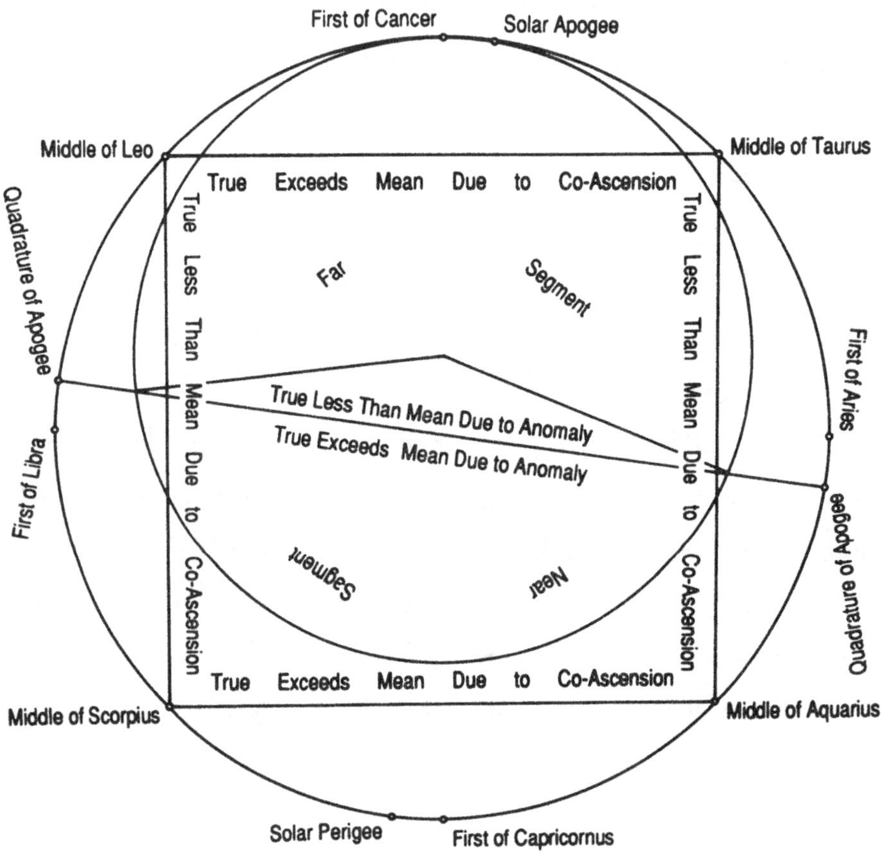

[Fig. T24]

The difference of the anomaly changes due to the motion of the apogee, but over a long period.

[8] This then is the explanation of the difference in the lengths of the days. Finding the [actual] amounts for all times pertains to the practical handbooks. This difference is called the **equation of the nychthemeron**. When there is a complete revolution [of the sun through the ecliptic], the true and mean days become equal, and this ceases to be a consideration.

/1/ in the end part] β = at the end] α, M.

وهذه صورة القطع على أنّ الأوج في / أواخر / الجوزاء :

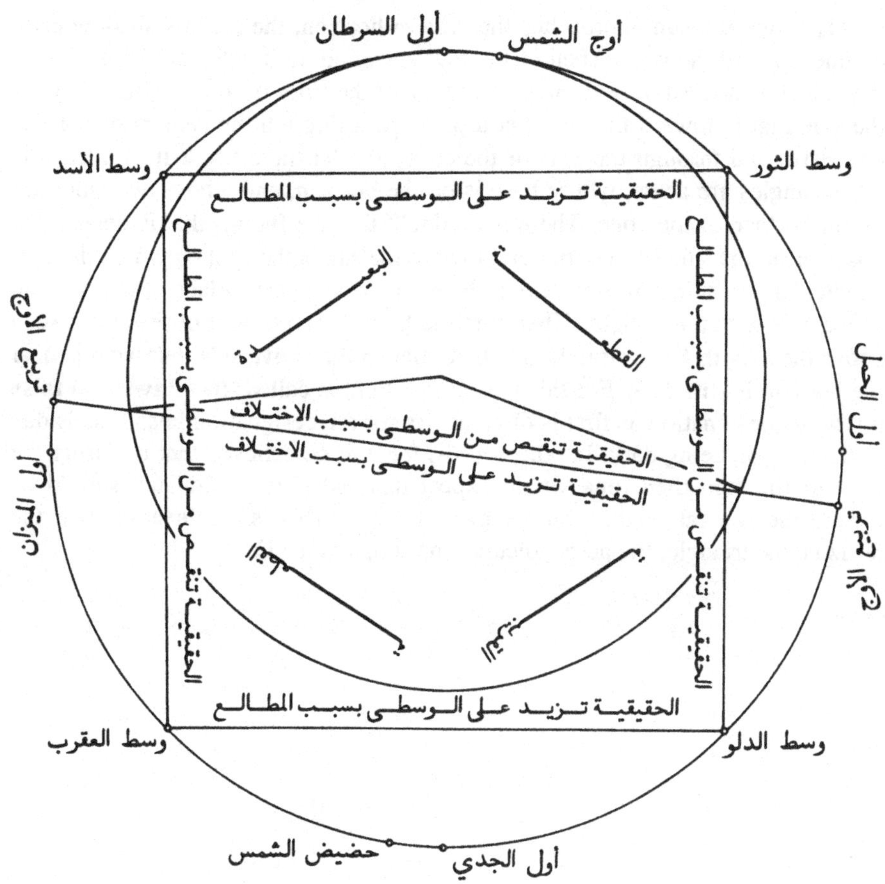

[شكل ٢٤]

ويتغيّر تفاوت الاختلاف بسبب حركة الأوج ولكنّ في مدة طويلة .

[٨] فهذا بيان التفاوت في مقادير الأيّام . ووجود المقادير في كل وقت يتعلّق بكتب العمل ، ويُسمّى هذا التفاوت تعديل الأيّام بلياليها . وإذا تمّ الدور تساوت الأيّام الحقيقية والوسطى وسقط هذا الاعتبار .

/1/ ... [= β آخر] M , α .

CHAPTER NINE

On Dawn and Dusk

[1] When the sun approaches the eastern horizon, the Earth's shadow cone inclines toward the west. Then of the rays surrounding it, what is visible first is that which is nearest the eye, and the nearest of the sides of the cone to the eye is the side that is toward the sun. Let a plane pass through the centers of the sun and Earth and through the axis of the cone, and let there occur in it a triangle whose angles are acute, whose base is on the horizon, and whose two sides are on the surface of the cone. There is no doubt that the [point] that is nearest the observer on the side [of the triangle] toward the sun is the spot on which the perpendicular extending from the eye falls on that side, not the place of intersection of the side with the horizon. Therefore the first observed light of the sun is seen above the / Earth / as a straight line that falls on the above side, but what is near the horizon is still dark. For this reason that light is called **first dawn** and **false dawn**. Its designation as first is obvious. As for the designation false, this is due to the horizon being dark; in other words, for it to be believed that it is truly the light of the sun, then [one would expect that] what is illuminated should be toward the sun rather than farther away from it. This is an illustration of the horizon, the triangle, the perpendicular, the sun, and the Earth:

/11/ Earth] β, M = horizon] α.

الفصل التاسع
في الصبح والشفق

[١] إذا قربت الشمس من الأفق الشرقي مال مخروط ظلّ الأرض نحو المغرب ؛ فيكون المرئي من الشعاع المحيط به أوّلاً ما هو أقرب إلى البصر ، والأقرب من جوانب المخروط إلى البصر هو الجانب الذي يلي الشمس . وليمرّ سطح بمركزي الشمس والأرض وبسهم المخروط ، وليحدث منه مثلّث حادّ الزوايا قاعدته على الأفق وضلعاه على سطح المخروط . ولا شكّ أنّ الأقرب من الضلع الذي يلي الشمس إلى الناظر يكون موقع العمود الخارج من النظر الواقع على ذلك الضلع ، لا موضع اتّصال الضلع بالأفق . فإذن أوّل ما يُرى نور الشمس يُرى فوق / الأرض / كخط مستقيم منطبق على الضلع المذكور ، ويكون ما يقرب من الأفق بعدُ مظلماً . فلذلك يسمّى ذلك النور بالصبح الأول والصبح الكاذب . أمّا تسميته بالأول فظاهر ؛ وأمّا تسميته بالكاذب فلكون الأفق مظلماً ، أي لو كان يصدُق أنّه نور الشمس لكان المنير ما يلي الشمس دون ما يبعد منه . وهذه صورة الأفق والمثلّث والعمود والشمس والأرض :

/11/ ...] M , β = الأفق] α .

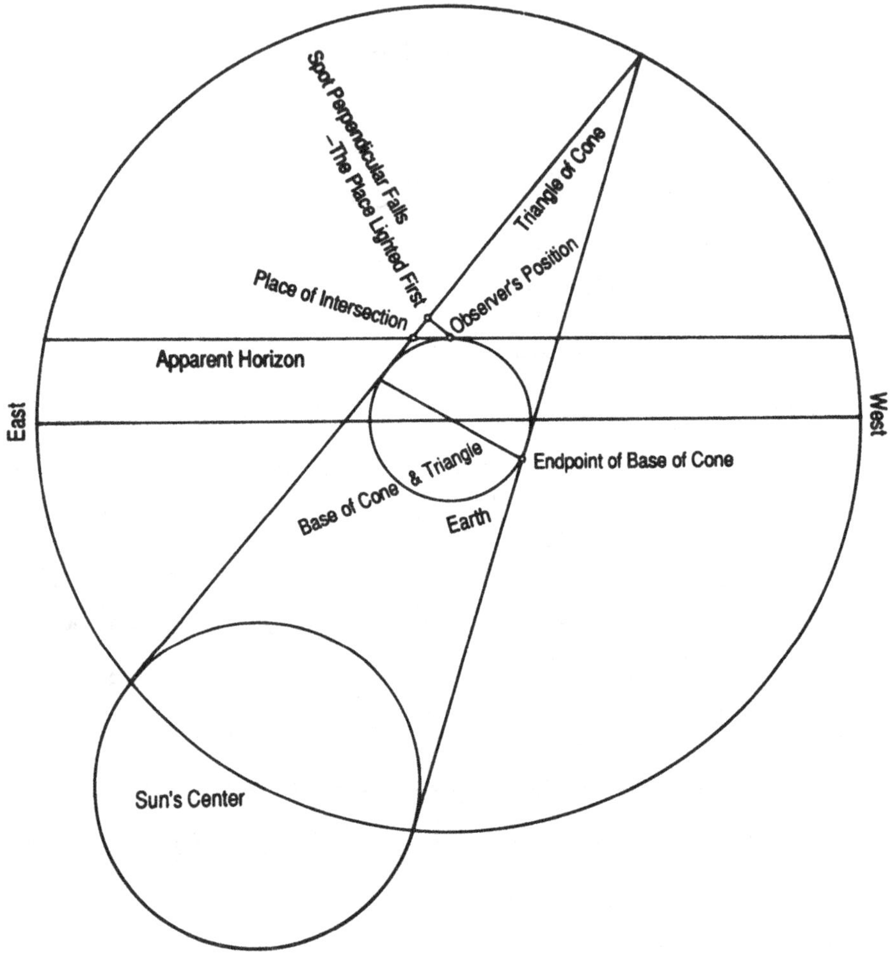

[Fig. T25]

[2] Then when the sun comes quite near, the light will spread; the horizon will then become lighted and the **dawn** will be true. **Dusk** will be the reverse of dawn. It has become known by trial and error that the depression of the sun below the horizon at the first rising of dawn and the final setting of dusk is 18°. Thus in the lands whose latitude is 48½ / degrees, / dusk will be continuous

---

/5/ degrees] $\beta = -\alpha, -M$.

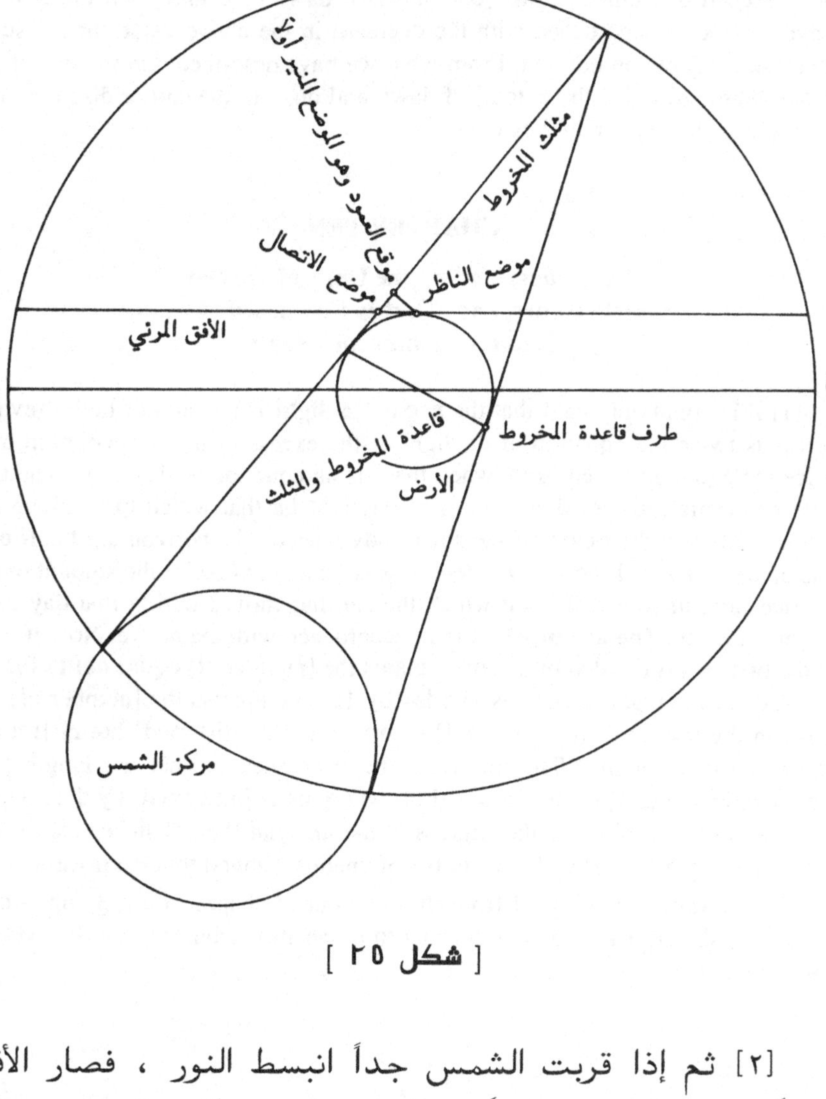

[شكل ٢٥]

[٢] ثم إذا قربت الشمس جداً انبسط النور ، فصار الأفق منيراً ويصير الصبح صادقاً . والشفق يكون بعكس الصبح . وقد عُرف بالتجربة أنّ انحطاط الشمس من الأفق عند أوّل طلوع الصبح وآخر غروب الشفق يكون ثمانية عشر جزءاً . ففي البلاد التي تكون عروضها ثمانية وأربعين / درجة / ونصفاً يتّصل الشفق

.−α , −M = β [… /5/

with dawn when the sun is at the summer solstice. In the [regions] whose latitudes exceed that amount, this [continuity of dawn and dusk] will occur for a longer period in accordance with the decrease in the above value for the sun's depression below the horizon. From what we have described, the reason for the stated delimitation [of the extent] of dawn and dusk in the earlier discussion of the spinning horizon should be clear.

CHAPTER TEN

On Understanding the Units of the Day, Namely Hours, and What Is Composed of Days, Namely Months and Years

[1] It is commonly held that the **arc of daylight** is the sum of half a revolution plus twice the equation of daylight or the excess of half a revolution over twice the equation of daylight (when there is an equation of daylight). The true state of affairs requires that the arc of daylight be that which turns along the equinoctial from the time half the solar body rises on the horizon until half of it sets on the horizon. It will be greater / or less / than the first by the amount of the co-ascension of [the arc] upon which the sun has moved during that day for a [given] locality. The **arc of night** is in accordance with the above. Now if each of the two arcs is divided by 15, one obtains the [number of] **equal hours** for the day and for the night. If each is divided by 12, one obtains the [number of] degrees in the **seasonal** [*lit.*, temporal] or **unequal** [*lit.*, distorted] **hours** [for the day and for the night]. The difference between them is that the length [*lit.*, length and shortness] of the days and of the nights is [measured] by the number of equal hours [or else] by the degrees in the unequal [hours] since it is the degrees in an equal [hour] and the number of unequal [hours] that are invariable.

[2] The **month** is derived from the illuminated shapes of the moon, which have been shown really to be according to its position from the sun. Its cycle is completed when

---

/13–14/ or less] $\beta = -\alpha$, $-M$.

بالصبح إذا كانت الشمس في المنقلب الصيفي . وفيما جاوزت عروضها ذلك المقدار يكون ذلك في زمان أكثر بحسب تناقص انحطاط الشمس عن الأفق القدر المذكور . ويتبيّن مما وصفنا السبب في تحديد الصبح والشفق المذكور فيما مرّ للآفق الرحوي .

الفصل العاشر
في معرفة أجزاء الأيام
وهي الساعات وما يتركب من الأيام
وهي الشهور والسنون

[١] المشهور أنّ قوس النهار هي مجموع نصف الدور وضعف تعديل النهار أو فضل نصف الدور على ضعف تعديل النهار إن كان تعديل نهار . والحقيقة تقتضي أن تكون قوس النهار هو[!] ما يدور من معدل النهار من وقت طلوع نصف جرم الشمس من الأفق إلى وقت غروب نصفه في الأفق ؛ وهو أزيد من الأول / أو أنقص / بقدر مطالع ما تسيره الشمس في ذلك اليوم لتلك البقعه . وقوس الليل بحسب ذلك . فإذا قُسم كل واحد من القوسين على خمسة عشر حصلت ساعات النهار والليل المستوية ؛ وإذا قُسم على اثني عشر حصلت أجزاء ساعاتهما الزمانية والمعوّجة . والفرق بينهما أنّ طول الأيام والليالي وقصرهما يكونان بعدد الساعات المستوية وأجزاء المعوّجة لأنّ أجزاء المستوية وعدد المعوّجة لا يختلفان .

[٢] وأما الشهر فمأخوذ من تشكّلات القمر النورية ، وقد تبيّن أنها إنما تكون بحسب أوضاعه من الشمس . ويتم دورُه إذا

/13–14/ أو أنقص] $-M = \beta$, $-\alpha$.

the excess of the moon's true motion over the sun's true motion becomes one revolution. Finding this is difficult. Compounding the difficulty is that it varies due to the irregularity of their motions. Those employing the [month] who [rely] on appearances take it to be from one day of conjunction to another, or from one night of [first] visibility of the crescent to another, or from some other shape to its like, depending on what they have conventionally adopted. Those employing the [month] who [rely] on calculation derive a cycle from the difference between the two mean motions; they find it to be in 29½ days plus a fraction. They then take one month to be 30, one month to be 29. / They add the accumulated fractions that exceed a half day [which is] 11 days in every 30 years; / thus in a period of 30 years, 11 months that should be 29 become 30 each. These are called **intercalary days**. Or else they add the intercalary days to the months in some other way. These months are **lunar**, of which there are true and of which there are mean.

[3] As for the **year**, it is derived from the return of the sun to its location on the ecliptic orb that results in a return of the [same] yearly condition determined by the seasons. This occurs in 365¼ days minus a fraction. During [a year], there are 12 complete mean lunar months plus an excess of 11 days minus a fractional part. Of those employing [the year] who do not take into account the lunar months, some take it to be from a day on which the sun reaches a certain point, such as the vernal equinox, to the same [point]. They take its months to be from the days on which [the sun] reaches points along the ecliptic analogous to the [initial point]. Or they reckon the months to be 30 each and add at the end [of the year] 5 or 6 [days]. The 5 are called the stolen or / the supplementary / [*i.e.* the epagomenal days] and the 6th intercalary. For these [people], their years are **true solar**, while their months are either true solar or conventional.

/8/ They add...in every 30 years] β = For the accumulated fractions that exceed a half day, they add 11 days in every 30 years] α, M. /22/ the supplementary] β = supplementary] α, M.

صار فضل حركة القمر على حركة الشمس الحقيقيتين دوراً ؛ ووجوده متعذر ومع تعذُره مختلف لاختلاف حركتيهما . فمستعمله من أهل الظاهر يأخذونه من يوم الاجتماع إلى يومه أو من ليلة رؤية الهلال إلى ليلتها أو من تشكّل آخر إلى مثله بحسب ما يصطلحون عليه . ومستعملوه من أهل الحساب يأخذون الدور من الفضل بين الحركتين الوسطيين ، فيجدونه في تسعة وعشرين يوماً ونصف وكسر ؛ فيأخذون لشهر ثلاثين ولشهر تسعة وعشرين . ويزيدون / الكسور / المجتمعة التي تزيد على نصف يوم في كل ثلاثين سنة أحد عشر يوماً فيصير أحد عشر شهراً مما يجب أن يكون تسعة وعشرين في مدة ثلاثين سنة ثلاثين ثلاثين ، وتسمى تلك الأيام كبائس ؛ أو يزيدون الكبائس في الشهور على وجه آخر . وهذه الشهور قمرية ، فمنها حقيقية ومنها وسطية .

[٣] وأما السنة فمأخوذة من عود الشمس إلى موضعها في فلك البروج المقتضي لعود حال السنة بحسب الفصول ؛ ويحصل ذلك في ثلاثمائة وخمسة وستين يوماً وربع يوم إلا كسراً . ويتم فيها من الشهور القمرية الوسطى اثنا عشر ، ويزيد عليها أحد عشر يوماً غير شيء من الكسور . ومستعملوها ، إن لم يعتبروا الشهور القمرية ، فربما يأخذونها من يوم تحلّ الشمس فيه نقطة بعينها – كالاعتدال الربيعي – إلى مثله . ويأخذون شهورها من الأيام التي تحلّ فيها أمثال تلك النقطة من البروج ؛ أو يعدّون الشهور ثلاثين ثلاثين ويزيدون في آخرها خمسة أو ستة ، وتسمى الخمسة المسترَقة / واللواحق / والسادس كبيسة . وهؤلاء سنوهم شمسية حقيقية وشهورهم إما شمسية حقيقية وإما اصطلاحية . وربما

/8/ ... [للكسور = β [M , α . /22/ ... [ولواحق = β [M , α .

Some others take [the year to begin] from a day agreed upon without regard to the position of the sun, and they have conventionally adopted months that are around 30 since the lunar months are approximately that. The fraction in excess of 365 is taken by some to be a quarter exactly, and they intercalate 1 day in every 4 years, while others drop it completely. These years are **conventional solar**. For those who do wish to take into account lunar months, they make the year solar and the months lunar. In every 3 years or in every 2 years, they add a month to the year to reconcile the previously mentioned 11 days minus a fraction in accordance with what they have conventionally adopted. One people has made every 12 lunar months a year, which they call a **lunar year**. Every people has an epoch to which they refer the years of their history; understanding the details of that does not pertain to this science.

CHAPTER ELEVEN

On the Degrees of Transit of the Stars on the Meridian and on Their [Degrees of] Rising and Setting

[1] When the poles of the ecliptic orb are on the meridian circle, this being when the two solstice points are also on it and the two equinox points are on the horizon, then the transit of the stars at that time is at their degrees in longitude since the meridian circle is their latitude circle. When the visible pole of the ecliptic orb is to the east of the meridian (this occurring, if the visible pole is northerly, during the transiting of the half of the ecliptic orb bisected by the autumnal equinox and the rising of the southern half [of the ecliptic], or, if it is southerly, [during] the transiting of the [corresponding] other half and the rising of the [corresponding] other half), then the star whose latitude is in the direction of the visible pole will transit the meridian circle after its degree [in longitude has done so] since its latitude circle extending from the pole meets the star before its [longitudinal] degree; thus when its [longitudinal] degree reaches the meridian, the star will be

يأخذونها من يوم يتّفق من غير ملاحظة موضع الشمس ، ويصطلحون على شهور تدور حول الثلاثين لكون الشهور القمرية قريبة منه . والكسر الزائد على ثلاثمائة وخمسة وستين ربما يأخذونه ربعاً تاماً ويكبسون في كل أربع سنين بيوم ، وربما يحذفونه مطلقاً ؛ وهذه السنون شمسية اصطلاحية . وإن أرادوا اعتبار الشهور القمرية جعلوا السنة شمسية والشهور قمرية ، وزادوا في كل ثلاث سنين أو في كل سنتين شهراً في السنة لاجتماع الأحد عشر يوماً غير كسر المذكور على حسب ما يصطلحون عليه . وقوم يجعلون كل اثني عشر من الشهور القمرية سنة ويسمونها سنين قمرية . ولكل قوم مبدأ ينسبون سني تأريخهم إليه ، ومعرفة تفاصيل ذلك غير متعلقة بهذا العلم .

الفصل الحادي عشر
في درجات ممرّ الكواكب بنصف النهار وطلوعها وغروبها

[١] إذا كان قطبا فلك البروج على دائرة نصف النهار ـ وذلك يكون عند كون نقطتي الانقلابين أيضاً عليها ونقطتي الاعتدالين على الأفق ـ فمرور الكواكب حينئذ يكون مع درجاتها الطولية لأنّ دائرة نصف النهار تكون دائرة عرضها . وإذا كان القطب الظاهر من فلك البروج شرقياً عن نصف النهار ـ وذلك يكون عند مرور النصف من فلك البروج الذي يتوسطه الاعتدال الخريفي وطلوع النصف الجنوبي منه إن كان القطب الظاهر شمالياً ، أو مرور النصف الآخر وطلوع النصف الآخر إن كان جنوبياً ـ فالكوكب الذي يكون عرضه في جهة القطب الظاهر يمرّ على دائرة نصف النهار بعد درجته لأنّ دائرة عرضه الخارج[!] من القطب تلاقي الكوكب قبل درجته ؛ فإذا وافى درجته نصف النهار كان الكوكب

away from [its degree] in the direction of the pole, *i.e.* it will still be to the east. The star whose latitude is in the direction opposite the visible pole transits [the meridian circle] before its [longitudinal] degree since the latitude circle mentioned above first meets the star's [longitudinal] degree that is on the meridian; it then meets the star, which has already transited and become westerly. When the visible pole is to the west (this occurring, if the pole is northerly, during the transiting of the half of the ecliptic orb bisected by the vernal equinox and the rising of the northern half [of the ecliptic], or, if it is southerly, [during] the transiting of the [corresponding] other half and the rising of the [corresponding] other half), then the star whose latitude is in the direction of the visible pole will transit before its degree [in longitude has done so] and that whose latitude is in the opposite direction will transit after it for exactly [the reasons] we have stated.

[2] The rising and setting of the stars on the horizons of the [Earth's] equator are similar to their transit across the meridian for the rest of the horizons. Thus a star that reaches the horizon along with the pole and the solstice rises or sets with its [longitudinal] degree; one that is in the direction of the visible pole rises in advance of its [longitudinal] degree and sets after it; and one that is in the direction of the invisible pole rises after its [longitudinal] degree and sets in advance of it. The northern pole will be visible there during the period of the rising of the half bisected by the vernal equinox and the transit of the southern half across the meridian from above [the horizon]; the southern pole will be visible during the period of the rising of the [corresponding] other half and the transit of the [corresponding] other half.

[3] As for the rising and setting of the stars in the remaining horizons, it is as we have described for the equator except for the transit and the rising of the halves of the ecliptic orb. For that will vary: it may be that one of the poles is visible and what transits or rises will be an arc that is smaller or larger than a half. In those horizons whose latitude exceeds the obliquity, one of the ecliptic poles will be permanently visible and the [above] rule concerning the stars may be uniformly applied without any variation [due to which ecliptic pole is visible].

منها في جهة القطب ، أعني يكون شرقياً بعدُ . والكوكب الذي يكون عرضه في خلاف جهة القطب الظاهر يمرّ عليها قبل درجته لأنّ دائرة العرض المذكورة تلاقي درجة الكوكب الكائنة على نصف النهار أولاً ، ثم تلاقي الكوكب وقد مرّ وصار غربياً قبل ذلك . وإذا كان القطب الظاهر غربياً ـ وذلك يكون عند مرور النصف من فلك البروج الذي يتوسطه الاعتدال الربيعي وطلوع النصف الشمالي منه إن كان القطب شمالياً ، أو مرور النصف الآخر وطلوع النصف الآخر إن كان جنوبياً ـ فالكوكب الذي يكون عرضه في جهة القطب الظاهر يمرّ قبل درجته والذي يكون عرضه في خلاف تلك الجهة يمرّ بعدها لِما ذكرنا بعينه .

[٢] وطلوع الكواكب وغروبها في آفاق خط الاستواء يكون كمرورها على نصف النهار في سائر الآفاق . فالكوكب الذي يوافي الأفق مع القطب والانقلاب يطلع أو يغرب مع درجته ، والذي يكون في جهة القطب الظاهر يطلع قبل درجته ويغيب بعدها ، والذي يكون في جهة القطب الخفي يطلع بعد درجته ويغيب قبلها . ويكون هناك القطب الشمالي ظاهراً مدة طلوع النصف الذي يتوسطه الاعتدال الربيعي ومرور النصف الجنوبي على نصف النهار من فوق ، والقطب الجنوبي ظاهراً مدة طلوع النصف الآخر ومرور النصف الآخر .

[٣] وأما طلوع الكواكب وغروبها في سائر الآفاق فكما وصفناه في خط الاستواء إلا في مرور الأنصاف وطلوع الأنصاف من فلك البروج ؛ فإنّ ذلك يختلف ، وربما يكون أحد القطبين ظاهراً والمارّة أو الطالعة قوس أصغر من النصف أو أكبر . وفي الآفاق التي يزيد عرضها على الميل الكلي يكون أحد قطبي البروج أبدي الظهور ، ويطّرد الحكم في الكواكب من غير اختلاف .

CHAPTER TWELVE

On Finding the Meridian Line and the *qibla* Bearing

[1] Two equal altitudes of the sun are observed on the same day from the two sides of its maximum altitude, and the directions of their shadows from the same gnomon are marked on level land. Then the angle occurring between them is bisected with a line. This line will then be in the plane of the meridian circle and is called the **meridian line**. The perpendicular to it is in line with the prime vertical circle.

[2] By another method, a gnomon is erected vertically to a plane of level land. A circle whose radius is equal to twice the gnomon is drawn, and the entry of the shadow into the circle and its emergence from it is observed before and after noon. The two places are marked and the arc that falls between them is bisected. The midpoint and the center are joined with a straight line; this then is the meridian line. The perpendicular to it that passes through the center of the circle is the east-west line. The two [lines] divide the circle into fourths. Each fourth is then divided into 90 equal parts in order to find the measures of the azimuths from the shadow lines falling on the circumference since [the number of] these parts between the east and west points and the shadow line is an azimuth. This circle is known as the **Indian**.

[3] As for the ***qibla*** **bearing**, let it be noted that the longitude of Mecca—may God Most High protect it—is 77⅙° from the Eternal Islands and 67⅙° from the coast of the western sea. Its latitude is 21⅔°. / For / every locality

---

/22/ For] β = Thus for] α, M.

الفصل الثاني عشر
في معرفة خط نصف النهار
وسمت القبلة

[١] يُرصد ارتفاعان متساويان للشمس في يوم واحد عن جنبتي غاية ارتفاعها ؛ ويُخط على أرض مستوية سمتا ظليهما عن مقياس واحد ؛ ثم تُنصّف الزاويةُ الحادثة بينهما بخط . فيكون ذلك الخط في سطح دائرة نصف النهار ويسمى خط نصف النهار. والقائم عليه عموداً يكون في سمت دائرة أول السموت .

[٢] وبوجه آخر يُقام مقياس قائم على سطح أرض مستوية ؛ وتُرسم دائرة نصف قطرها بقدر ضعف المقياس ؛ ويُرصد دخول الظل / في / الدائرة وخروجها عنها قبل نصف النهار وبعده ؛ ويُعلّم على الموضعين وتُنصّف القوسُ التي تقع بينهما ؛ ويوصَل بين المنتصف وبين المركز بخط مستقيم ، فهو خط نصف النهار . والقائم عليه عموداً المارّ بمركز الدائرة خط المشرق والمغرب . ويُربِّعان الدائرةَ ، ثم يُقسم كلُ ربع / بتسعين / قسماً متساوية ليعرف مقادير السموت من خطوط الظل الواقعة على المحيط لأنّ ما بين نقطتي المشرق والمغرب وخط الظل من تلك الأقسام سمت . وهذه الدائرة يُعرف بالهندية .

[٣] وأما سمت القبلة فليعلم أنّ طول مكّة _ حماها الله تعالى _ عن جزائر الخالدات سبع وسبعون جزءاً وسدس جزء ، وعن ساحل البحر الغربي سبع وستون جزءاً وسدس جزء . وعرضها أحد وعشرون جزءاً وثُلثا جزء . / وكل / بلدة يكون

/11/ ...] β = –M , –α . /15/ ...] β = تسعين] M , α . /22/ ...] β = فكل] M , α .

whose longitude is less than the longitude of Mecca, Mecca is to the east of it; for every locality whose longitude is greater than the longitude of Mecca, Mecca is to the west of it. If their longitudes are equal, then Mecca is on its meridian line—to the south if the latitude of Mecca is less than its latitude, to the north if it is greater. Every locality / whose latitude equals that of Mecca / is beneath the same day-circle as Mecca; then if its longitude is less than the longitude of Mecca, Mecca is to the left of the rising place of the equinox for that locality, and if its longitude is greater, Mecca is to the right of the setting place of the equinox.

[4] There are many ways to determine the *qibla* bearing, but it would not be appropriate to present them here. Let us instead limit ourselves to one simple method, which is [as follows]. The sun transits the zenith of Mecca when it is in degree 8 of Gemini and in [degree] 23 of Cancer at noontime there. The difference between its noon and the noon of other localities is measured by the difference between the two longitudes. Let this [latter] difference be taken and let an hour be assumed for each 15 degrees and 4 minutes for each degree. The resulting total is the interval in hours from noon [for that locality]. Let an observation be made on that day at that time—before noon if Mecca is to the east or after if it is to the west; the direction of the shadow at that time is the *qibla* bearing.

/4–5/ whose latitude equals that of Mecca] β = whose latitude and that of Mecca are equal] α, M.

طولها أقل من طول مكّة فمكّة شرقية عنها ؛ وكل بلدة يكون طولها أكثر من طول مكّة فمكّة غربية عنها . وإن تساوى طولاهما فمكّة على خط نصف نهارها ـ جنوبية إن كان عرض مكّة أقل من عرضها ، وشمالية إن كان أكثر . وكل بلدة / يساوي عرضها عرض مكّة / كانت مع مكّة تحت مدار واحد يومي ؛ فإن كان طولها أقل من طول مكّة فمكّة عن يسار مشرق الاعتدال لتلك البلدة ، وإن كان طولها أكثر فمكّة عن يمين مغرب الاعتدال .

[٤] ولمعرفة سمت القبلة طرق كثيرة لا يليق إيرادها هاهنا . فلنقتصر على وجه سهل وهو أنّ الشمس تكون مارّة بسمت مكّة عند كونها في الدرجة الثامنة من الجوزاء والثالثة والعشرين من السرطان وقت انتصاف النهار هناك . والفضل بين نصف نهارها وبين نصف نهار سائر البُلدان يكون بقدر التفاوت بين الطولين . فليؤخذ التفاوت وتؤخذ لكل خمسة عشر جزءاً ساعة ولكل جزء أربع دقائق ؛ فيكون ما اجتمع ساعات البعد عن نصف النهار . وليُرصد في ذلك اليوم ذلك الوقت قبل نصف النهار إن كانت مكّة شرقية أو بَعده إن كانت غربية . فسمت الظل ساعتئذ يكون سمت القبلة .

/4–5/ [...] = β تساوى عرضها وعرض مكة] M , α .

BOOK IV

On Finding the Measurements of the Distances and the Bodies

Seven Chapters

CHAPTER ONE

On the Measure [*misāḥa*] of the Earth

[1] In this book one has need of preliminary propositions other than those that have been stated. Among these is what was proven by Archimedes regarding the measure of a circle and a sphere, namely: that the circumference of every circle is approximately equal to $3\frac{1}{7}$ times its diameter; that the surface enclosed by the radius times half the circumference is equal to the area [*taksīr*] of the circle; that the surface enclosed by the diameter of the sphere times the circumference of the largest circle occurring in it is equal to the surface enclosing the sphere; and that each portion on the surface of the sphere bounded by two great circles [*i.e.* a lune] is equal to the surface enclosed by the diameter times the maximum inclination between them.

[2] Having presented these lemmas, we say: if a person travels along the meridian line on level land in an amount whereby the local latitude increases, or decreases, by one degree, the amount that he has traveled is a one-degree unit of a great circle that occurs upon the Earth. The great circle is 360 times that amount; the diameter of the Earth is 1 part in $3\frac{1}{7}$ parts, this being the total circumference of that great circle. Many people have undertaken to determine this among whom is a group of scientists during the reign of al-Ma'mūn, may God be pleased with him. By his decree, they came to the Plain of Sinjār and obtained the measure of 1° of the 360° along the meridian line; they found it to be $22\frac{2}{9}$ parasangs, each

الباب الرابع
في معرفة مقادير الأبعاد والأجرام
سبعة فصول

الفصل الأول
في مساحة الأرض

[١] يُحتاج في هذا الباب إلى مصادرات غير ما ذُكر . من ذلك ما بيّنه أرشميدس في مساحة الدوائر والأكر وهو : أنّ محيط كل دائرة مثل ثلاثة أمثال قطرها ومثل سُبع قطرها بالتقريب ؛ وأنّ السطح الذي يحيط به نصف القطر في نصف المحيط مساوٍ لتكسير الدائرة ؛ وأنّ السطح الذي يحيط به قطر الكرة في محيط أعظم دائرة تقع فيها مساوٍ للسطح المحيط بالكرة ؛ وأنّ كل قطعة من سطح الكرة تحيط بها دائرتان عظيمتان فهي مساوية لسطح يحيط به القطر في غاية الميل بينهما .

[٢] وبعد تقديم هذه المقدمات نقول : إذا سار سائر على خط نصف النهار في أرض مستوية بقدر ما يزيد جزء واحد في عرض البلد أو ينقص فالقدر الذي ساره يكون حصّة درجة واحدة من الدائرة العظيمة التي تقع على الأرض . والدائرة العظيمة تكون ثلاثمائة وستين مرة مثل ذلك القدر ، وقطر الأرض يكون جزءاً من ثلاثة أجزاء وسُبع جزء وهي مجموع محيط تلك العظيمة . وقد قام بتحقيق ذلك قوم كثير منهم طائفة من الحكماء في عهد المأمون – رضي الله عنه – حضروا بأمره برّية سِنجار وحصلوا مقدار الجزء الواحد من ثلاثمائة وستين جزءاً من خط نصف النهار ؛ فوجدوه اثنين وعشرين فرسخاً وتُسعي فرسخ على أنّ كل

parasang being 3 miles, each mile being 4000 cubits, each cubit being 24 digits, and each digit being the measure of six average barleycorns laid side by side. When the parasangs plus the fractional [part] are multiplied by 360, one obtains the size of the circumference of the great circle on the Earth, which is 8000 parasangs. If this amount is divided by $3\frac{1}{7}$, one obtains for the size of the diameter 2545 / and a half / parasangs, approximately. Its radius is then 1273 parasangs, approximately. This is the standard with which the distances are measured; similarly the sphere of the Earth is the body with which the [other] bodies are measured.

[3] If the diameter is multiplied by the circumference of the great circle, one obtains the surface area of the Earth, which is twenty thousand thousands three hundred sixty thousand [20,360,000] parasangs. A fourth of this is the area of the populated quarter; the length of the / populated / quarter is half the circumference, while its width is a fourth of it. As for the amount that is [actually] inhabited, which is between the equator and the location whose latitude is equal to the complement of the obliquity, its length is likewise 4000 parasangs and its width, obtained by multiplying the parasangs in 1° by $(66 + \frac{1}{4} + \frac{1}{6})°$, is 1476 parasangs, / approximately. / Its area, obtained by multiplying this by the parasangs in the diameter, is / three thousand thousands seven hundred fifty-six thousand four hundred twenty [3,756,420] parasangs, / which is nearly $(\frac{1}{6} + (\frac{1}{6}$ of $\frac{1}{10}))$ of the entire surface of the Earth. If one wishes to know this in miles, he should multiply the linear parasangs by 3 and the square ones by 9. Likewise if he wishes the amount in cubits, digits or barleycorns, he should multiply [the parasangs] by the number of them in a linear or a square parasang. This then is how to find the measure of the Earth.

/7/ and a half] β = –α, –M. /12/ populated] β = –α, –M. /17/ approximately] β = –α, –M. /18–19/ three thousand thousands seven hundred fifty-six thousand four hundred twenty [3,756,420] parasangs] β = three thousand thousands seven hundred sixty-five thousand four hundred twenty [3,765,420] parasangs] α, M.

فرسخ ثلاثة أميال ، وكل ميل أربعة آلاف ذراع ، وكل ذراع أربعة وعشرون أصبعاً ، وكل أصبع مقدار ست شعيرات مضمومة بُطون بعضها إلى بعض من الشعيرات المعتدلة . فإذا ضُرب الفراسخ مع الكسر في ثلاثمائة وستين حصل مقدار محيط الدائرة العظمى من الأرض وهو ثمانية آلاف فرسخ . وإذا قُسِمَ هذا المبلغ على ثلاثة وسُبع حصل مقدار قطرها ألفين وخمسمائة وخمسة وأربعين فرسخاً / ونصف فرسخ / بالتقريب . فيكون نصف قطرها ألفاً ومائتين وثلاثة وسبعين فرسخاً تقريباً ، وهو المقدار الذي تقدّر به الأبعادُ كما أنّ كرة الأرض هي الجرم الذي تقدّر به الأجرام .

[٣] وإذا ضرب القطر في محيط الدائرة العظمى حصل تكسير سطح الأرض وهو عشرون ألف ألف وثلاثمائة وستون ألف فرسخ . وربع ذلك تكسير الربع المسكون ، ويكون طول الربع / المسكون / نصف المحيط وعرضه ربعه . وأما القدر المعمور ، وهو ما بين خط الاستواء والموضع الذي عرضه بقدر تمام الميل ، فيكون طوله أيضاً أربعة آلاف فرسخ وعرضه الحاصل من ضرب فراسخ الجزء الواحد في ستة وستين جزءاً وربع وسدس جزء ألف وأربعمائة وستة وسبعون فرسخاً / تقريباً / . وتكسيره الحاصل من ضرب ذلك في فراسخ القطر / ثلاثة آلاف ألف وسبعمائة وستة وخمسون ألفاً وأربعمائة وعشرون فرسخاً / ، وهو قريب من سدس جميع سطح الأرض وسدس عُشره . وإن أراد مريد أن يعرف ذلك بالأميال ضرب الفراسخ الطولية في ثلاثة والتكسيرية في تسعة ؛ وكذلك إن أراد مقاديرها بالذُرعان والأصابع والشعيرات ضربها في أعدادها لفرسخ طولي أو تكسيري . فهذه معرفة مساحة الأرض .

/7/ ...] β = M– , α– . /12/ ...] β = M– , α– . /17/ ...] β = M– , α– .

/18–19/ ...] β = ثلاثة آلاف ألف وسبعمائة وخمسة وستين ألفاً وأربعمائة وعشرين فرسخاً] M , α .

[4] Abū al-Rayḥān has another method for determining the measure of the Earth that is found by observing the depression of the horizon from the peak of a high mountain whose height is possible to ascertain. We shall not present it here, however, inasmuch as it contains geometrical proofs.

[5] As for what we promised to show in the first part of this book, namely how to find the ratio of a mountain whose height is half a parasang to the diameter of the Earth, the way we do this is to double the parasangs in the diameter, which comes to 5090 parasangs. The ratio of half a parasang to the diameter is the same as the ratio of one to this amount. Then we take the barleycorns in a cubit, which is 144, and we divide the above amount by it; the result is then 35. The ratio of one part of this, which is $\frac{1}{5}$ of $\frac{1}{7}$ of the width of a barleycorn, to a cubit is the same as the ratio of half a parasang to the diameter.

CHAPTER TWO

On Finding the Distances of the Moon from the Center of the World

[1] The distances of the moon and the other wandering stars from the center of the World are known for any time based upon the radii of their orbs being 60 parts as is stated in calculating their true positions by the method of geometry. The ratio of one to the other, [however], is not known; thus finding that is required. One needs to assume a standard with which all of them may be measured, and so the radius of the Earth has been made that [standard].

[2] In order to find the lunar distances with this standard, Ptolemy observed the moon at the time of its minimum / altitude / on the meridian circle, and he found its apparent altitude to be exactly $(39 + (\frac{1}{2} \text{ of } \frac{1}{6}))^\circ$. Its true altitude by calculation for that time and place was $40\frac{1}{5}^\circ$. Thus he found the difference between them to be 1 degree and 7 minutes, which is the lunar parallax.

/20/ altitude] β = altitudes] α, M.

[٤] ولأبي الريحان طريق آخر في معرفة مساحة الأرض تعرف برصد انحطاط الأفق عن رأس جبل مرتفع يمكن الوقوف على ارتفاعه ؛ وإنما لم نورده هاهنا لاشتماله على براهين هندسية .

[٥] وأما ما وعدنا بيانه في صدر هذا الكتاب ، وهو معرفة نسبة جبل يكون ارتفاعه نصف فرسخ إلى قطر الأرض ، فالوجه فيه أن نضعّف فراسخ القطر فيصير خمسة آلاف وتسعين فرسخاً ؛ وتكون نسبة نصف فرسخ إلى القطر كنسبة الواحد إلى هذا القدر . ثم نأخذ شعيرات الذراع وهي مائة وأربع وأربعون ونقسم ذلك المبلغ عليها ، فتخرج خمسة وثلاثون . وتكون نسبة جزء منها وهو خُمس سُبع عرض شعيرة إلى ذراع كنسبة نصف فرسخ إلى القطر .

الفصل الثاني
في معرفة أبعاد القمر
من مركز العالم

[١] كان أبعاد القمر وغيره من الكواكب السيّارة من مركز العالم معلومة في كل وقت بحسب كون أنصاف أقطار أفلاكها ستين جزءاً على ما يُذكر في حساب تقويماتها بطريق الهندسة ؛ ولم تكن نسبة البعض إلى البعض معلومة ، فطلب معرفة ذلك . واحتيج إلى فرض مقدار يقدّر به الجميع ، فجُعل ذلك نصف قطر الأرض .

[٢] ولمعرفة أبعاد القمر بذلك المقدار رصد بطلميوس القمر في وقت كان في أقل / ارتفاعه / على دائرة نصف النهار ؛ فوجد ارتفاعه المرئي بالتدقيق تسعة وثلاثين جزءاً ونصف سدس جزء . وكان ارتفاعه الحقيقي بالحساب لذلك الوقت في تلك البقعة أربعين جزءاً وخُمس جزء ؛ فوجد التفاوت بينهما جزءاً وسَبع دقائق ، وهو اختلاف منظر القمر .

/20/ ...] β = ارتفاعاته] M , α .

[3] It may be shown in the science of geometry that if the sizes of two angles and a side of a triangle with straight sides are known, the sizes of its remaining sides and angles are known. If one pictures the figure for parallax, which is below, then in the triangle in which one of the angles is the parallax

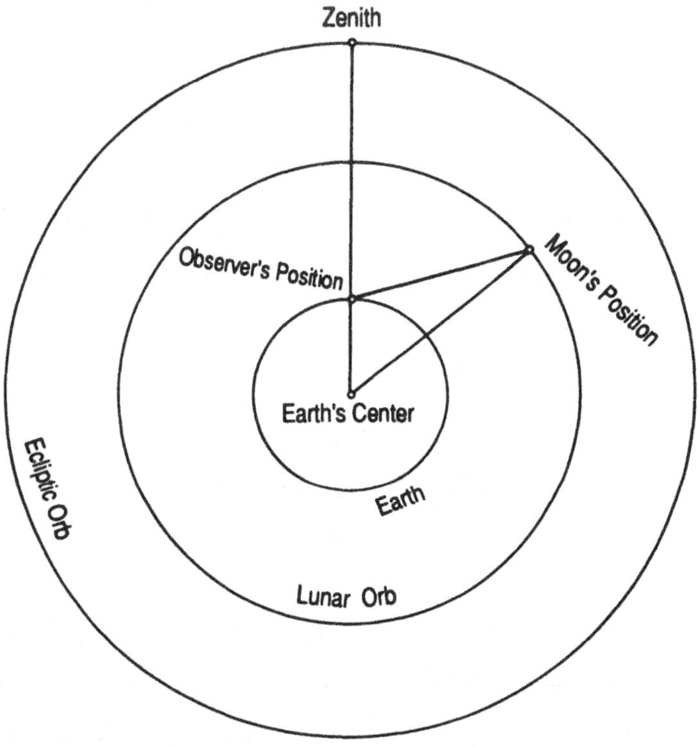

[Fig. T26]

at which is the position of the moon, the second the complement of the true altitude at which is the center of the Earth, and the third that at which is the position of the observer, there are two known angles, *i.e.* the parallax and the complement of the altitude. If the side that is the radius of the Earth is assumed to be 1, two angles and a side become known, and it is possible to find the remaining angle and the two remaining sides of the [triangle]. There results / by calculation / for the size of the side that is the distance of the moon from the center of the Earth (39 + ½ + ¼) parts, the radius of the Earth assumed to be 1 part.

/9/ by calculation] β = from calculation] α, M.

[٣] وقد تبيّن في علم الهندسة أنه إذا كانت مقادير زاويتين وضلع من مثلث مستقيم الأضلاع معلومة كانت مقادير الباقية من أضلاعه وزواياه معلومة . وإذا صُوّر شكل اختلاف المنظر ــ وهو هذا ــ كان ، في المثلث الذي إحدى زواياه اختلاف المنظر وهي

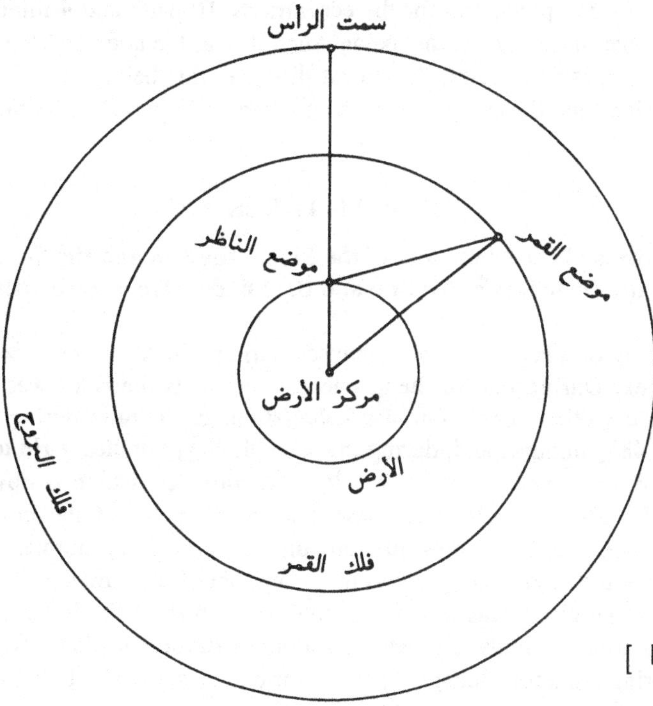

[شكل ٢٦]

التي عندها موضع القمر والثانية تمام الارتفاع الحقيقي وهي التي عندها مركز الأرض والثالثة التي عندها موضع الناظر ، زاويتان معلومتان أعني اختلاف المنظر وتمام الارتفاع . وإذا فرض الضلع الذي هو نصف قطر الأرض واحداً صارت زاويتان وضلع معلومة وأمكن معرفة الزاوية الباقية والضلعين الباقيين منه . وقد خرج / بالحساب / مقدار الضلع الذي هو بعد القمر عن مركز الأرض تسعة وثلاثين جزءاً ونصف وربع جزء على أنّ نصف قطر الأرض جزء واحد .

/9/ [...] من الحساب = β [M , α .

[4] By calculating the true positions with the scale whereby the inclined radius is 60, the epicyclic radius is 5¼, and the eccentricity is 10 parts and 19 minutes, the distance of the moon from the center of the World for the above time is (40 + ¼ + ⅙) parts. When the same quantity is known in two units of measure, it is possible to convert anything that is measured in one of these two units to the other unit since they will all be according to the ratio [of the units]. Thus Ptolemy converted the above quantities into a unit of measure by which the radius of the Earth is 1: for the inclined radius the result is then 59, for the epicyclic radius 5⅙ parts, and for the eccentricity 10 parts and 9 minutes. The farthest distance of the moon, this being when it is at the apex and the epicycle is at the apogee, is 64⅙ parts; its nearest distance, this being when it is at the epicyclic perigee and the epicycle is at the perigee, is 33 parts and 33 minutes.

CHAPTER THREE

On the Sizes of the Diameters of the Moon, the Sun and the Shadow, and the Distances of the Sun and the Shadow from the Earth

[1] Ptolemy observed two lunar eclipses during which the moon was at the epicyclic apex. During one of them, one-quarter of its diameter was eclipsed and, during the other, half of it. By calculation, its latitude during the first eclipse was 48½ minutes and, during the second, 40⅔ minutes. / He took / the difference between them, namely (7 + ½ + ⅓) minutes, which is obviously a quarter of the diameter. He thus knew that the diameter of the moon at its farthest distance was four times this [amount], namely 31⅓ minutes, and that the latitude for the second eclipse was in the amount of the radius of the shadow since the shadow circle was passing through the center of the lunar disk. This [radius] was approximately 2 times the radius of the moon plus / ⅗ times its radius. / During numerous lunar eclipses at various distances, he found this same ratio between them.

/19/ He took] β = He thus took] α, M. /24/ ⅗ times its radius] β = ⅗ times its diameter] α, M.

[٤] وكان بحساب التقاويم بالقدر الذي يكون نصف قطر المائل ستين ونصف قطر التدوير خمسة وربعاً وما بين المركزين عشرة أجزاء وتسع عشرة دقيقة بُعدُ القمر عن مركز العالم في ذلك الوقت أربعين جزءاً وربع وسدس جزء . وإذا عُرف مقدار واحد بتقديرين أمكن أن يحوّل كل ما يقدّر بواحد من ذينك التقديرين إلى التقدير الآخر لكون الجميع على نسبتهما . فحوّل بطلميوس المقادير المذكورة إلى التقدير الذي به نصف قطر الأرض واحد ؛ فخرج نصف قطر المائل تسعة وخمسين ، ونصف قطر التدوير خمسة أجزاء وسدس ، وما بين المركزين عشرة أجزاء وتسع دقائق . ويكون أبعد بعد القمر ، وذلك عند كونه في الذروة والتدوير في الأوج ، أربعة وستين جزءاً وسدس جزء ؛ وأقرب بعده ، وذلك عند كونه في حضيض التدوير والتدوير في الحضيض ، ثلاثة وثلاثين جزءاً وثلاثاً وثلاثين دقيقة .

الفصل الثالث
في مقادير أقطار القمر والشمس والظل وأبعاد الشمس والظل عن الأرض

[١] رصد بطلميوس خسوفين للقمر كان القمر فيهما في ذروة التدوير ؛ وقد انخسف من قطره في أحدهما ربعه وفي الآخر نصفه . وكان بالحساب عرضه في الخسوف الأول ثمانية[!] وأربعين دقيقة ونصف وفي الثاني أربعين دقيقة وثُلثي دقيقة . / وأخذ / الفضل بينهما ، وهو سبع دقائق ونصف وثُلث دقيقة ، ويكون لا محالة ربع القطر ؛ فعرف أنّ قطر القمر في أبعد بعده أربعة أمثال ذلك ، وهو أحد[!] وثلاثون دقيقة وثُلث ، وأنّ العرض في الخسوف الثاني هو مقدار نصف قطر الظل لكون دائرة الظل مارّة بمركز صفحة القمر ، وهو بالتقريب مِثْلاً نصف قطر القمر ومثل ثلاثة أخماس / نصف / قطره . وقد وجد في خسوفات كثيرة في أبعاد مختلفة النسبة بينهما هذه النسبة .

/19/ ...] β = فأخذ] M, α . /24/ ...] β = –M, –α .

[2] Furthermore, he found the diameter of the sun under most circumstances to be equal in appearance to the diameter of the moon at its farthest distance. He thus determined that the diameter of the sun at its mean distance is equal to the diameter of the moon at its farthest distance.

[3] He then set forth in the plane passing through them the figures of the two luminaries, the Earth, the two shadow cones, and the moon according to this illustration:

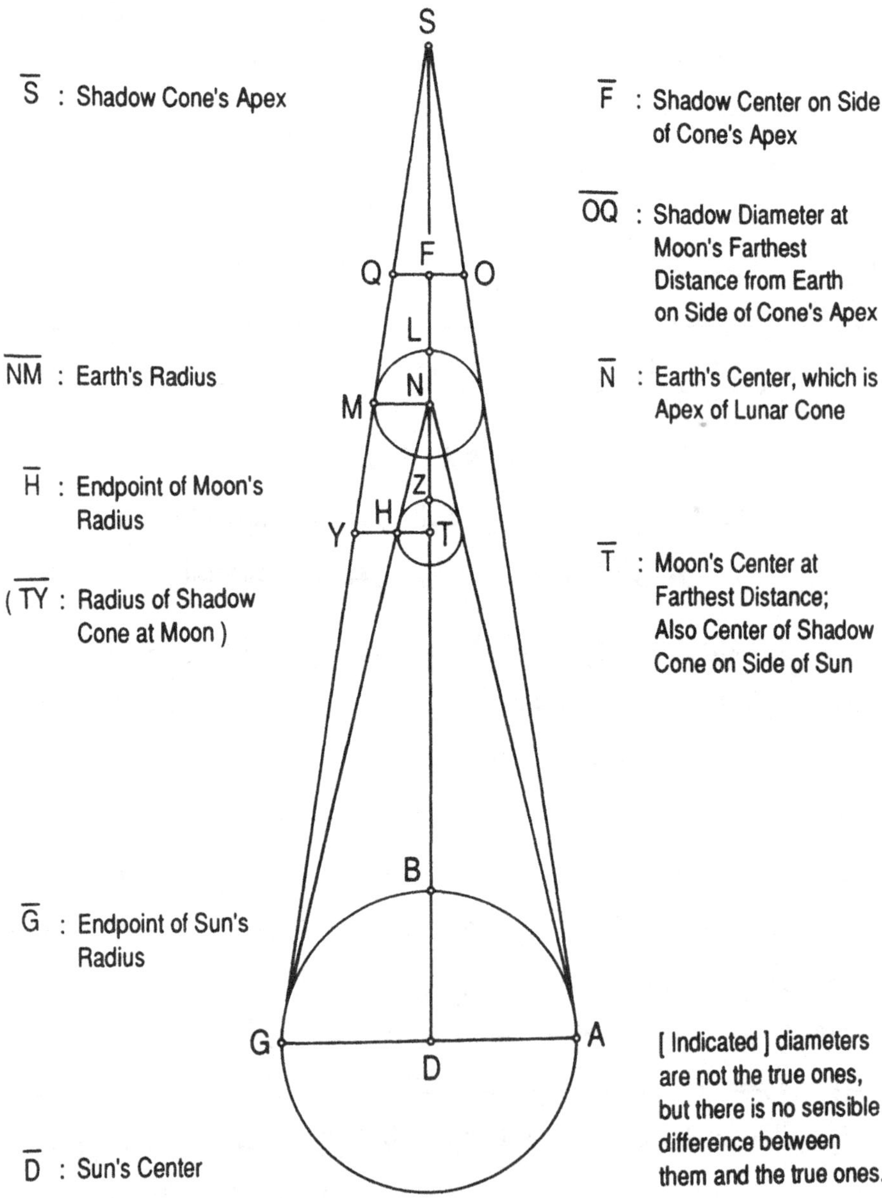

[Fig. T27]

[٢] وأيضاً وَجَد قطر الشمس في أكثر الأحوال مساوياً في النظر لقطر القمر في بعده الأبعد ؛ فحكم بأنَ قطر الشمس في بعدها الأوسط مساوٍ لقطر القمر في بعده الأبعد .

[٣] ثم وضع في السطح المارّ بالنيّرين والأرض ومخروطي الظل والقمر أشكالها على هذه الصورة :

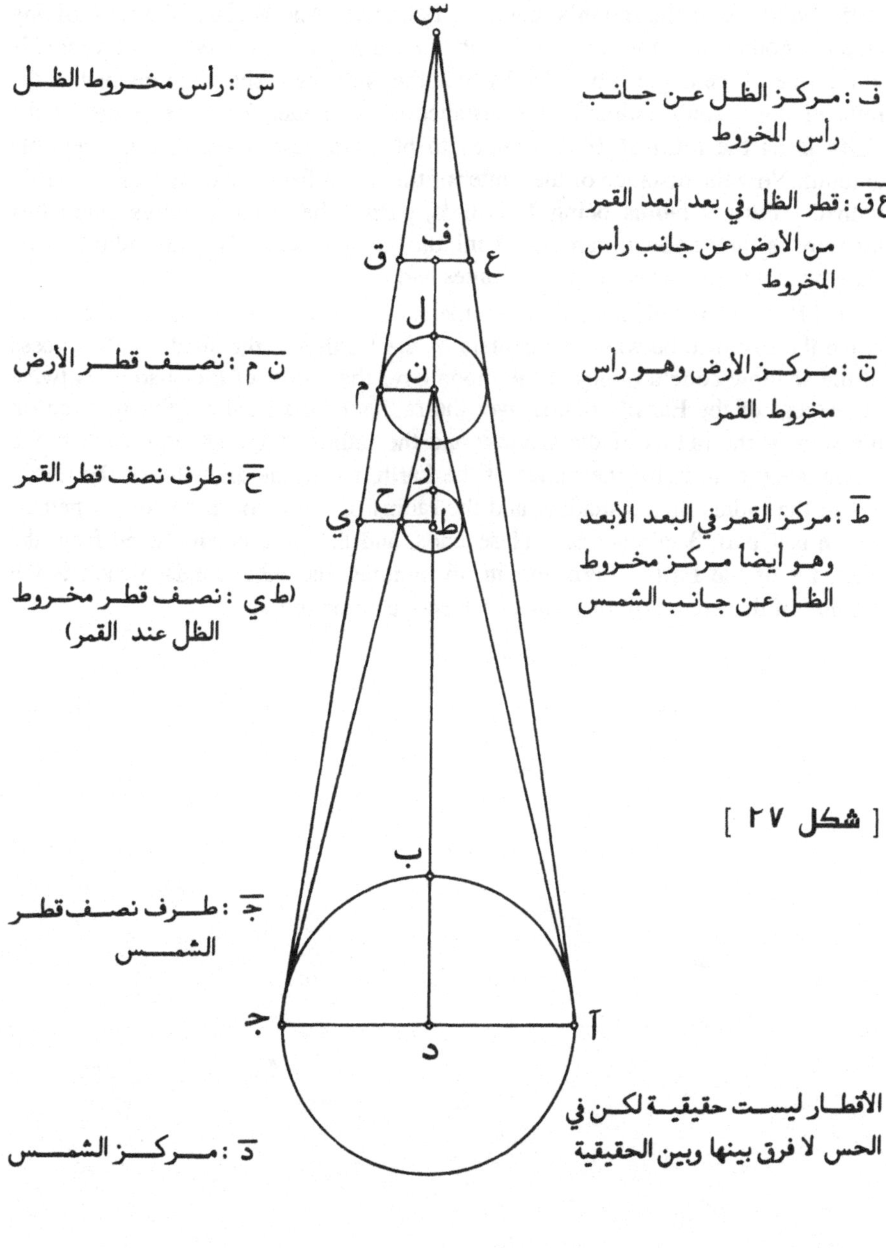

[شكل ٢٧]

He assumed the moon to be at its farthest distance and a diameter of the shadow on the other side to be at the moon's [other] farthest distance; the distance between the centers of the shadow and the Earth and [the distance] between the centers of the moon and the Earth would then be equal, each one of them being $64\frac{1}{6}$. In the triangle that occurs in the moon's cone between the centers of the moon and the Earth and the endpoint of the moon's radius, the angle that is at the center of the Earth, which is in the amount of the moon's radius, and the angle that is at the center of the moon, which is right, are known. And since the angles of every triangle are equal to two right [angles], the third angle, which is at the endpoint of the moon's diameter, is known. And because the ratio of any side to another is as the ratio of the sine of the angle subtended by the first side to the sine of the angle subtended by the other side, as is shown in geometry, the ratio of the moon's radius to the distance of its center from the center of the Earth is as the ratio of $16\frac{2}{5}$ minutes to 60 parts less a small, imperceptible amount. Now the distance of the center of the moon from the center of the Earth, with the Earth's radius being 1, is $64\frac{1}{6}$ parts. The moon's radius using this measure is then known, namely 17 minutes and 33 seconds. The radius of the shadow using this measure is 45 minutes and 38 seconds.

[4] Because the distance between the centers of the moon and the shadow is twice the distance between the centers of the Earth and the shadow, the excess of the shadow cone's radius at the moon over the radius of the shadow is twice the excess of the Earth's radius over the radius of the shadow. For that reason the sum of the radius of the shadow and the radius of the shadow cone at the moon is equal to twice the radius of the Earth, *i.e.* to the diameter of the Earth. When the radius of the shadow and the radius of the moon are added together, which is 1 part, 3 minutes and 11 seconds, and this sum is subtracted from the diameter of the Earth, there remain 56 minutes and 49 seconds, which is the amount of the excess of the radius of the cone at the moon

وفَرَض القمر في بعده الأبعد وقطرَ الظل عن الجانب الآخر في بعد القمر الأبعد ؛ فيكون البعد بين مركزي الظل والأرض وبين مركزي القمر والأرض متساويين ، كل واحد منهما أربعة وستون وسدس . وتكون في المثلث الذي حدث في مخروط القمر بين مركزي القمر والأرض وطرف نصف قطر القمر الزاويةُ التي على مركز الأرض ، وهي بقدر نصف قطر القمر ، والزاويةُ التي على مركز القمر ، وهي قائمة ، معلومتين . ولكون زوايا كل مثلث مساوية لقائمتين تصير الزاوية الثالثة ، وهي التي على طرف قطر القمر ، معلومة . ولأنّ نسبة كل ضلع إلى آخر تكون كنسبة جيب الزاوية التي يوترها الضلع الأول إلى جيب الزاوية التي يوترها الضلع الآخر – على ما بُيّن في الهندسة – تكون نسبة نصف قطر القمر إلى بعد مركزه عن مركز الأرض كنسبة ست عشرة دقيقة وخُمسي دقيقة إلى ستين جزءاً إلاّ شيء قليل غير محسوس . وكان بعد مركز القمر عن مركز الأرض – على أنّ نصف قطر الأرض واحد – أربعة وستين جزءاً وسدس جزء ؛ فنصف قطر القمر بذلك المقدار يكون معلوماً ، وهو سبع عشرة دقيقة وثلاث وثلاثون ثانية ، ويكون نصف قطر الظل بذلك المقدار خمساً وأربعين دقيقة وثمانياً وثلاثين ثانية .

[٤] ولأنّ البعد بين مركزي القمر والظل ضعف البعد بين مركزي الأرض والظل تكون زيادة نصف قطر مخروط الظل الذي عند القمر على نصف قطر الظل ضعف زيادة نصف قطر الأرض على نصف قطر الظل ؛ ويكون لذلك مجموع نصفي قطر الظل وقطر مخروط الظل عند القمر مساوياً لضعف نصف قطر الأرض ، أعني لقطر الأرض . وإذا جمع نصف قطر الظل ونصف قطر القمر ، وهما جزء وثلاث دقائق وإحدى عشرة ثانية ، ونقص المجموع من قطر الأرض ، بقيت ست وخمسون دقيقة وتسع وأربعون ثانية ، وهي مقدار فضل نصف قطر المخروط عند القمر

over the radius of the moon. The ratio of the Earth's radius to this is as the ratio of the distance between the centers of the Earth and the sun to the distance between the centers of the two luminaries, which is as the ratio of 1 to 56 minutes and 49 seconds. Thus when the distance of the sun from the center of the Earth is 1, the distance between the two luminaries is 56 minutes and 49 seconds, and there / remain / 3 minutes and 11 seconds for the distance of the moon from the Earth; with the radius of the Earth being 1, this distance is 64⅙ parts. Then by this [measure], the distance of the sun from the center of the Earth at its mean distance is 1210 times the Earth's radius.

[5] Furthermore the ratio of the Earth's radius to the radius of the shadow, which is 45 minutes and 38 seconds, is as the ratio of the distance of the cone's apex from the center of the Earth to its distance from the center of the shadow. Therefore when the distance of the cone's apex from the center of the Earth is 1, its distance from the center of the shadow is 45 minutes and 38 seconds, and there remain (14 + ⅕ + ⅙) minutes for the distance of the shadow's center from the center of the Earth, which is, with the radius of the Earth being 1, 64⅙. Then by this [measure], the distance of the cone's apex from the center of the / Earth / is (203 + ½ + ⅓) times the Earth's radius.

/6/ remain] β = are] α, M. /19/ Earth] α, β = shadow] M (see commentary).

على نصف قطر القمر . وتكون نسبة نصف قطر الأرض إليه كنسبة البعد بين مركزي الأرض والشمس إلى البعد بين مركزي النيّرين ، وهي كنسبة الواحد إلى ست وخمسين دقيقة وتسع وأربعين ثانية . فإذن إذا كان بعد الشمس عن مركز الأرض واحداً كان البعد بين النيّرين ستاً وخمسين دقيقة وتسعاً وأربعين ثانية ؛ / ويبقى / بعد القمر عن الأرض ثلاث دقائق وإحدى عشرة ثانية ، وكان هذا البعد – على أنّ نصف قطر الأرض واحد – أربعة وستين جزءاً وسدس جزء . فبحسب ذلك يكون بعد الشمس عن مركز الأرض في / بعده / الأوسط ألفاً ومائتين وعشرة أمثال نصف قطر الأرض .

[٥] وأيضاً نسبة نصف قطر الأرض إلى نصف قطر الظل ، وهو خمس وأربعون دقيقة وثمان وثلاثون ثانية ، كنسبة بعد رأس المخروط عن مركز الأرض إلى بعده عن مركز الظل . فلذلك إذا كان بعد رأس المخروط عن مركز الأرض واحداً كان بعده عن مركز الظل خمساً وأربعين دقيقة وثمانياً وثلاثين ثانية ؛ ويبقى بُعد مركز الظل عن مركز الأرض أربع عشرة دقيقة وخُمس وسدس دقيقة ، وكان – على أنّ نصف قطر الأرض واحد – أربعة وستين وسدساً . فبحسب ذلك يكون بعد رأس المخروط عن مركز / الأرض / مائتين وثلاثة أمثال ونصف وثُلث مثل لنصف قطر الأرض .

/6/ ...] β = وكان] M , α . /9/ بعده] β , α = بعدها] M . /19/ الأرض] β , α = الظل] M (انظر شرحنا) .

CHAPTER FOUR

On the Volume of the Two Luminaries

[1] It has been established in the science of optics that for any two bodies that are equal in appearance but are at different distances, the ratio of the nearer of the two to the one farther away with respect to the size of the diameter of the body is as the ratio of the distance of the nearer to the distance of the farther. For that reason, the ratio of the moon's radius, which is 17 minutes and 33 seconds, to the sun's radius is as the ratio of the distance of the moon from the Earth, which is $64\frac{1}{6}$, to the distance of the sun from the Earth, which is 1210. Thus the radius of the sun is also known, namely $5\frac{1}{2}$, with the Earth's radius being 1. If the diameter of the moon is assumed to be 1, the diameter of the Earth becomes $3\frac{2}{5}$ and the diameter of the sun $18\frac{4}{5}$.

[2] Euclid has proven that the ratio of one sphere to [another] sphere is as the ratio of the cube of the [first] diameter to the cube of the [other] diameter. Then when the above quantities are multiplied by themselves twice so they become cubed, one finds that the sun is $(166 + \frac{1}{4} + \frac{1}{8})$ times the Earth and 6644 times / and $\frac{2}{3}$ times / the moon, and that the Earth is $39\frac{1}{4}$ / and ($\frac{1}{2}$ of $\frac{1}{10}$) / times the moon.

/17/ and $\frac{2}{3}$ times] $\beta = -\alpha$, –M. /18/ and $\frac{1}{2}$ of $\frac{1}{10}$] $\beta = -\alpha$, –M.

الفصل الرابع
في مقدار جرم النيّرين

[١] ثبت في علم المناظر أنّ كل جرمين متساويين في الرؤية ومختلفين في البعد تكون نسبة أقربهما إلى أبعدهما في مقدار قطر الجرم كنسبة بعد الأقرب إلى بعد الأبعد . ولذلك تكون نسبة نصف قطر القمر ، الذي هو سبع عشرة دقيقة وثلاث وثلاثون ثانية ، إلى نصف قطر الشمس كنسبة بعد القمر عن الأرض ، الذي هو أربعة وستون وسدس ، إلى بعد الشمس عن الأرض ، الذي هو ألف ومائتان وعشرة . فيكون نصف قطر الشمس أيضاً معلوماً ، وهو خمسة ونصف على أنّ نصف قطر الأرض واحد . وإن فرض قطر القمر واحداً صار قطر الأرض ثلاثة وخُمسَين وقطر الشمس ثمانية عشر وأربعة أخماس .

[٢] وقد بيّن أقليدس أنّ نسبة الكرة إلى الكرة تكون كنسبة مكعّب القطر إلى مكعّب القطر . فإذا ضربت هذه المقادير في أنفسها مرتين لتصير مكعّبة علم أنّ الشمس مائة وستة وستون مثلاً وربع وثمن مثل / الأرض / وستة آلاف وستمائة وأربعة وأربعون مثلاً / وثلثي مثل / القمر ، وأنّ الأرض تسعة وثلاثون مثلاً وربع / ونصف عشر / مثل / القمر / .

/16/ ...] β = للأرض] M , α . /17/ وثلثي مثل] β = –M , –α . /18/ ونصف عشر] β = –M , –α . /18/ القمر] M , β = للقمر] α .

CHAPTER FIVE

On the Rest of the Distances of the Sun and the Distances and Body [Sizes] of the Two Lower Planets

[1] The known distance for the sun that was given above was only made on the assumption that it was at its mean distance. Its elongation from there at the two other distances is according to the amount between the centers of the two [orbs of the sun], which is, based on Ptolemy's observations, $2\frac{1}{2}$ of those parts by which the radius of its eccentric orb is 60; therefore, it is 1 part in 24 of its mean distance. When we divide the known distance of the sun, which is 1210, by 24, the result is 50 and a fraction, which is the size of the eccentricity. The farthest distance of the sun is then approximately 1260 times the radius of the Earth, and its nearest distance is 1160 times it.

[2] Since there is no void between the orbs of the planets, and there is no known body other than their orbs, the farthest distance of each planet has been made the nearest distance of the planet above it so that the adopted distances are the minimum possible. Thus the nearest distance of the sun is the farthest distance of Venus.

[3] As for Venus, one finds in calculating true positions that its eccentricity is $1\frac{1}{4}$ parts, and the radius of its epicycle is $43\frac{1}{6}$ of those parts by which the radius of / its deferent / is 60. Thus its farthest distance is $(104 + \frac{1}{4} + \frac{1}{6})$ parts, and its nearest distance is $(15 + \frac{1}{3} + \frac{1}{4})$ of those parts, which is approximately $(\frac{1}{10} + (\frac{1}{2} \text{ of } \frac{1}{10}))$ of the farthest distance.

/19/ its deferent] β, M = the deferent] α.

الفصل الخامس
في سائر أبعاد الشمس
وأبعاد السفليين وجرميهما

[١] البعد المعلوم للشمس المذكور إنما فرض عند كونها في بعدها الأوسط . ويكون تباعدها عنه في البعدين الآخرين بقدر ما بين مركزيهما ؛ وكان ذلك ، بحسب أرصاد بطلميوس ، جزأين / ونصفاً / من الأجزاء التي بها نصف قطر فلكها الخارج المركز ستون . فإذن هو جزء من أربعة وعشرين من بعدها الأوسط . وإذا قسمنا بُعد الشمس المعلوم ، وهو ألف ومائتان وعشرة ، على أربعة وعشرين خرج خمسون وكسر ، وهو مقدار خروج المركز . فيكون بعد الشمس الأبعد ألفاً ومائتين وستين مثلاً لنصف قطر الأرض بالتقريب ، وبُعدها الأقرب ألفاً ومائة وستين مثلاً له .

[٢] ولمّا لم يكن بين أفلاك الكواكب خلاء ولا جرم معلوم غير أفلاكها ، جُعل البعد الأبعد لكل كوكب البعد الأقرب للكوكب الذي فوقه لتكون الأبعاد المأخوذة هي التي لا يمكن أن تكون أقل منها ؛ فيكون البعد الأقرب للشمس البعد الأبعد للزهرة .

[٣] أما الزهرة فقد عُلم في حساب التقاويم أنّ ما بين مركزيها جزء وربع ، ونصف قطر تدويرها ثلاثة وأربعون وسدس من الأجزاء التي بها نصف قطر / حاملها / ستون . فيكون بعدها الأبعد مائة وأربعة أجزءا وربع وسدس ، وبعدها الأقرب خمسة عشر جزءاً وثُلث وربع بتلك الأجزاء وهو عُشر البعد الأبعد ونصف عُشره بالتقريب .

/7/ ...] β = ونصف] M , α . /19/ ...] M , β = الحامل] α .

[4] Moreover Mercury's eccentricity is 3 parts, which is equal to the distance between each of the centers of its orbs and the one adjacent to it. The radius of its epicycle is 22½ of those parts by which the radius of its deferent is 60. Its farthest distance is 91½ parts, and its nearest distance is 33 parts 4 minutes. This is known only by successive approximation [*istiqrā'*] because its nearest distance is not directly opposite its farthest distance. Its nearest distance is therefore (⅕ + ⅙) of its farthest distance or, [with respect] to the farthest distance of Venus, 11 : 200 parts, which is nearly 1 : 18 parts of it.

[5] The farthest distance of the moon to the nearest distance of the sun was found above also to be nearly 1 : 18 parts. So it seemed likely to them that their two orbs were between the orbs of the two luminaries since there was no way to leave this distance between the orbs unoccupied. This is the basis for our statement earlier [in the book] that the distance of the sun from the Earth is consistent with Venus and Mercury being below [the sun].

[6] To return to what we were doing: when we take (⅒ + (½ of ⅒)) of Venus's farthest distance, one obtains 174 times the radius of the Earth, which is then Venus's nearest distance and Mercury's farthest distance. It was previously stated that the height of the shadow cone is 203 plus a fraction times the radius of the Earth. It is then known that the Earth's shadow will disappear in the orb of Venus between its near and mean distances. Furthermore it is clear from the [above] that the thickness of Venus's orb is 1000 less 14 times the Earth's radius, and that the thickness of Mercury's orb plus what is contained inside it is 348 times, which is approximately ⅓ of [Venus's orb]. We then take (⅕ + ⅙) of Mercury's farthest distance with the result being 64 times the Earth's radius, which is the nearest of Mercury's distances and the farthest of the moon's distances, in accord with the result of the earlier calculation.

[٤] وأيضاً ما بين مركزي عطارد ثلاثة أجزاء ويساويه البعد بين كل مركز من مراكز أفلاكه وبين الذي يليه ، ونصف قطر تدويره اثنان وعشرون جزءاً ونصف بالأجزاء التي بها نصف قطر حامله ستون . وبعده الأبعد أحد وتسعون جزءاً ونصف ، وبُعد أقربه ثلاث[!] وثلاثون جزءاً وأربع دقائق . وإنما عُرف ذلك بالاستقراء لأنّ بعده الأقرب لا يقابل بعده الأبعد . فيكون بعده الأقرب خُمساً وسدساً من بعده الأبعد أو أحد عشر جزءاً من مائتي جزء هي أجزاء بعد الزهرة الأبعد ، وهي قريبة من جزء من ثمانية عشر منه .

[٥] ووُجد بعد القمر الأبعد من بعد / الشمس الأقرب / أيضاً قريباً من جزء من ثمانية عشر كما مرّ . فغلب على ظنونهم كون فلكيهما بين فلكي النيّرين إذ لا وجه لتعطيل هذا البعد بين الأفلاك . وهذا هو الوجه لقولنا فيما مرّ أنّ بعد الشمس من الأرض يناسب كون الزهرة وعطارد تحتها .

[٦] ونعود إلى ما كُنّا فيه . فإذا أخذنا العُشر ونصف العُشر من بعد الزهرة الأبعد حصل مائة وأربعة وسبعون مثلاً لنصف قطر الأرض ، فهو البعد الأقرب للزهرة والبعد الأبعد لعطارد . وقد مرّ أنّ ارتفاع مخروط الظلّ مائتان وثلاثة أمثال نصف قطر الأرض وكسر ، فعُلم أنّ ظلّ الأرض ينعدم في فلك الزهرة بين بعديه الأقرب والأوسط . وأيضاً يتبيّن منه أنّ ثخن فلك الزهرة ألف مثلاً لنصف قطر الأرض غير أربعة عشر مثلاً ، وأنّ ثخن فلك عطارد بما في ضمنه ثلاثمائة وثمانية وأربعون مثلاً وهو قريب من ثُلثه . ثم أخذنا الخُمس والسدس من بعد عطارد الأبعد فحصل أربعة وستون مثلاً لنصف قطر الأرض ، وهو أقرب أبعاد عطارد وأبعد أبعاد القمر موافقاً لما خرج من الحساب الأول .

/10/ ...] M , β = أقرب الشمس] α .

[7] Turning now to the body [size] of Venus and of Mercury: they have stated that the diameter of Venus at its mean distance is approximately equal to $\frac{1}{10}$ the diameter of the sun, and that the diameter of Mercury to the diameter of the sun is as 1 : 15. Then the [mid-distance] between Venus's two [extreme] distances is taken, and one obtains 667, which is its mean distance. Its ratio to the sun's mean distance is as the ratio of Venus's diameter to $\frac{1}{10}$ the sun's diameter. Venus's mean distance to the sun's mean distance is as 1 : 1;49 [*lit.*, one to one and forty-nine minutes], which is the size of Venus's diameter to $\frac{1}{10}$ the sun's diameter. When one multiplies 1;49 by 10, the result is 18$\frac{1}{6}$. Thus the diameter of Venus to the diameter of the sun is as 1 : 18$\frac{1}{6}$ parts. When 2 parts in 11 is taken of this, one obtains 3$\frac{3}{10}$ parts; thus the diameter of Venus to the diameter of the Earth is as 1 : 3$\frac{3}{10}$ parts. / If / the two quantities are cubed, it becomes approximately 1 : 35;56 [*lit.*, one to thirty-five and fifty-six minutes]; therefore, the body of the Earth is approximately 36 times the body of Venus.

[8] Furthermore Mercury's mean distance occurring between its two [extreme] distances is 119 times the radius of the Earth. To the sun's mean distance this is approximately as 1 : 10$\frac{1}{6}$ parts, which is the size of Mercury's diameter to ($\frac{1}{3}$ of $\frac{1}{5}$) the diameter of the sun. Multiplied by 15, the result is 153. Thus the size of Mercury's diameter to the sun's diameter is 1 : 153. When 2 parts in 11 is taken of this, [the result] is approximately 28; thus the size of Mercury's diameter to the Earth's diameter is as 1 : 28 parts. The cube of 28 is 21,952. Thus the Earth's body is approximately equal to 22,000 times Mercury's body.

/13/ If] β = When] α, M.

[٧] وأما جرم الزهرة وعطارد فذكروا أنّ قطر الزهرة في بعدها الأوسط يكون مثل عُشر قطر الشمس تقريباً وأنّ قطر عطارد من قطر الشمس يكون كواحد من خمسة عشر . فأخذ ما بين بعدي الزهرة فحصل ستمائة وسبعة وستون ، وهو بعدها الأوسط . وتكون نسبتها إلى بعد الشمس الأوسط كنسبة قطر الزهرة إلى عُشر قطر الشمس . وبُعد الزهرة الأوسط من بعد الشمس الأوسط كواحد من واحد وتسع وأربعين دقيقة ، وهي قدر قطر الزهرة من عُشر قطر الشمس . وإذا ضرب واحد وتسع وأربعون دقيقة في عشرة بلغ ثمانية عَشر وسدساً ، فيكون قطر الزهرة من قطر الشمس كواحد من ثمانية عشر جزءاً وسدس جزء . وإذا أخذ منها جزءان من أحد عشر حصل ثلاثة أجزاء وثلاثة أعشار جزء ؛ فقطر الزهرة من قطر الأرض كواحد من ثلاثة أجزاء وثلاثة أعشار . / وإن / كُعّب المقداران يصير واحداً من خمسة وثلاثين وست وخمسين دقيقة بالتقريب ؛ فإذن جرم الأرض ستة وثلاثون مثلاً / لجرم الزهرة بالتقريب / .

[٨] وأيضاً بُعد عطارد الأوسط الكائن بين بعديه مائة وتسعة عشر مثلاً لنصف قطر الأرض . وهو من بعد الشمس الأوسط كواحد من عشرة أجزاء وسدس بالتقريب ، وهو قدر قطر عطارد من ثُلث خُمس قطر الشمس . ضرب في خمسة عشر بلغ مائة وثلاثة وخمسين . فقدر قطر عطارد من قطر الشمس واحد من مائة وثلاثة وخمسين . وإذا أخذ منه جزءان من أحد عشر كان ثمانية وعشرين بالتقريب ؛ فقدر قطر عطارد من قطر الأرض كجزء من ثمانية وعشرين . ومكعّب ثمانية وعشرين أحد وعشرون ألفاً وتسعمائة واثنان وخمسون ؛ فجرم الأرض مثل جرم عطارد اثنان وعشرون ألف مرة بالتقريب .

/13/ ...] β = وإذا] M , α . /15/ ...] M , β = بالتقريب لجرم الزهرة] α .

CHAPTER SIX

On the Distances of the Upper Planets and Their Body [Sizes]

[1] Ptolemy found the eccentricity of Mars to be 6 parts and the radius of its epicycle 39½ parts, based on the deferent radius being 60. Its farthest distance is then 105½ parts, and its nearest distance is 14½ parts, which is to the farthest distance as 1 : 7, approximately. Then multiplying the farthest distance of the sun, which is 1260, by 7 results in 8820 times the radius of the Earth, which is then the farthest distance of Mars.

[2] They have stated that the diameter of Mars at its mean distance is [with respect] to the diameter of the sun as 1 : 20 parts. They then took its mean distance, *i.e.* the mid-distance between its two [extreme] distances; this is equal to 5040 times the radius of the Earth, which is 4⅙ times the sun's mean distance. When one takes (½ of ⅒) the diameter of the sun, one obtains 16½ minutes; multiplying by 4⅙ results in 1;9 [*lit.*, one and nine minutes], which is the diameter of Mars when the diameter of the Earth is 1. By taking its cube, which is then 1;31 [*lit.*, one and thirty-one minutes], one finds that the volume of Mars is approximately equal to 1½ times the volume of the Earth.

[3] It is evident that the thickness of Mars's orb is 7560 times the Earth's radius and that the diameter of the sun's sphere is 2520 times it. Therefore the thickness of Mars's orb is 3 times the breadth of [the combination of] the sun's orb and the orbs and the elements [contained] inside it. This is an elucidation of what we stated in the chapter on the configurations [*hay'āt*] of the orbs of the upper planets.

الفصل السادس
في أبعاد الكواكب العلوية وأجرامها

[١] وجد بطلميوس ما بين مركزي المرّيخ ستة أجزاء ونصف قطر تدويره تسعة وثلاثين جزءاً / ونصفاً / على أنّ نصف قطر الحامل ستون . فيكون بعده الأبعد مائة وخمسة أجزاء ونصف وبعده الأقرب أربعة عشر جزءاً ونصف ، وهو من بعده الأبعد كواحد من سبعة تقريباً . فضَرْب أبعد بعد الشمس ، وهو ألف ومائتان وستون ، في سبعة بلغ ثمانية آلاف وثمانمائة وعشرين مثلاً لنصف قطر الأرض ، فهو بعد المرّيخ الأبعد .

[٢] وذكروا أنّ قطر المرّيخ في بعد أوسطه يكون من قطر الشمس كجزء من عشرين . فأخذوا بعده الأوسط _ أعني منتصف ما بين بعديه _ فكان خمسة آلاف وأربعين مثلاً لنصف قطر الأرض ، وهو أربع مرات وسدس مرة مثل بعد الشمس الأوسط . وإذا أخذ نصف عُشر قطر الشمس خرج ست عشرة دقيقة ونصف ؛ ضرب في أربعة وسدس بلغ واحداً وتسع دقائق ، وهو قطر المرّيخ إذا كان قطر الأرض واحداً . أخذ مكعّبه فكان واحداً وأحداً [!] وثلاثين دقيقة ، فعلم أنّ جرم المرّيخ مثل جرم الأرض مرة ونصف بالتقريب .

[٣] وقد ظهر أنّ ثخن فلك المرّيخ سبعة آلاف وخمسمائة وستون مثلاً لنصف قطر الأرض ، وقطر كرة الشمس يكون ألفين وخمسمائة وعشرين مثلاً له . فثخن فلك المرّيخ ثلاثة أمثال غلظ فلك الشمس مع ما فيه من الأفلاك والعناصر . وهذا بيان ما ذكرناه في باب هيئات أفلاك الكواكب العلوية .

/4/ ...] ونصف = β [M , α .

[4] Turning to Jupiter: by calculation Ptolemy found its eccentricity to be $(2 + \frac{1}{2} + \frac{1}{4})$ parts and the radius of its epicycle $11\frac{1}{2}$ parts, based on the radius of its deferent being 60. Thus its farthest distance is $74\frac{1}{4}$ parts, and its nearest distance is $(45 + \frac{1}{2} + \frac{1}{4})$ parts. So the first is $(1 + \frac{1}{4} + \frac{1}{5} + \frac{1}{6})$ times the second. When one takes the farthest distance of Mars times $(1 + \frac{1}{4} + \frac{1}{5} + \frac{1}{6})$, the result is 14,259 times the Earth's radius, which is then Jupiter's farthest distance.

[5] They have stated that [Jupiter's] diameter is equal to ($\frac{1}{2}$ of $\frac{1}{6}$) the diameter of the sun when both are at their mean distances. When one then takes the mid-distance of its two [extreme] distances, it is 11,540 times / the Earth's radius, / which is equal to $(9 + \frac{1}{3} + \frac{1}{5})$ times the sun's mean distance. When one takes ($\frac{1}{2}$ of $\frac{1}{6}$) the diameter of the sun, it is $27\frac{1}{2}$ minutes. If [this] is then multiplied by $(9 + \frac{1}{3} + \frac{1}{5})$, the result is $(4 + \frac{1}{5} + \frac{1}{6})$ units. The diameter of the Earth to the diameter of Jupiter is therefore as $1 : (4 + \frac{1}{5} + \frac{1}{6})$ units. When these two [quantities] are cubed, [one finds] the volume of Jupiter to be equal to / $83\frac{1}{4}$ / times the volume of the Earth.

[6] As for Saturn: by calculation Ptolemy found its eccentricity to be $(3 + \frac{1}{4} + \frac{1}{6})$ parts and the radius of its epicycle $6\frac{1}{2}$ parts, with the parts such that the radius of its deferent is 60 parts. Thus its farthest distance is $(69 + \frac{2}{3} + \frac{1}{4})$ parts, and its nearest distance is $(50 + (\frac{1}{2}$ of $\frac{1}{6}))$ parts; the farthest [distance] is equal to $1\frac{2}{5}$ the nearest. Then multiplying Jupiter's farthest distance by $1\frac{2}{5}$ results in 19,963 times the Earth's radius, which is Saturn's farthest distance.

/11/ the Earth's radius] β, M(?) = the Earth] α. /17/ $83\frac{1}{4}$] $\beta = 82\frac{1}{4}$] α, M.

[٤] وأما المشتري فقد وجد بطلميوس بالحساب ما بين مركزيه جزأين ونصف وربع جزء ونصف قطر تدويره أحد عشر جزءاً ونصف على أنّ نصف قطر حامله ستون . فيكون بعده الأبعد أربعة وسبعين جزءاً وربع جزء وبعده الأقرب خمسة وأربعين جزءاً ونصف وربع جزء ؛ ويكون الأول من الثاني مثله ومثل ربعه وخُمسه وسدسه . وإذا أخذ مثل بعد المرّيخ الأبعد ومثل ربعه وخُمسه وسدسه بلغ أربعة عشر ألفاً ومائتين وتسعة وخمسين مثلاً لنصف قطر الأرض ، فهو البعد الأبعد للمشتري .

[٥] وذكروا أنّ قطره مثل نصف سدس قطر الشمس إذا كانا في بعديهما الأوسطين . فإذا أخذ منتصف بعديه كان أحد عشر ألفاً وخمسمائة وأربعين مثلاً / لنصف قطر الأرض / ، وهو تسع مرات مثل بعد الشمس الأوسط وثُلث وخُمس مرة . وإذا أخذ نصف سدس قطر الشمس كان سبعاً وعشرين دقيقة / ونصفاً / . فإذا ضرب في تسعة وثُلث وخُمس بلغ أربعة وخُمس وسدس واحدٍ ، فقطر الأرض من قطر المشتري كواحد من أربعة وخُمس وسدس واحدٍ . وإذا كُعّبا كان جرم المشتري مثل جرم الأرض / ثلاثة[!] / وثمانين مرة وربع مرة .

[٦] وأما زحل فقد وجد بطلميوس بالحساب ما بين مركزيه ثلاثة أجزاء وربع وسدس جزء ونصف قطر تدويره ستة أجزاء ونصف بالأجزاء التي بها نصف قطر حامله ستون جزءاً . فيكون بعده الأبعد تسعة وستين جزءاً وثُلثي جزء وربعه وبعده الأقرب خَمسين جزءاً ونصف سدس جزء ؛ والأبعد مثل الأقرب ومثل خُمسيه . فضرب بعد المشتري الأبعد في واحد وخُمسين بلغ تسعة عشر ألفاً وتسعمائة وثلاثة وستين مثلاً لنصف قطر الأرض ، وهو البعد الأبعد لزحل .

/11/...] β, M(?) = للأرض] α. /13/...] β = ونصف] α, M. /17/...] β = اثنين] α, M.

[7] They have stated that [Saturn's] diameter is to the sun's diameter as 1 : 18 when both are at their mean distances. When one takes the mid-distance of its two [extreme] distances, it is 17,111 times the Earth's radius, this then being Saturn's mean distance, and it is approximately equal to 14 times the sun's mean distance. When one takes 1 part in 18 of the sun's diameter, it is $18\frac{1}{3}$ minutes. When multiplied by 14, the result is approximately $4\frac{1}{4}$ parts. The Earth's diameter is then to Saturn's diameter as 1 : $4\frac{1}{4}$ parts, approximately. When these two [quantities] are cubed, [one finds] the volume of Saturn to be approximately equal to 77 times the volume of the Earth.

CHAPTER SEVEN

On the Distance of the Fixed Stars and Their Body [Sizes] and a Concluding Discussion Regarding This Section

[1] The farthest distance of Saturn was made the distance of the fixed stars from the Earth—inasmuch as [the amount] by which it exceeds it is not known—in order that the [assigned] boundary not be beyond what exists.

[2] They have stated that the diameter of average-size stars of first magnitude in relation to the diameter of the sun is about ($\frac{1}{2}$ of $\frac{1}{10}$) of it. Their distance is approximately $16\frac{1}{2}$ times the mean distance of the sun. 1 : 20 parts of the sun's diameter is $16\frac{1}{2}$ minutes; when this is multiplied by $16\frac{1}{2}$, the result is $(4 + \frac{1}{3} + \frac{1}{5})$ units. Thus the diameter of average stars of first magnitude is equal to $(4 + \frac{1}{3} + \frac{1}{5})$ times the diameter of the Earth. When the two [quantities] are cubed, [one finds] their volume to be approximately equal to 93 times the Earth's volume.

[٧] وذكروا أنّ قطره من قطر الشمس كواحد من ثمانية عشر عند كونهما في بعديهما الأوسطين . وإذا أخذ منتصف بعديه كان سبعة عشر الفاً ومائة وأحد عشر مثلاً لنصف قطر الأرض ، فهو بعد زحل الأوسط ، وهو أربع عشرة مرة مثل بعد الشمس الأوسط تقريباً . وإذا أخذ جزء من ثمانية عشر من قطر الشمس كان ثماني عشرة دقيقة / وثُلثاً / . فإذا ضرب في أربعة عشر بلغ أربعة أجزاء وربع جزء بالتقريب ، فقطر الأرض من قطر زحل كجزء واحد من أربعة أجزاء وربع / تقريباً / . وإذا كُعّبا كان جرم زحل مثل جرم الأرض سبعاً وسبعين مرة بالتقريب .

الفصل السابع
في بعد الثوابت وأجرامها
وتمام القول في هذا الباب

[١] جُعل أبعد بعد زحل بعد الثوابت من الأرض إذ لم تكن الزيادة عليه معلومة لئلاّ يكون المحدود أكثر من الموجود .

[٢] وذكروا أنّ قطر أوسط كواكب القدر الأول جرماً يكون من قطر الشمس بالقياس قريباً من نصف عُشره . وكان بعدها ستة عشر مثلاً ونصفاً لبعد الشمس الأوسط بالتقريب . والجزء من عشرين من قطر الشمس ست عشرة دقيقة ونصف . وإذا ضرب في ستة عشر ونصف بلغ أربعة وثُلث وخُمس واحدٍ . فقطر أوسط كواكب القدر الأول أربع مرات مثل قطر الأرض ومثل ثُلثه وخُمسه . وإذا كُعّبا كان جرمه ثلاثاً وتسعين مرة بالتقريب مثل جرم الأرض .

/6/ ... [وثُلث = β , M [α . /8/ ... [بالتقريب = β , M [α .

[3] One should divide this amount by 6, and this sixth is made the difference between the mean of any magnitude and the mean of the magnitude that follows it. The sixth is divided by 3, and this third of the sixth is made the difference between the largest and the mean [values] for any magnitude or between the mean and the smallest [values]. Thus the largest of the fixed stars is $98\frac{1}{6}$ times the Earth, and the smallest is $10\frac{1}{3}$ times it.

[4] It has become clear from this study that the largest of these bodies is the sun; then there are the fixed stars of first magnitude, then Jupiter, then Saturn, then the remaining fixed stars, then Mars, then the Earth, then Venus, then the moon, and then Mercury, which is the smallest star.

[5] Whoever wishes to convert the distances to parasangs, miles, and other [units] may do so. We have converted two of these distances to parasangs. The first is the nearest, which is the nearest distance of the moon from the center of the Earth, *i.e.* the radius of the world of generation and corruption; it is 42,709 parasangs. As for [the distance] from the surface of the Earth to the part of the moon's orb closest to us, it is 41,436 parasangs. The second is the most distant, which is the distance of the fixed stars from the center of the Earth; it is twenty-five thousand thousands four hundred twelve thousand eight hundred ninety-nine [25,412,899] parasangs.

[6] Let us end the book here, praising God Most High and praying for His prophet, the chosen one. God is for us sufficient, how wonderful is the Entrusted One!

[٣] وينبغي أن يقسم هذا القدر على ستة ويجعل السدس التفاضل بين أوسط كل قدر وأوسط القدر الذي يليه . ويقسم السدس على ثلاثة ويجعل ثُلث السدس التفاضل بين أكبر كل قدر وبين أوسطه أو بين أوسطه وأصغره . فيكون أكبر الثوابت ثمانية وتسعين مثلاً وسدس مثل للأرض ، وأصغرها عشرة أمثالها وثُلث مثلها .

[٤] وقد بان من هذا البحث أنّ أعظم هذه الأجرام الشمس ؛ ثم كواكب القدر الأول من الثوابت ، ثم المشتري ، ثم زحل ، ثم باقي الكواكب الثابتة ، ثم المرّيخ ، ثم الأرض ، ثم الزهرة ، ثم القمر ، ثم عطارد ، وهو أصغر الكواكب .

[٥] ومَن أراد أن يحوّل الأبعاد إلى الفراسخ والأميال وغيرهما فله ذلك . ونحن حوّلنا بُعدين منها إلى الفراسخ . الأول أقربها ، وهو بعد القمر الأقرب من مركز الأرض ، أعني نصف قطر عالم الكون والفساد ؛ فكان اثنين وأربعين ألفاً وسبعمائة وتسع[!] فراسخ . وأما من سطح الأرض إلى ما هو أقرب إلينا من فلك القمر فأحد وأربعون ألفاً وأربعمائة وستة وثلاثون فرسخاً . والثاني أبعدها ، وهو بعد الثوابت عن مركز الأرض ؛ فكان خمسة وعشرين ألف ألف وأربعمائة واثني عشر ألفاً وثمانمائة وتسعة وتسعين فرسخاً .

[٦] ولنختم الكتاب هاهنا حامدين لله تعالى ومصلّين على نبيّه المصطفى . والله حسبنا ونعم الوكيل .

Part III

Commentary Figures

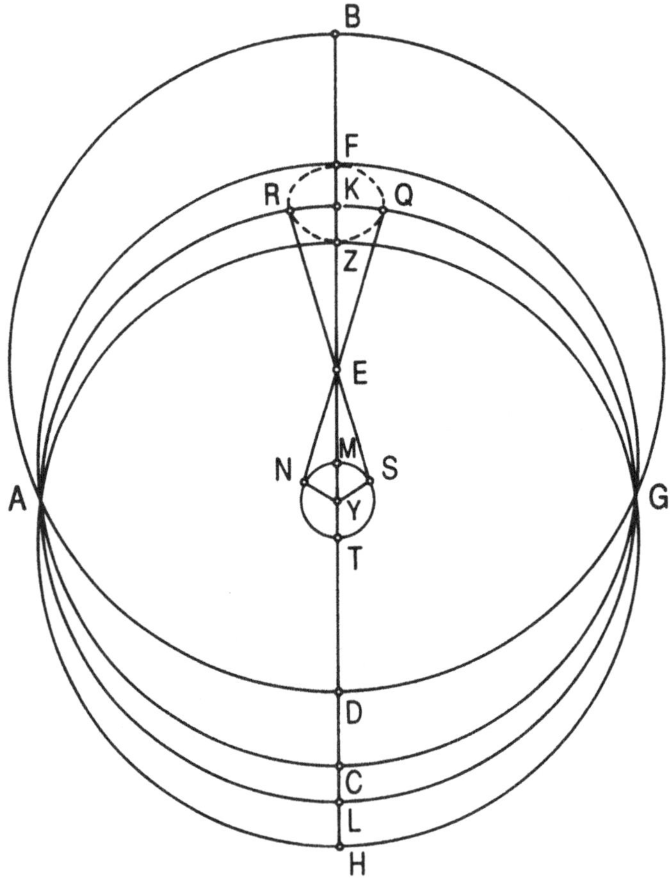

Fig. C1

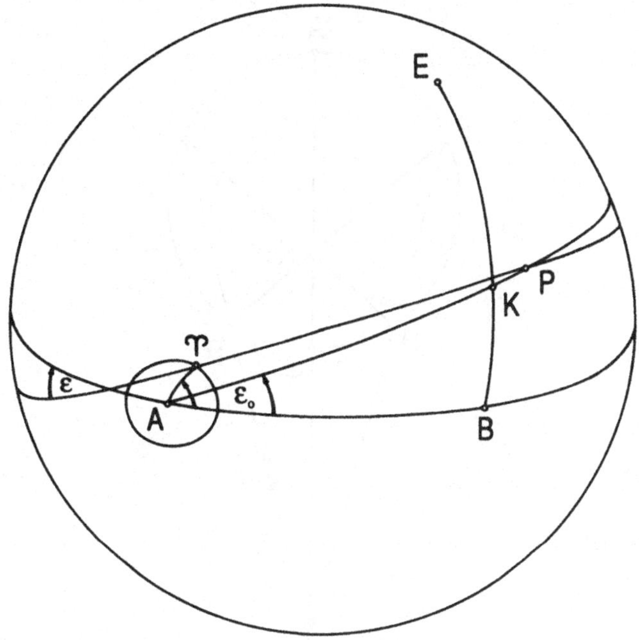

Fig. C2

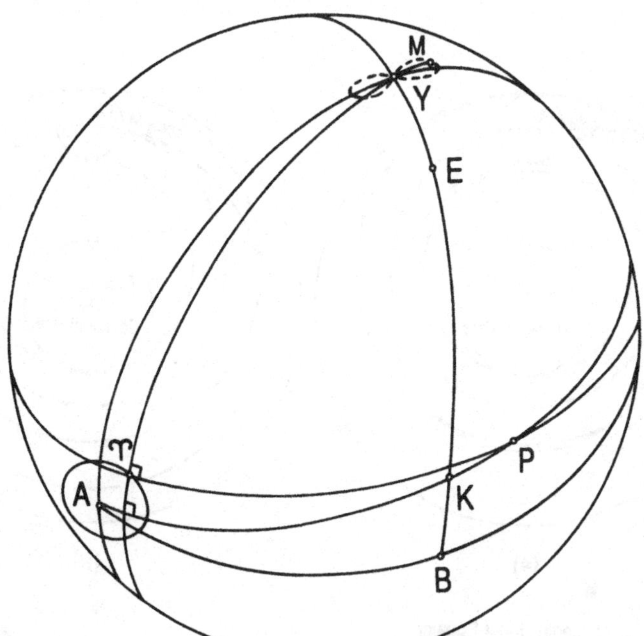

Fig. C3

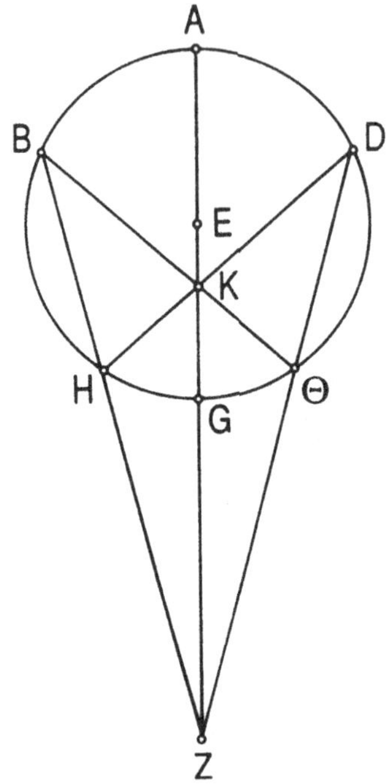
A
B
D
E
K
H
G
Θ
Z

Fig. C4

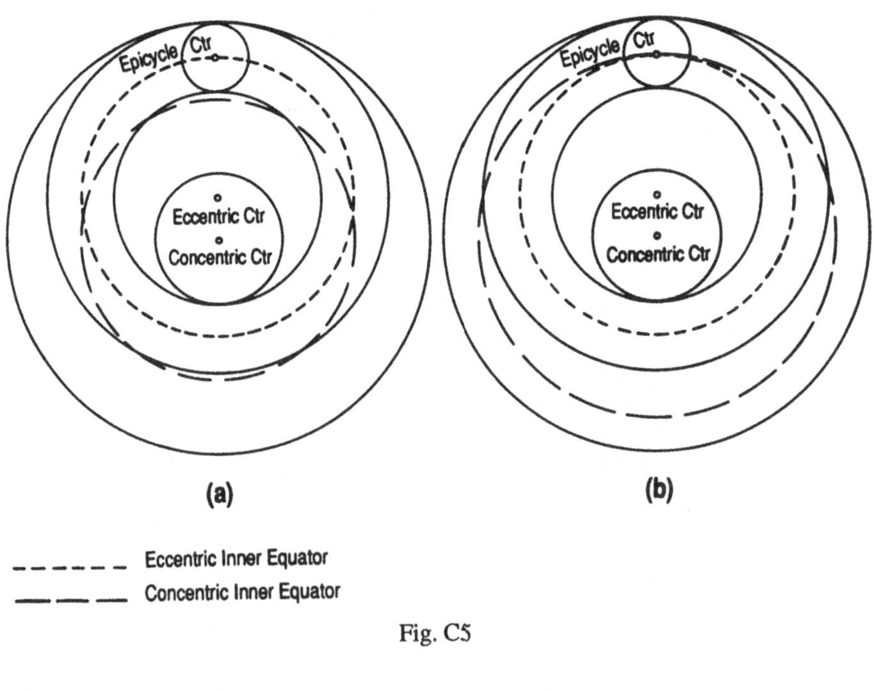
Epicycle
Ctr
Eccentric Ctr
Concentric Ctr
(a)
Epicycle
Ctr
Eccentric Ctr
Concentric Ctr
(b)
Eccentric Inner Equator
Concentric Inner Equator

Fig. C5

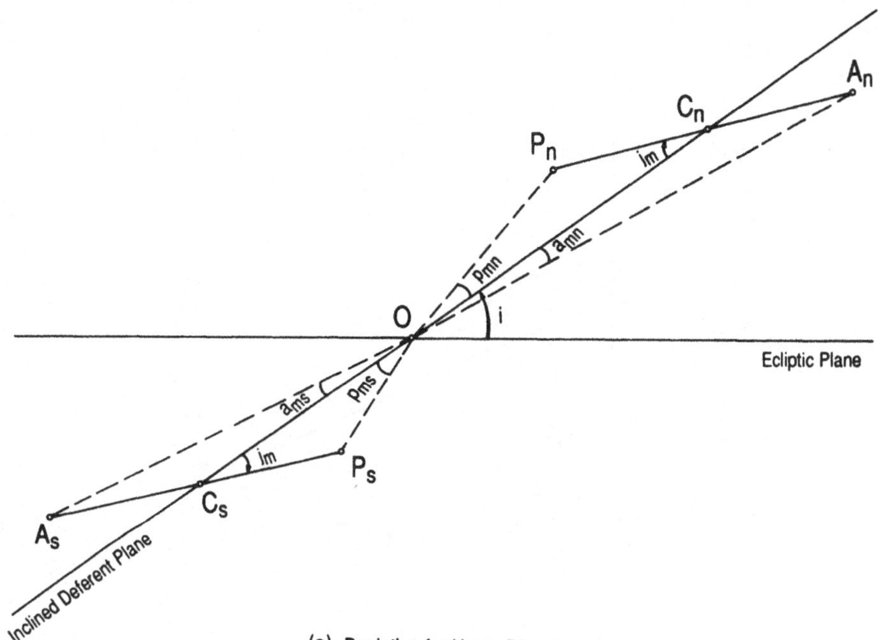

(a) Deviation for Upper Planets

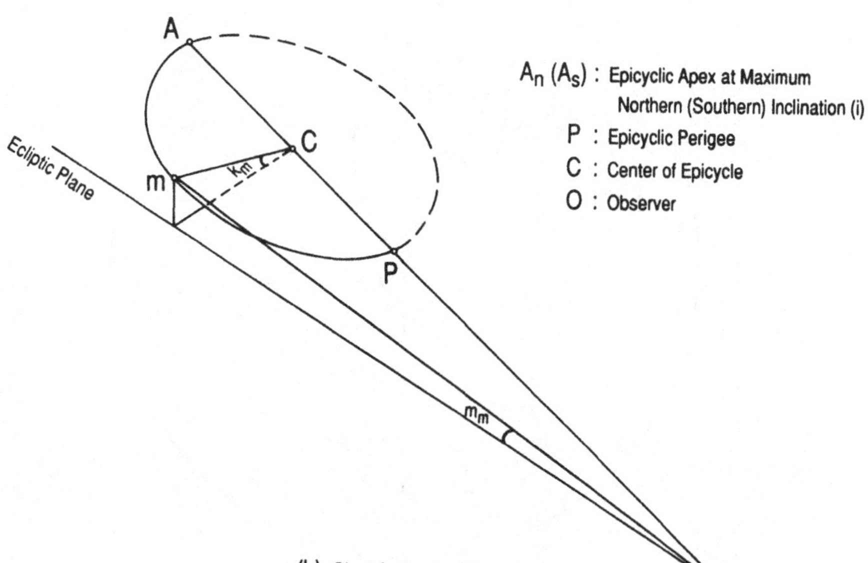

(b) Slant for Lower Planets

Fig. C6

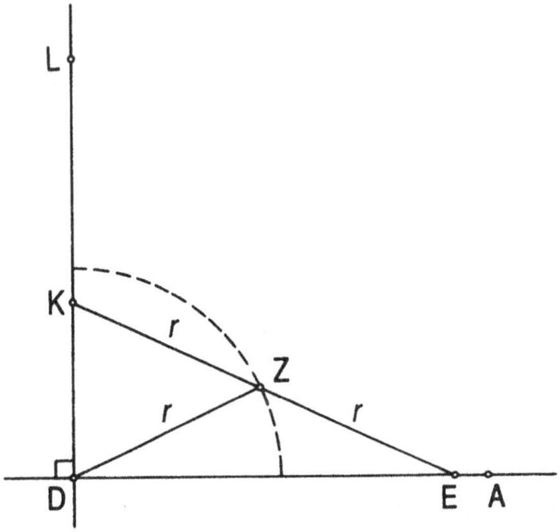

Fig. C7

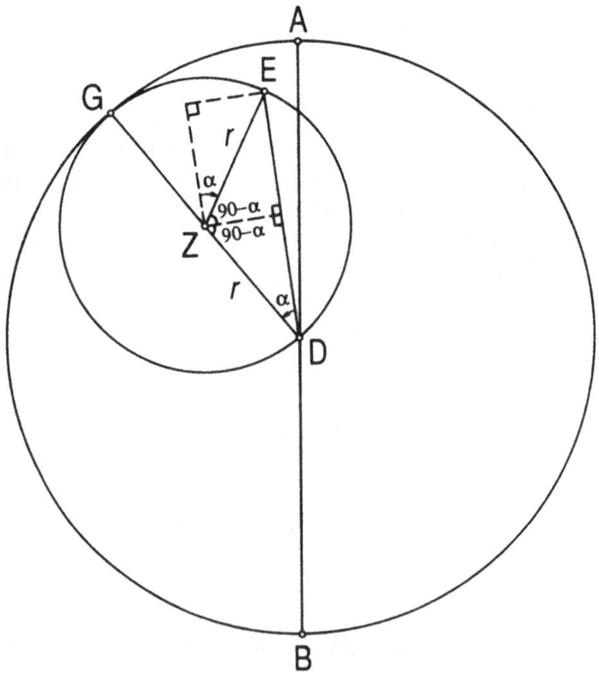

Fig. C8

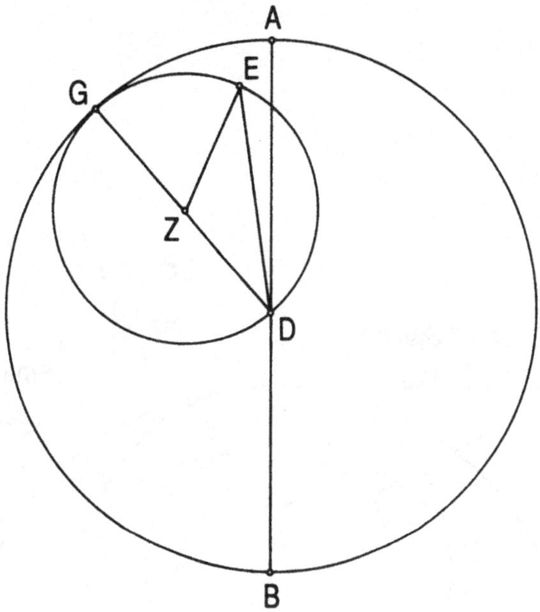

Fig. C9

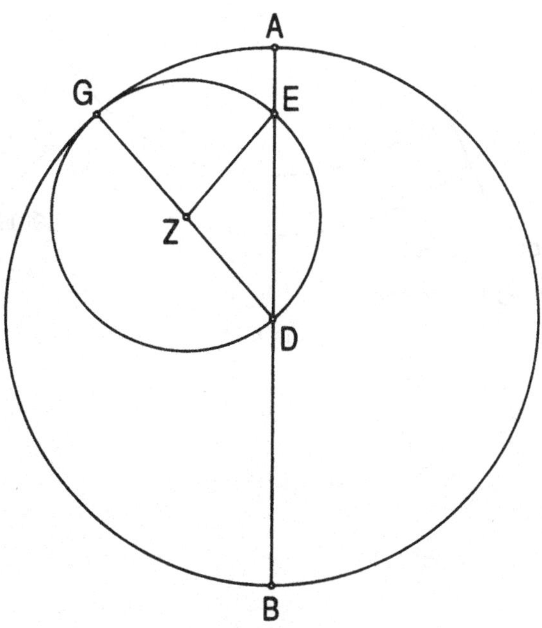

Fig. C10

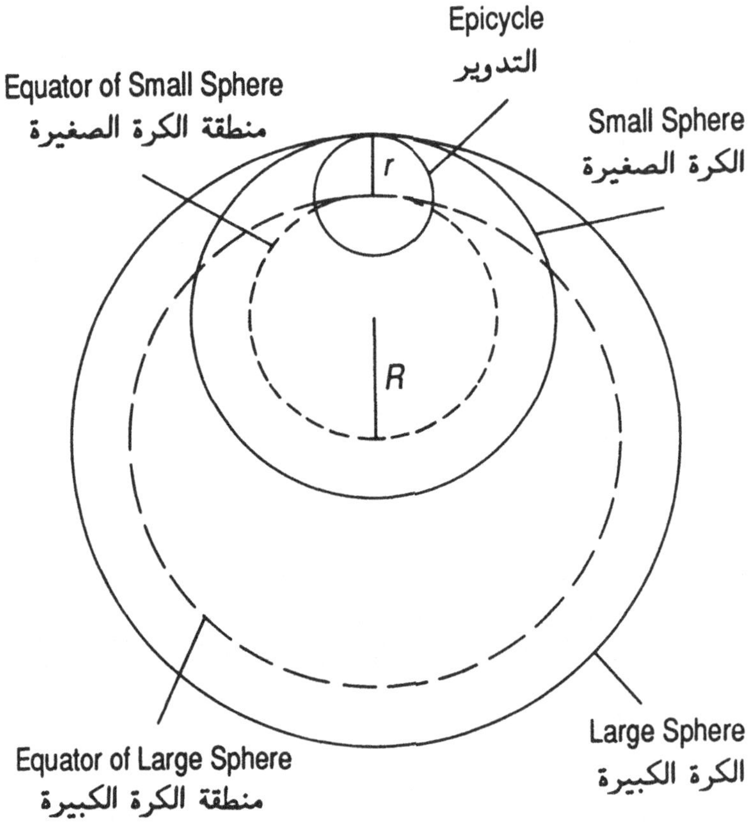
Epicycle
التدوير
Equator of Small Sphere
منطقة الكرة الصغيرة
Small Sphere
الكرة الصغيرة
r
R
Equator of Large Sphere
منطقة الكرة الكبيرة
Large Sphere
الكرة الكبيرة

Fig. C11

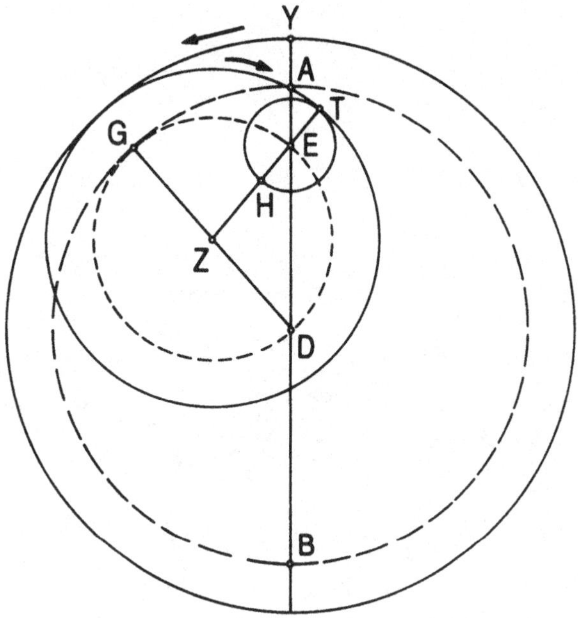

Fig. C12

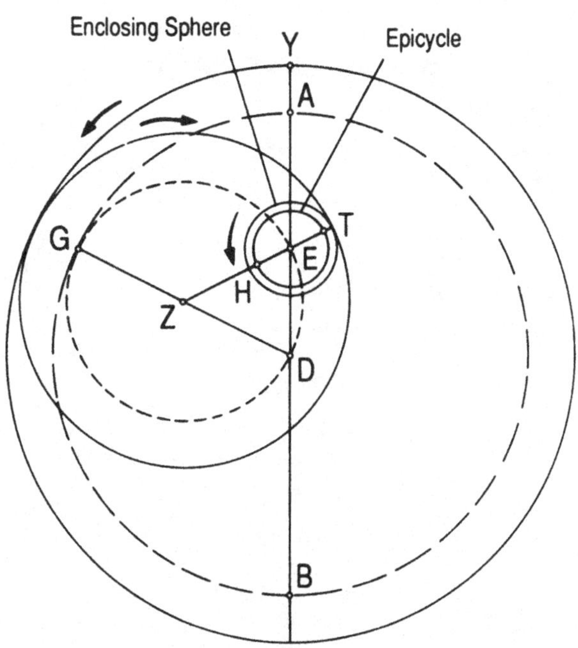

Fig. C13

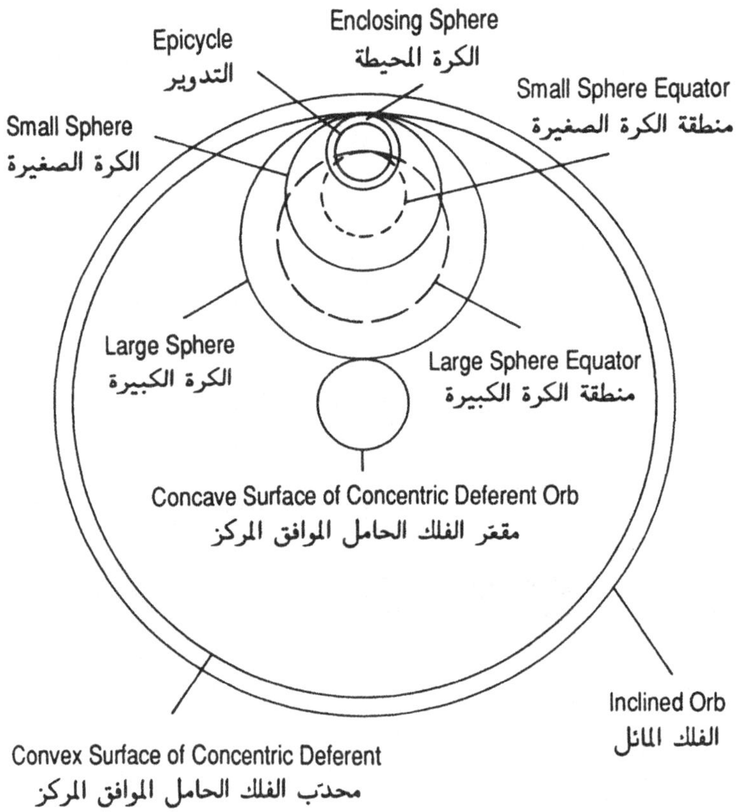
Enclosing Sphere
الكرة المحيطة
Epicycle
التدوير
Small Sphere Equator
منطقة الكرة الصغيرة
Small Sphere
الكرة الصغيرة
Large Sphere
الكرة الكبيرة
Large Sphere Equator
منطقة الكرة الكبيرة
Concave Surface of Concentric Deferent Orb
مقعّر الفلك الحامل الموافق المركز
Inclined Orb
الفلك المائل
Convex Surface of Concentric Deferent
محدّب الفلك الحامل الموافق المركز

Fig. C14

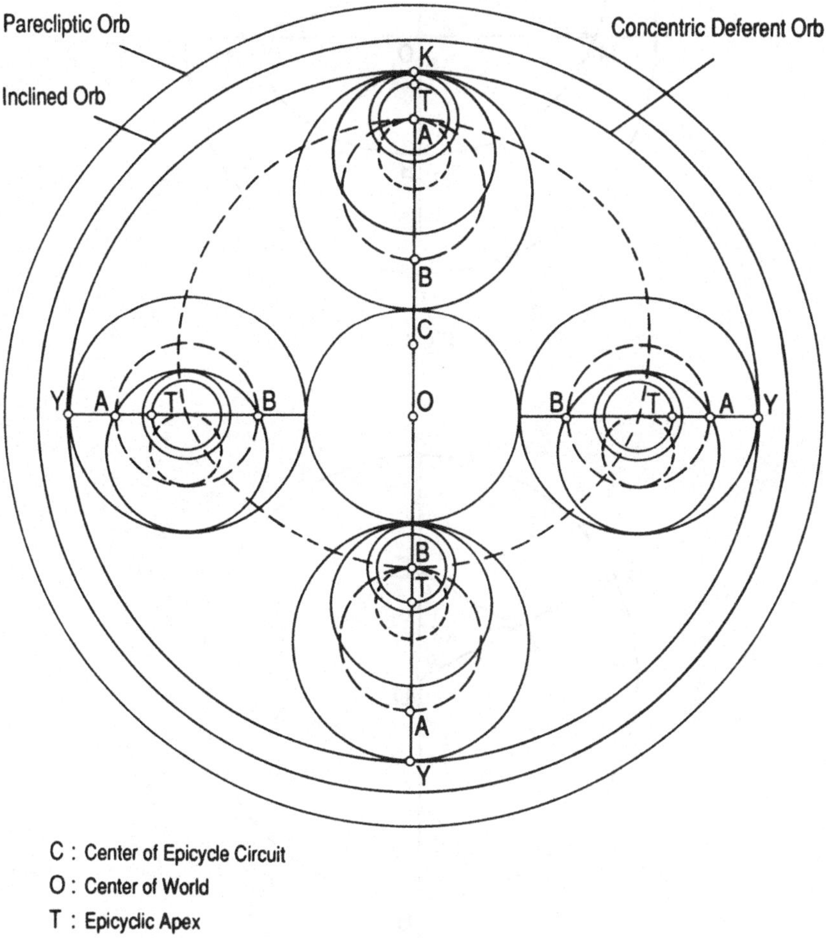
Parecliptic Orb
Concentric Deferent Orb
Inclined Orb
K
T
A
B
C
O
Y
A
T
B
B
T
A
Y
B
T
A
Y
C : Center of Epicycle Circuit
O : Center of World
T : Epicyclic Apex

Fig. C15

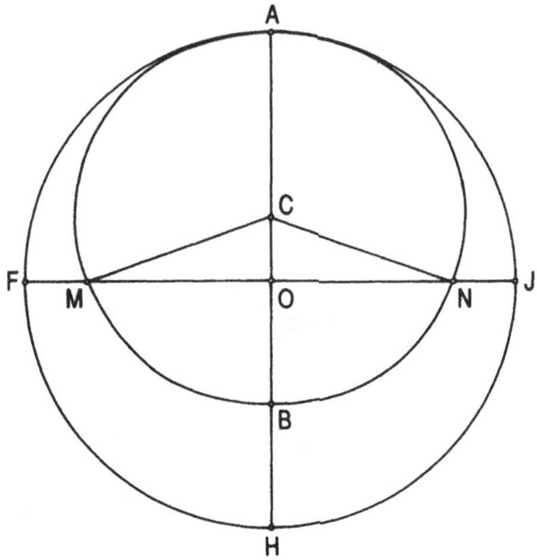

Fig. C16

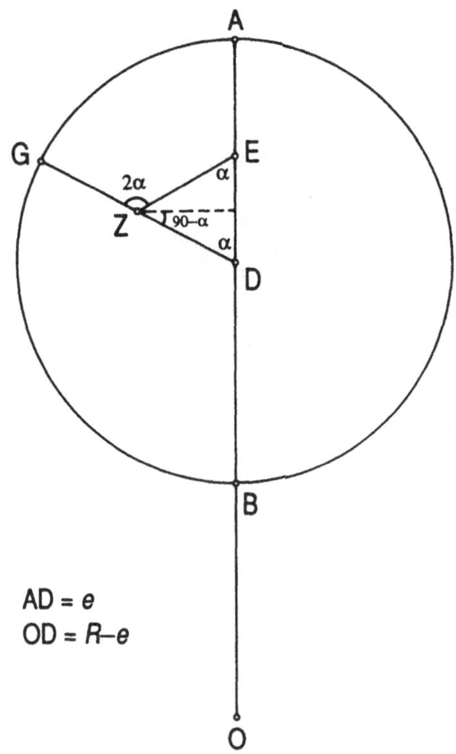

Fig. C17

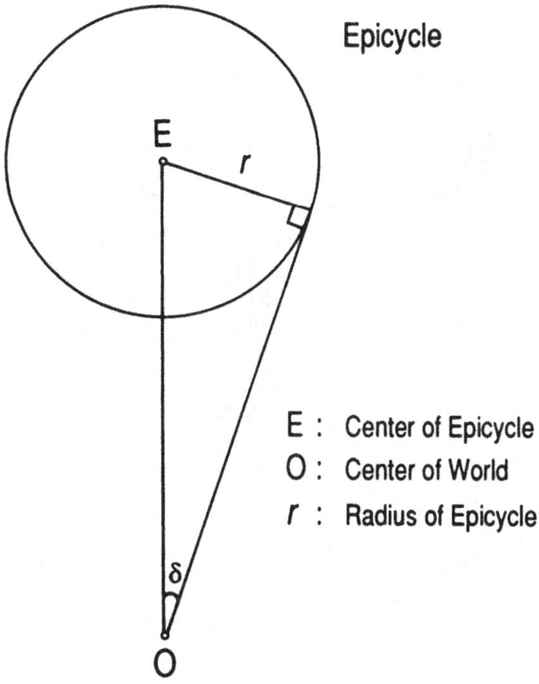
Epicycle
E
r
δ
O
E : Center of Epicycle
O : Center of World
r : Radius of Epicycle

Fig. C18

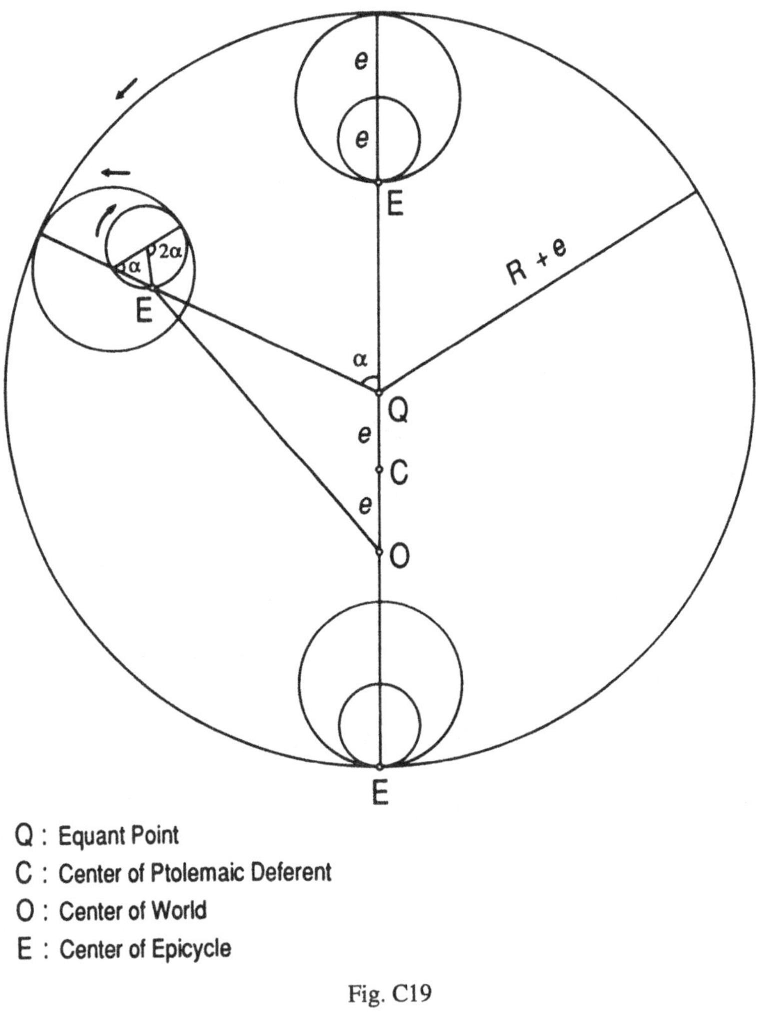
e
e
E
R + e
2α
α
E
α
Q
e
C
e
O
E
Q : Equant Point
C : Center of Ptolemaic Deferent
O : Center of World
E : Center of Epicycle

Fig. C19

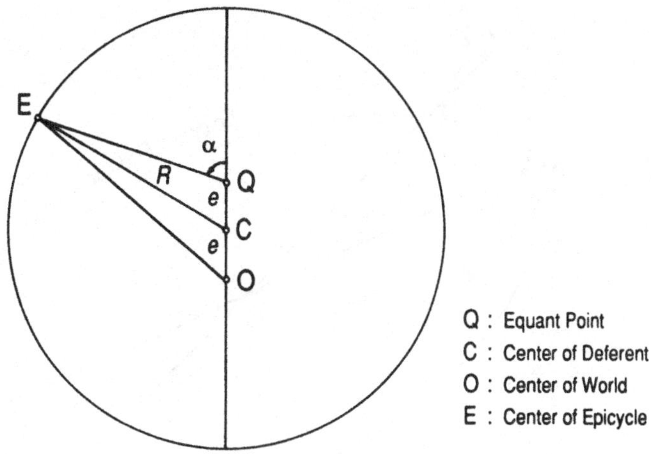

Fig. C20

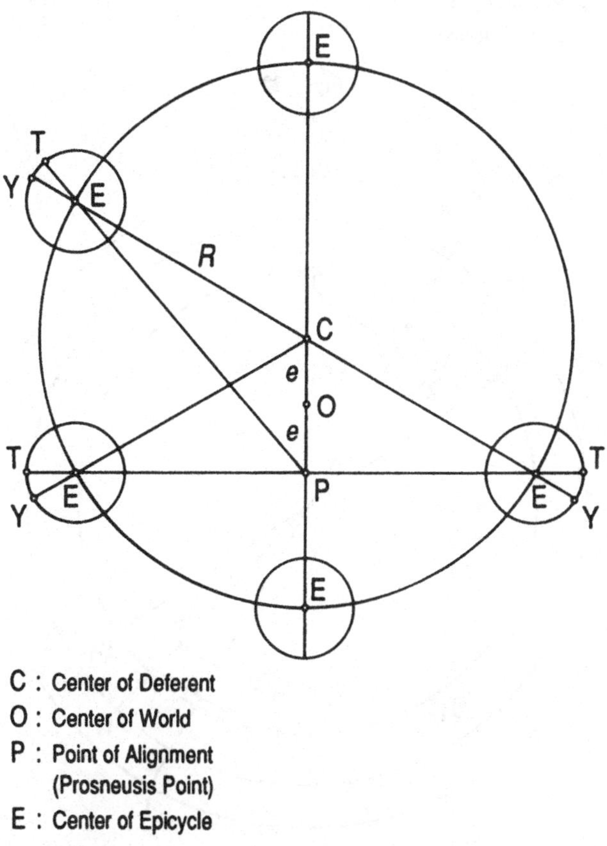

Fig. C21

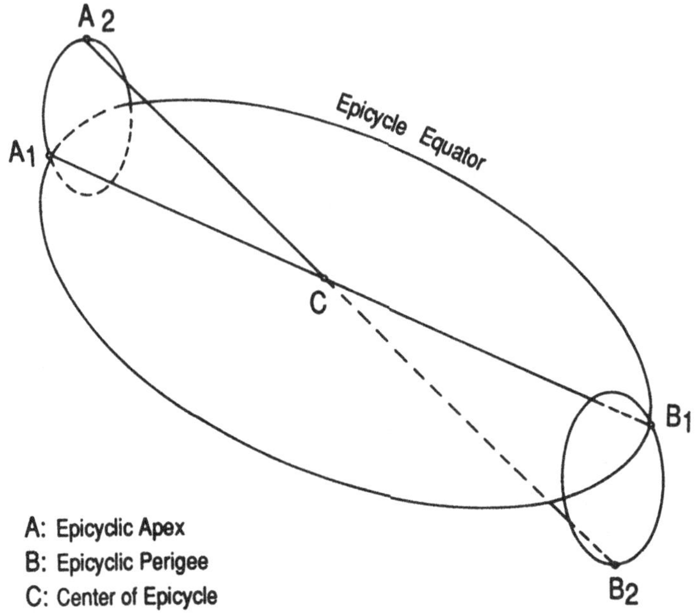

Fig. C22

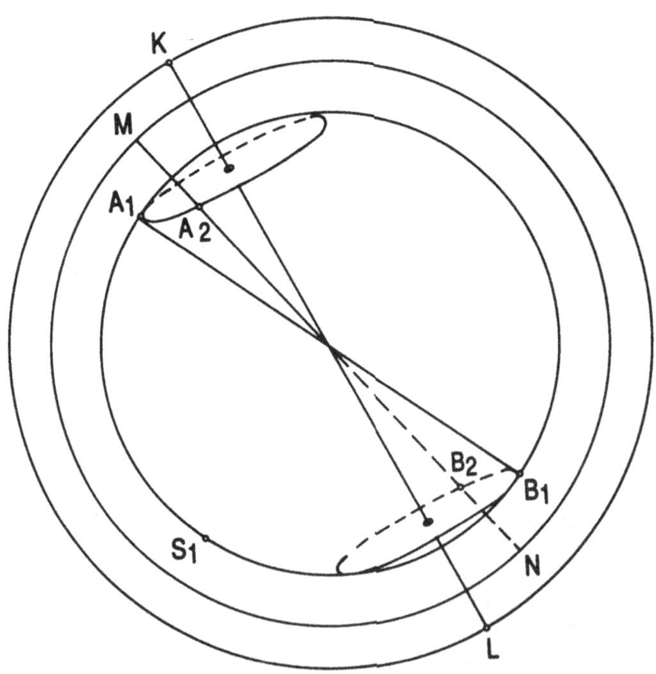

Fig. C23

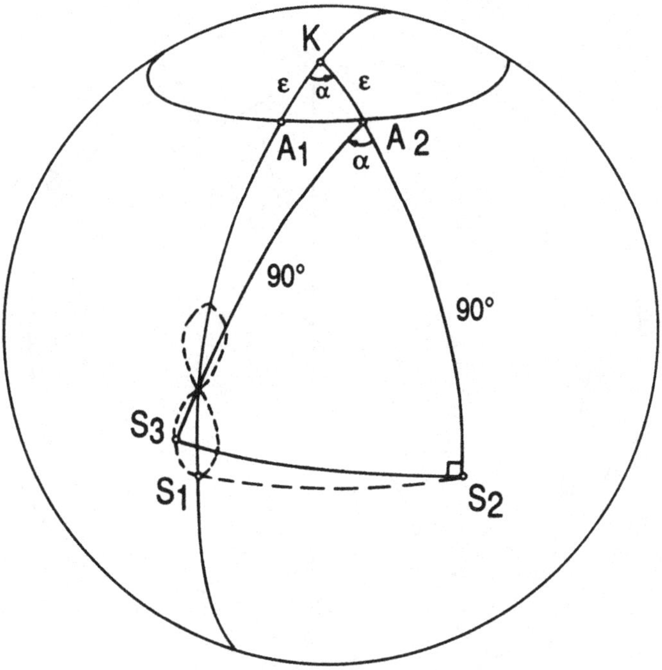

Fig. C24

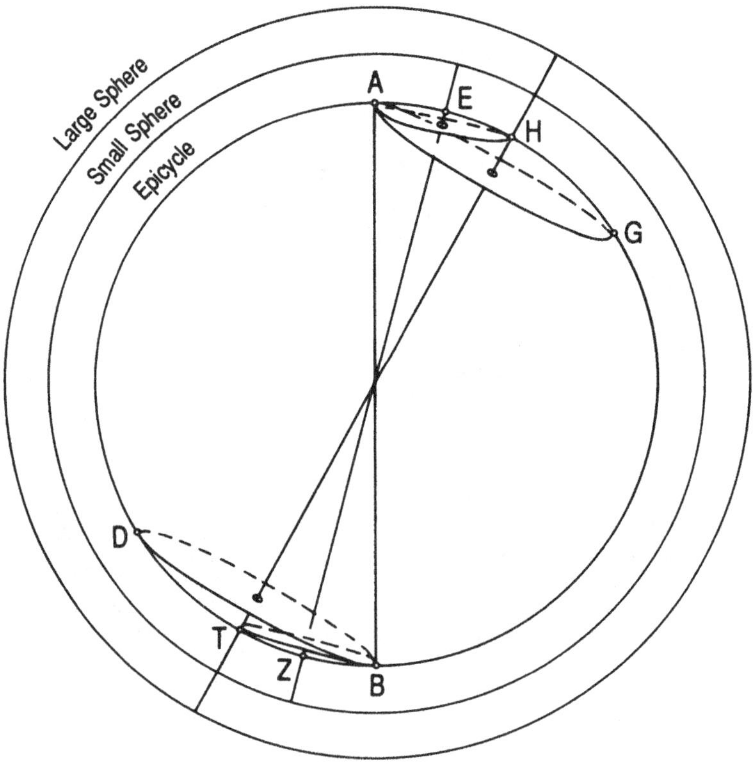

Fig. C25

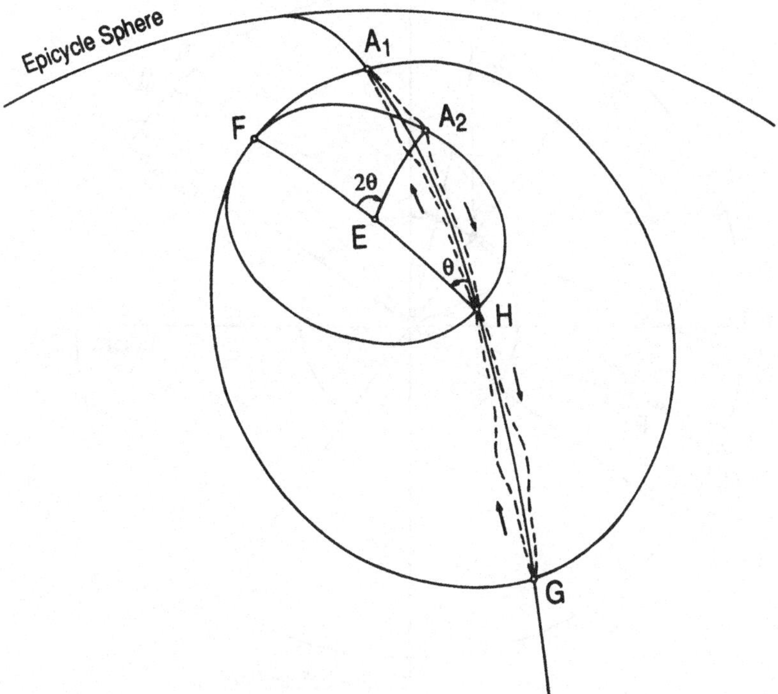

Fig. C26

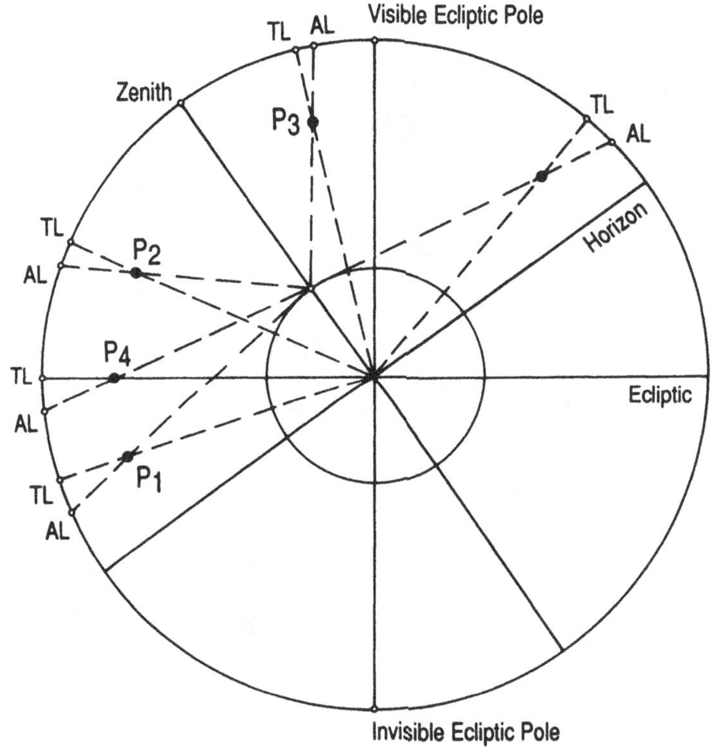

P : Planet
TL : True Latitude
AL : Apparent Latitude

Fig. C27

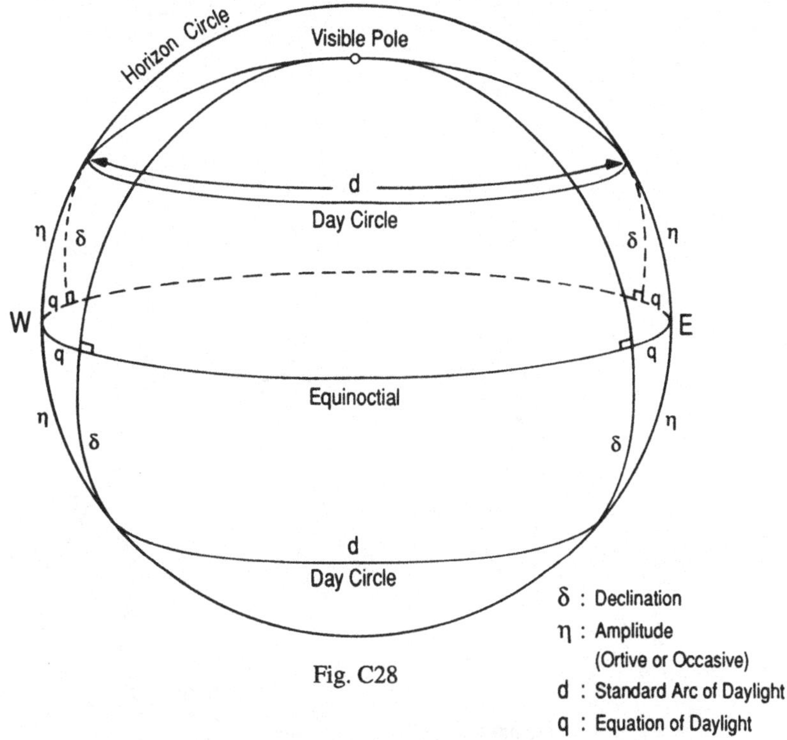

Fig. C28

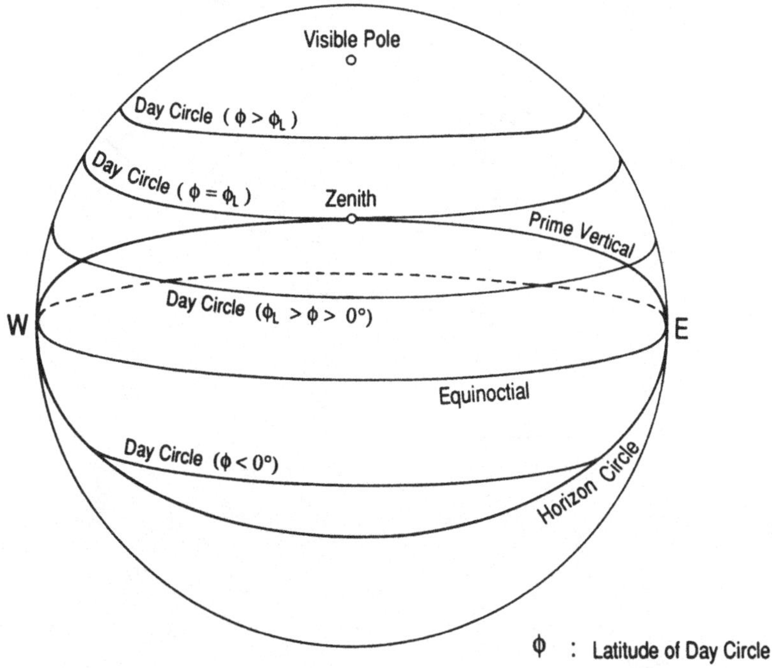

Fig. C29

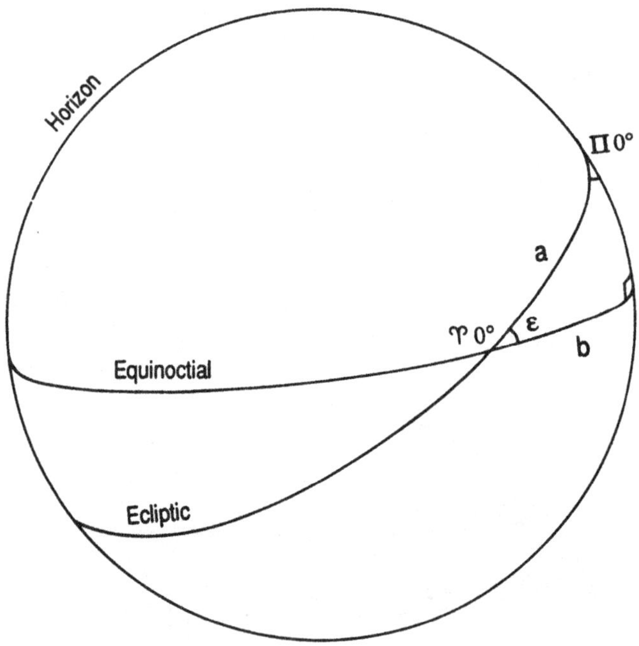

Fig. C30

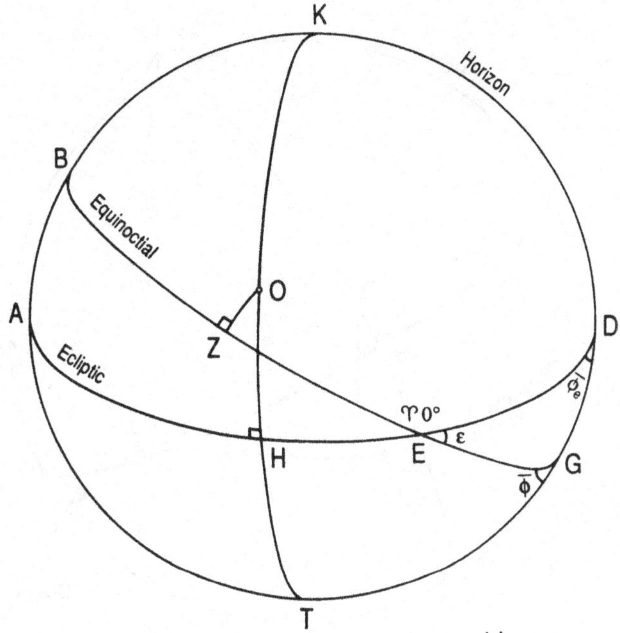

(a)

O : Zenith
ZO : ϕ (Local Latitude)
HO : ϕ_e (Local Ecliptic Latitude)

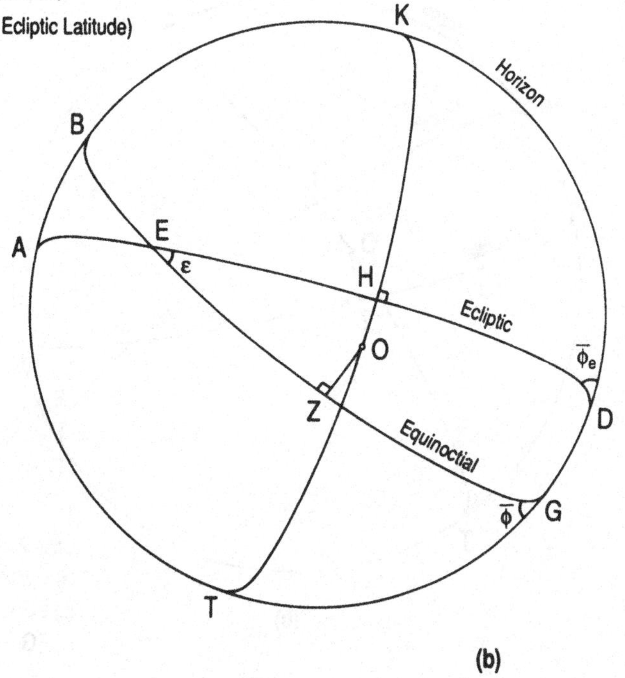

(b)

Fig. C31

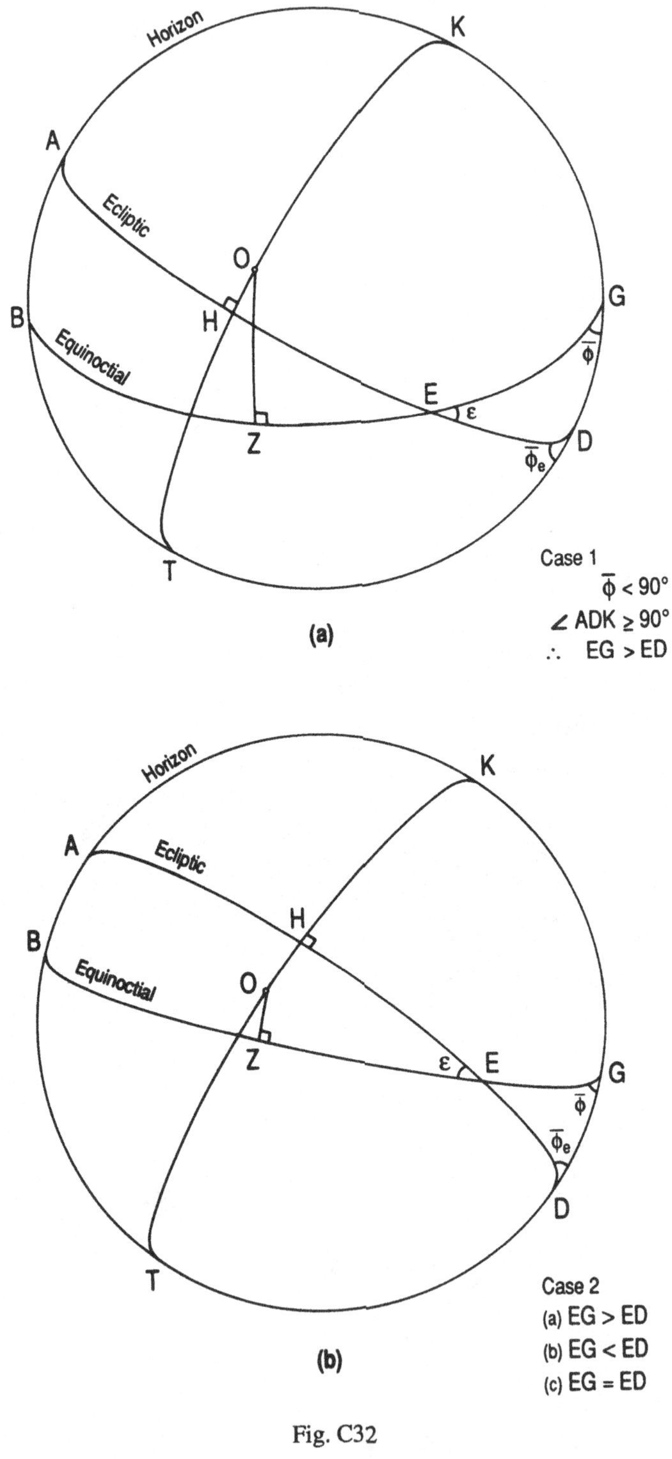
Horizon
K
A
Ecliptic
O
B
H
G
Equinoctial
$\bar{\phi}$
E
ε
Z
$\bar{\phi}_e$
D
T
Case 1
$\bar{\phi} < 90°$
$\angle$ ADK $\geq$ 90°
$\therefore$ EG > ED
(a)
Horizon
K
A
Ecliptic
H
B
Equinoctial
O
Z
ε
E
G
$\bar{\phi}$
$\bar{\phi}_e$
D
T
Case 2
(a) EG > ED
(b) EG < ED
(c) EG = ED
(b)

Fig. C32

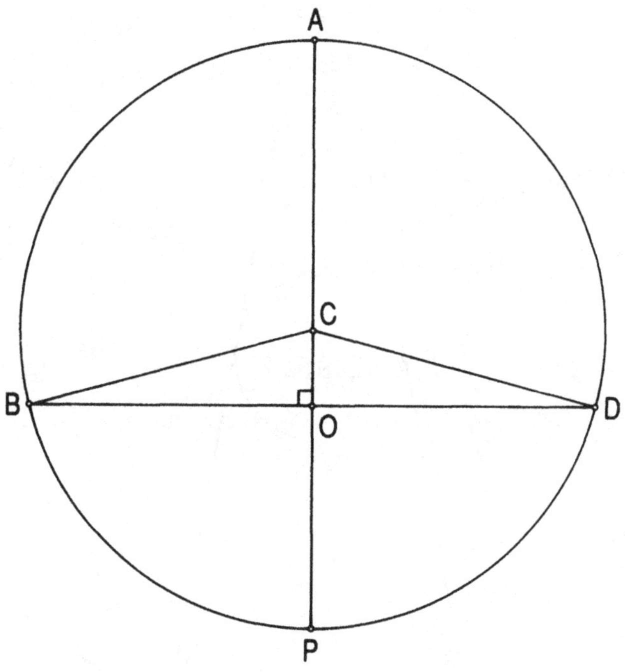

C : Center of Eccentric
O : Earth
B&D : Mean Distances

Fig. C33

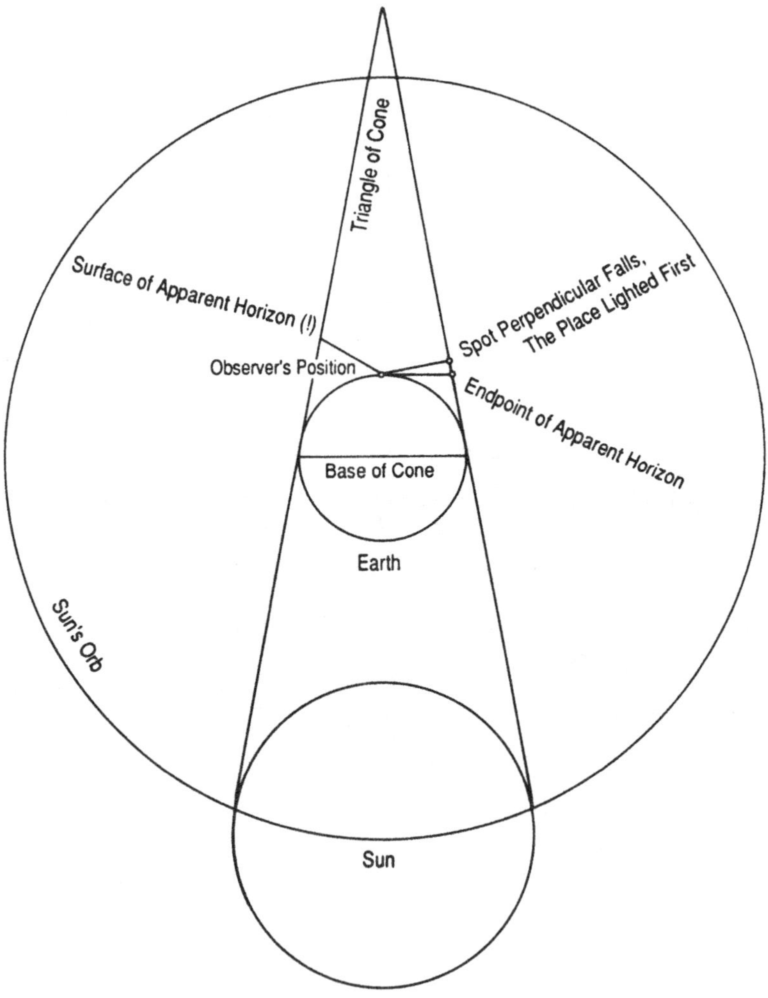
Triangle of Cone
Surface of Apparent Horizon (!)
Spot Perpendicular Falls,
The Place Lighted First
Observer's Position
Endpoint of Apparent Horizon
Base of Cone
Earth
Sun's Orb
Sun

Fig. C34a

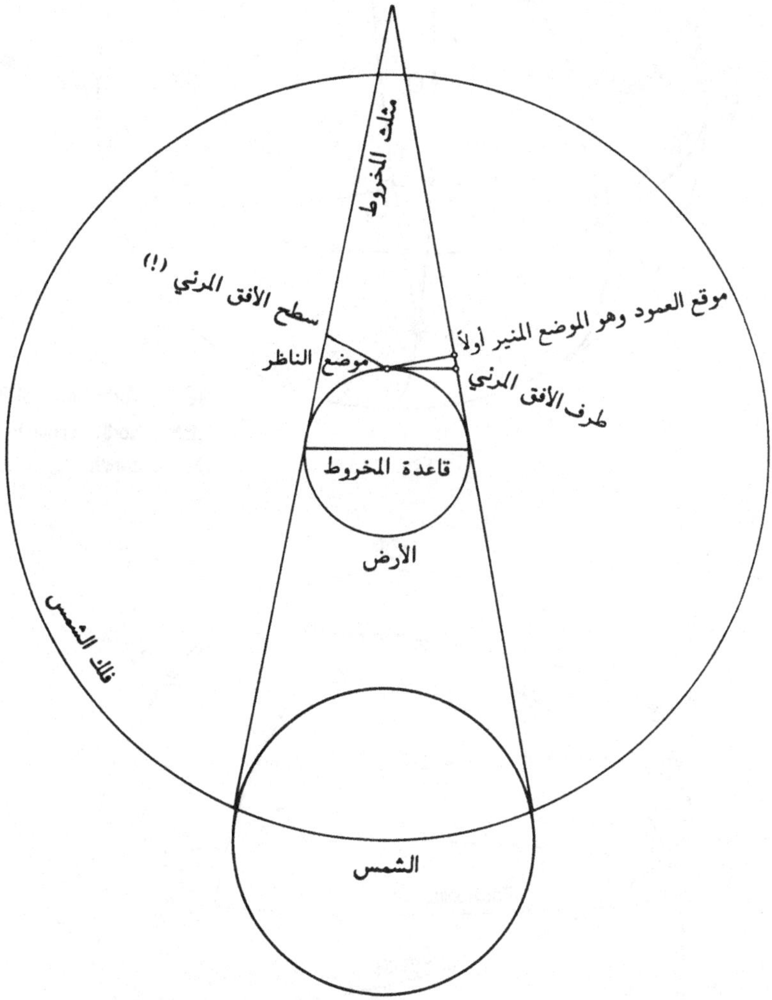
مثلث المخروط
سطح الأفق المرئي (!)
موضع الناظر
موقع العمود وهو الموضع المنير أولاً
طرف الأفق المرئي
قاعدة المخروط
الأرض
فلك الشمس
الشمس

Fig. C34b

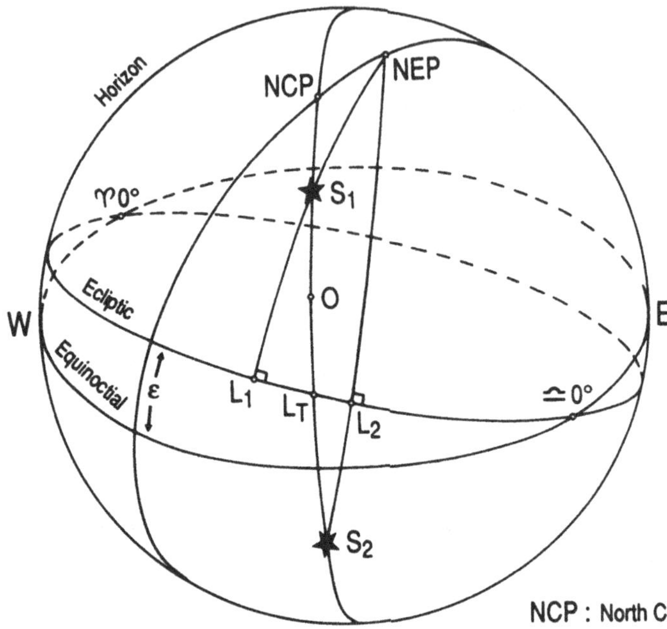

Fig. C35

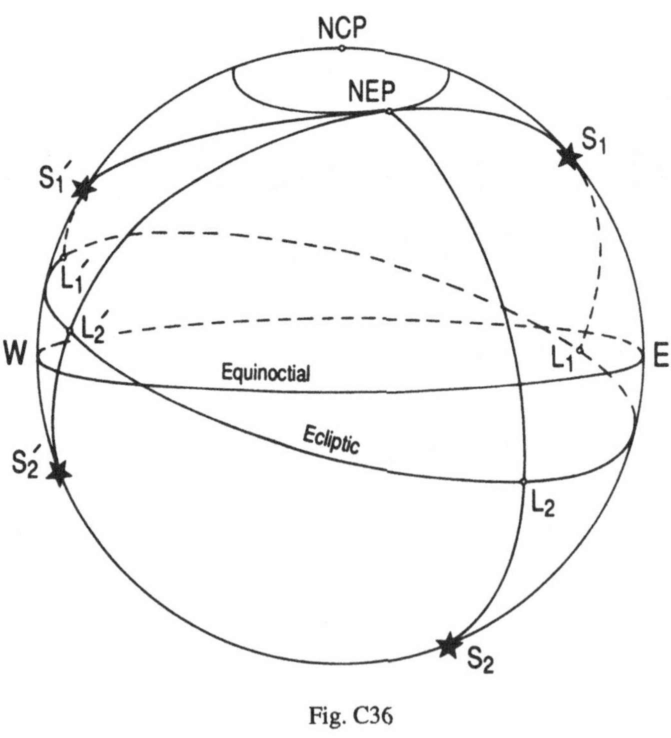

Fig. C36

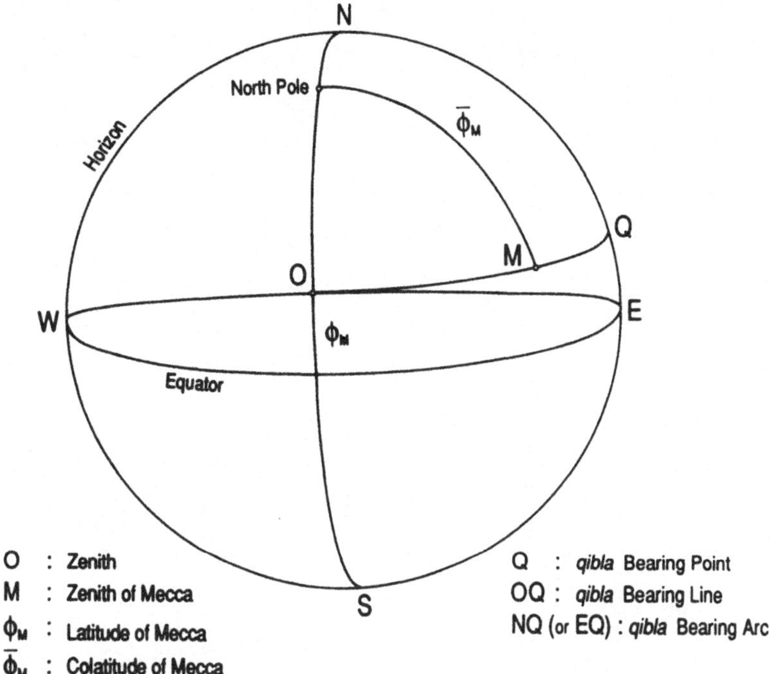

Fig. C37

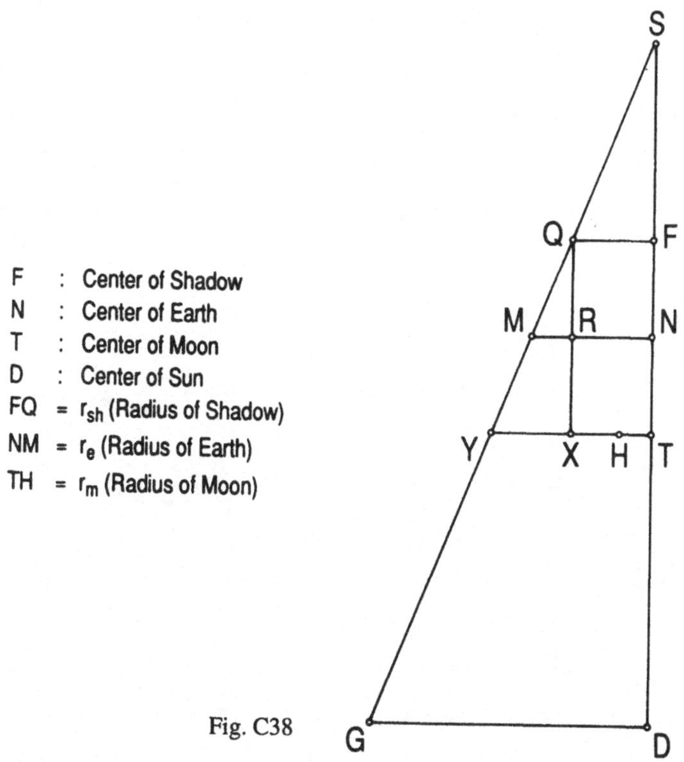

Fig. C38

Made in the USA
Monee, IL
13 September 2026

62976343R00214